T0406447

NanoScience and Technology

The series NanoScience and Technology is focused on the fascinating nano-world, mesoscopic physics, analysis with atomic resolution, nano and quantum-effect devices, nanomechanics and atomic-scale processes. All the basic aspects and technology-oriented developments in this emerging discipline are covered by comprehensive and timely books. The series constitutes a survey of the relevant special topics, which are presented by leading experts in the field. These books will appeal to researchers, engineers, and advanced students.

Krystian Mistewicz

Low-Dimensional Chalcohalide Nanomaterials

Energy Conversion and Sensor-Based Technologies

Krystian Mistewicz
Institute of Physics—Center for Science
and Education
Silesian University of Technology
Katowice, Slaskie, Poland

ISSN 1434-4904 ISSN 2197-7127 (electronic)
NanoScience and Technology
ISBN 978-3-031-25138-2 ISBN 978-3-031-25136-8 (eBook)
https://doi.org/10.1007/978-3-031-25136-8

This Springer imprint is published by the registered company Springer Nature Switzerland AG
The registered company address is: Gewerbestrasse 11, 6330 Cham, Switzerland

I dedicate this book to my beloved wife Dominika.

Preface

The energy conversion and sensor-based technologies play the crucial roles in development of modern science and industry. A global energy consumption increases intensively. A large amount of the pollutants and waste are released to the natural environment due to a production of the electric energy from the non-renewable resources. Thus, there is a strong need for construction of devices and systems able to harvest mechanical, thermal, and solar energies and their conversion into the electric output. Furthermore, the new reliable sensors are necessary for the environmental pollution monitoring. The worldwide research on novel catalytic materials is in progress in order to ensure the effective and large-scale methods of environmental remediation.

This book provides a first review on recent achievements in fabrication, characterization, and applications of low-dimensional chalcohalide nanomaterials. Such comprehensive overview of the chalcohalide nanomaterials and chalcohalide based devices has not been presented so far. These compounds possess exceptional features, including semiconducting, ferroelectric, piezoelectric, pyroelectric, electrocaloric, thermoelectric, photovoltaic, ferroelectric-photovoltaic, photoelectrochemical, photocatalytic, and piezocatalytic properties. Some of the chalcohalide compounds are known as photoferroelectrics, since they exhibit both photoactive and ferroelectric characteristics.

This book presents various methods of chalcohalide nanomaterials fabrication, such as mechanical milling of bulk crystals, liquid-phase exfoliation, vapor-phase growth, hydro/solvothermal methods, synthesis under ultrasonic irradiation, microwave synthesis, laser/heat-induced crystallization, electrospinning, successive ionic layer adsorption and reaction. The recently developed technologies for an incorporation of the chalcohalide nanomaterials into functional structures and devices are elaborated, i.e., solution processing for thin films preparation, spin-coating deposition of polymer composites, solution casting, films deposition via drop-casting, high pressure compression of nanowires into the bulk samples, pressure-assisted sintering, and electric field-assisted alignment of nanowires. The piezoelectric and triboelectric nanogenerators based on the chalcohalide nanomaterials are described herein as the devices for mechanical energy harvesting and vibration detection. The pyroelectric nanogenerators are characterized in term of their possible application for waste heat recovery. The new approach in electrical energy storage is proposed by using chalcohalide-based supercapacitors. The antimony and bismuth chalcohalide solar cells, photodetectors as well as ionizing radiation detectors are presented. The conductometric, photoconductive, impedance, and quartz crystal microbalance sensors are analyzed in detail. Applications of the chalcohalide nanomaterials for photo- and piezocatalytic degradation of toxic pollutants are summarized. This book discusses also major challenges in synthesis, examination, and applications of low-dimensional chalcohalide nanomaterials. At the end of the book, the future trends in this field of research are provided.

Katowice, Poland Krystian Mistewicz

Acknowledgements This book was supported by the Silesian University of Technology (Gliwice, Poland) through the Rector's habilitation grant No. 14/010/RGH21/0008.

Contents

Chapter 1
Introduction

1.1 General Introduction to Chalcohalide Materials

The chalcohalides can be defined in general as a broad group of the inorganic materials which contain both chalcogen and halogen atoms. The chalcogens are the chemical elements originating from group 16 of the periodic table [1] according to the modern notation established by the International Union of Pure and Applied Chemistry (IUPAC). This group of the chemical elements is also known as the oxygen family [2]. It consists of oxygen (O), sulfur (S), selenium (Se), tellurium (Te), and polonium (Po). The halogens belong to the group 17 of the periodic table [3] which includes fluorine (F), chlorine (Cl), bromine (Br), iodine (I), and astatine (At). The chalcohalide family of materials comprises a large number of various chemical compounds, including binary [4, 5], ternary [6, 7], quaternary [8–10], pentanary [11, 12] or much more complex materials [13, 14]. Among them there are the pnictogen chalcohalides [15], known also as an $A^{15}{}_{x}B^{16}{}_{y}C^{17}{}_{z}$ group of semiconductors [16–18], where A^{15} refers to the certain element from group 15 of the periodic table (arsenic (As), antimony (Sb) or bismuth (Bi)), and B^{16}, C^{17} denote chalcogen and halogen, respectively. The symbols *x*, *y*, and *z* can take different values depending on the material stoichiometry. The $A^{15}{}_{x}B^{16}{}_{y}C^{17}{}_{z}$ ternary chalcohalides consist of one cation and two anions [19]. They exhibit the ns^2 electronic configuration, what results in numerous compositional combinations of these compounds as well as their unique physical and chemical properties [20, 21]. As similar to methylammonium lead triiodide (MAPI), the chalcohalides containing post-transition elements with an ns^2 electronic configuration are anticipated to possess the extraordinary charge-carrier transport properties [22]. The chalcohalide semiconductors are considered as the potential "defect-tolerant" materials [23] with high charge-carrier mobilities due to the presence of lone pairs of electrons and strongly dispersive characters for valence and conduction bands [22]. Furthermore, the bandgap of the metal chalcohalides vary in the very wide range (Table 1.1), since halide and chalcogenide coexist as anions in these materials [24]. As presented in Table 1.1, the material

K. Mistewicz, *Low-Dimensional Chalcohalide Nanomaterials*, NanoScience and Technology, https://doi.org/10.1007/978-3-031-25136-8_1

morphology and method of its preparation are also crucial factors which can influence the value of the band gap. In Ref. [25] the novel computer-aided hierarchical screening procedure was developed and used to search through a large space of over 161,000 compositions of metal chalcohalides to identify promising candidate photoactive semiconductors. Due to possible tuning of the band gaps values of the $A^{15}{}_{x}B^{16}{}_{y}C^{17}{}_{z}$ semiconductors, these compounds are especially attractive for application in the optoelectronic devices [21, 26], solar cells [20, 27, 28], photodetectors [29], ionizing radiation detectors [30, 31], infrared glasses [32], infrared gradient refractive index lenses [33], photonic crystals [34, 35], and photocatalysts for an environmental remediation [36].

Figure 1.1 presents a typical crystal structure of $A^{15}{}_{x}B^{16}{}_{y}C^{17}{}_{z}$ chalcohalides with elemental ratio x:y:z = 1:1:1. Antimony sulfoiodide (SbSI) was chosen as the most commonly investigated representative of the ternary pnictogen chalcohalides. The majority of the $A^{15}B^{16}C^{17}$ compounds, excluding oxyhalides and arsenic selenoiodide (AsSeI), grow into double chains. The chains are held together by weak van der Waals interaction. The one-dimensional crystal structure favors charge transport along the c-axis of the crystals [29]. The anisotropic morphology of $A^{15}B^{16}C^{17}$ chalcohalides results in high internal electric fields enhancing charge carriers separation, what is beneficial for both photodetector and photocatalytic applications [29].

A plenty of the ternary $A^{15}{}_{x}B^{16}{}_{y}C^{17}{}_{z}$ compounds were confirmed to exhibit ferroelectricity [61]. As described above, these materials also show photosensitive properties. Thus, they are named as photoferroelectrics [62]. The enormous, above-bandgap open-circuit voltage can be generated in the photoferroelectric semiconductor under light illumination due to polarization-related charge-separation mechanism [63], what makes photoferroelectrics suitable for application in solar cells [64, 65]. Moreover, the chalcohalides possess several interesting electro-optic properties [30], including the shift of the optical absorption edge by applying an electric field to the material [66]. Table 1.2 shows that the crystal structure of the chalcohalide compound depends significantly on the temperature of the material which determines its phase (ferroelectric or paraelectric). The fundamentals of the ferroelectric materials and their basic properties will be discussed further in the next section of this chapter.

The values of Curie temperatures (T_C) of ternary and quaternary pnictogen chalcohalides can be found in very wide range (Table 1.3). The chemical composition and morphology of the material influence the phase transition temperature. Even slight change in components of the pnictogen chalcohalides may result in significant variation of the Curie temperature. For instance, substitution of iodine with chlorine in SbSI leads to rise of T_C (compare T_C values for SbSI and $SbSCl_{0.2}I_{0.8}$ in Table 1.3). In addition, the method of material fabrication can be also important factor determining the phase transition temperature, what is clearly seen in the case of antimony sulfobromide (SbSBr). More detailed discussion of the engineering of tuning of the Curie temperature will be given in the Sect. 1.4 of this chapter.

The dielectrics are electrical insulators that can be polarized under an external applied electric field [90]. They are characterized by the dielectric permittivity which

Table 1.1 The comparison of the values of the band gaps (E_g) for selected $A^{15}{}_xB^{16}{}_yC^{17}{}_z$ ternary chalcohalides with different morphologies (used abbreviations: C—computed value of the band gap; D—direct band gap; E—the value of the band gap determined experimentally; I—indirect band gap)

Material	Morphology of the material	E_g, eV	Character of the band gap	Method of E_g determination	References
BiTeI		0.207	I	C	[37]
		0.478		C	[38]
BiTeBr		0.595		C	[38]
$Bi_{13}S_{18}Cl_2$	Nanocrystals	0.76	I	E	[39]
$Bi_{13}S_{18}Br_2$	Nanocrystals	0.80	I	E	[39]
$Bi_{13}S_{18}I_2$	Nanocrystals	0.81	I	E	[39]
$Sn_2SbS_2I_3$		1.08	D	C	[40]
BiSI	Thin film	1.57	D	E	[41]
	Nanorods	1.6	I	E	[30]
	Nanorods	1.61		E	[42]
SbSeI	Nanorods	1.68		E	[43]
	Bulk crystal	1.7	I	E	[44]
	Nanowires	1.71		E	[45]
		1.86	I	C	[19]
Tl_6SeI_4	Bulk crystal	1.86	I	E	[46]
SbSI	Microrods	1.53	I	C	[47]
	Microrods	1.8	I	E	[48]
	Nanowires	1.81	I	E	[49]
	Microrods	1.84	I	E	[47]
	Nanowires	1.86	I	E	[50]
	Nanowires	1.91	I	E	[51]
	Thin film	1.96		E	[52]
BiSBr		1.89	D	C	[53]
AsSeI	Film	1.91		E	[54]
BiOI	Nanosheets	1.77	I	E	[55]
	Nanoparticles	1.94	D	E	[56]
BiSCl		1.98	D	C	[53]
$Bi_4O_5Br_2$	Nanosheets	2.0		E	[57]
$Bi_4O_5I_2$	Nanosheets	2.08	I	E	[55]
Bi_5O_7I	Nanosheets	2.32	I	E	[55]
Bi_3O_4Br	Nanorings	2.3	I	E	[58]
	Nanoplates	2.4	I	E	[58]
BiOBr		2.75		E	[57]

(continued)

Table 1.1 (continued)

Material	Morphology of the material	E_g, eV	Character of the band gap	Method of E_g determination	References
Bi_3O_4Cl	Nanosheets	2.7		E	[59]
	Powder	2.79	I	E	[60]
BiOF		3.94	D	C	[53]
	Nanoparticles	3.99	D	E	[56]

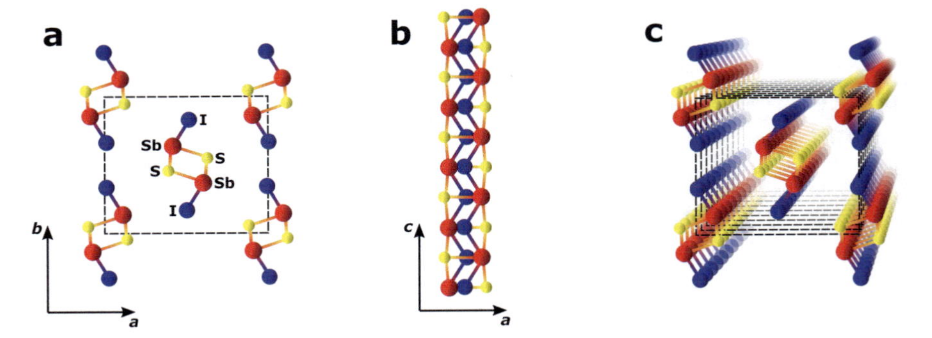

Fig. 1.1 The crystal structure of SbSI projected on the *a*-*b* plane (**a**), *a*-*c* plane (**b**), and three dimensional view of SbSI chains (**c**). The antimony, sulfur, and iodine are represented by the red, yellow, and blue spheres, respectively

Table 1.2 The crystal data of the various ternary pnictogen chalcohalides

Material	Crystal structure	Phase	Space group	Lattice parameters			References
				a, Å	*b*, Å	*c*, Å	
AsSI	Monoclinic		$P2_1$	8.61	4.22	9.95	[67]
BiSCl	Orthorhombic	Paraelectric	*Pnam*	7.7508	9.9920	3.9955	[68]
BiSeCl	Orthorhombic		*Pnam*	8.0787	9.3046	4.2182	[69]
BiSeI	Orthorhombic		*Pnam*	8.7	10.5	4.1	[70]
BiSI	Orthorhombic		*Pnma*	8.44	4.13	10.26	[71]
SbSBr	Orthorhombic	Ferroelectric	$Pna2_1$	8.168	9.700	3.942	[72]
		Paraelectric	*Pnam*	8.212	9.720	3.963	[72]
SbSeBr	Orthorhombic	Paraelectric	*Pnam*	8.3	10.2	4.0	[73]
SbSel	Orthorhombic	Paraelectric	*Pnam*	8.6862	10.3927	4.1452	[74]
SbSI	Orthorhombic	Ferroelectric	$Pna2_1$	8.525	10.137	4.097	[16]
		Paraelectric	*Pnam*	8.522	10.130	4.088	[16]
SbTel	Monoclinic		*C2/m*	13.7008	4.2418	9.2005	[75]

Table 1.3 The comparison of the curie temperature (T_C) values for selected ternary and quaternary pnictogen chalcohalides

Material	Morphology of the material	Fabrication technology of the material	T_C, K	References
SbSBr	Single crystal	Vapor transport reaction	21	[76]
	Single crystal	Vapor transport reaction	22.8	[77]
	Single crystal	Bridgman method	39	[78]
	Single crystal	Bridgman method	93	[79]
BiSBr	Single crystal	Bridgman method	103	[79]
BiSI	Single crystal	Bridgman method	113	[79]
$SbSBr_{0.5}I_{0.5}$	Single crystal	Bridgman method	207	[78]
SbSeI	Single crystal		223	[80]
$Sb_{0.9}Bi_{0.1}SI$	Single crystal	Sublimation method	~250	[81]
$Sb_{0.9}As_{0.1}SI$	Single crystal	Bridgman method	273	[79]
SbSI	Nanowires	Sonochemical synthesis	291	[50]
	Film	Pulsed laser deposition	292	[82]
	Single crystal	Vapor phase growth	292–293	[83]
	Single crystal	Bridgman method and vapor transport technique	293	[84]
	Thin film	Flash evaporation	293	[85]
	Thin film	Electron beam evaporation	294	[86]
	Single crystal		295	[87]
	Ceramics	Bridgman method	331	[88]
$SbS_{0.8}O_{0.2}I$	Single crystal	Bridgman method	338	[79]
$SbSCl_{0.2}I_{0.8}$	Single crystal	Bridgman method	340	[89]

value reflects the resistance of the dielectric against polarization by an external electric field. The piezoelectrics are group of dielectrics that exhibit change of the electric polarization under strain or stress applied to the material. Reverse effect can be also observed in the case of piezoelectric material. A mechanical deformation occurs when piezoelectric is subjected to an external electric field. The pyroelectrics belong to the piezoelectric compounds. They display change of polarization due to temperature variation. As depicted in Venn diagram (Fig. 1.2), every ferroelectric material possesses both pyroelectric and piezoelectric properties. Therefore, ferroelectric chalcohalides are considered as suitable materials for use in piezoelectric [91–94] and pyroelectric nanogenerators [45, 95] for mechanical and thermal energy harvesting, respectively. First principles study of electronic and thermoelectric properties of ternary chalcohalide semiconductors [6] indicated that these materials have also a great potential for application in thermoelectric generators based on the Seebeck effect [96].

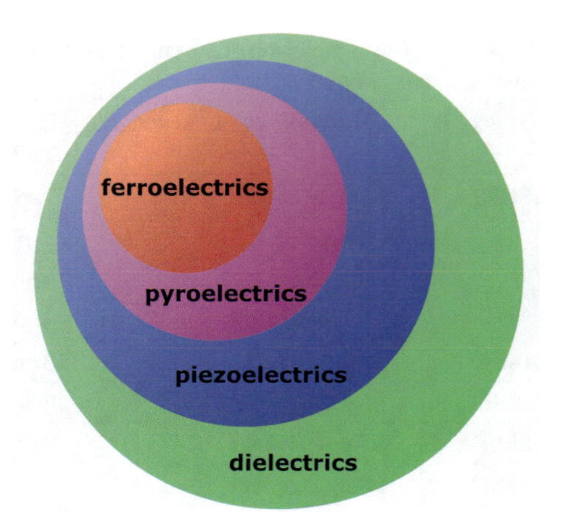

Fig. 1.2 The Venn diagram presenting relationships between ferroelectric, pyroelectric, piezoelectric, and dielectric materials

Antimony sulfoiodide is one of the most interesting members of the ternary pnictogen chalcohalide family of materials. In 1824, SbSI was synthesized for the first time [97]. However, SbSI received an attention of the scientific community very lately, i.e. in the early 60s of the twentieth century when photoconductivity [98] and ferroelectricity [87] were discovered in this material. The first-order phase transition in SbSI at temperature of about 290 K is of ion displacive type [78] what is similar to the barium titanate ($BaTiO_3$). According to [99], in the ferroelectric phase of SbSI, antimony and sulfur shift in the *c* direction with respect to iodine by approximately 0.2 Å and 0.05 Å, respectively. The large dielectric constant of $5 \cdot 10^4$ was reported for the bulk crystals of SbSI at the Curie temperature [87]. It was examined in Ref. [100] that illumination leads to the change in the dielectric constant of SbSI which is positive below the Curie temperature and negative above T_C. Numerous studies of bulk crystals of SbSI revealed their other exceptional features, such as piezoelectric [101, 102], electromechanical [66], pyroelectric [84], pyro-optic [103], and electro-optical properties [66]. Further investigations of SbSI were focused on the thin films of this material [85, 104–106]. Recently, a great interest in the fabrication and characterization of SbSI nanostructures has been observed. A low dimensional antimony sulfoiodide has been prepared so far in the form of nanocrystals [107, 108], nanowires [49, 51, 109], nanorods [110–115], and nanoneedles [116]. The methods of fabrication and selected applications of SbSI nanostructures will be discussed in detail in the next chapters of this book.

1.2 Fundamentals of the Ferroelectric Materials and Their Basic Properties

A ferroelectric is recognized as material exhibiting nonzero electric polarization below the Curie temperature (T_C), even when no electric field is applied to the material. This electric polarization, known as spontaneous polarization, can be totally reversed under a sufficiently large external electric field [117]. According to other definition of ferroelectric, it is an insulating system with multiple discrete stable or metastable states of spontaneous polarization [118]. The switching between accessible states can be achieved by applying of an external electric field what leads to the variation in relative energy of the states. Generally, the electric polarization (***P***) is described by the well-known equation

$$\vec{P} = \frac{\sum_{i=1}^{n} \overrightarrow{p_{ei}}}{V} \tag{1.1}$$

where $\boldsymbol{p}_e$ means electric dipole moment, V is a volume. As presented in Fig. 1.3, the electric polarization originates from three main contributions, i.e. electronic, ionic, and dipole reorientation. When an electric field is applied to the dielectric material, electron clouds are deformed, what leads to formation of electric polarization. In the case of ionic crystal, the anions and cations are shifted toward positive and negative electric potentials, respectively. The electric dipoles in a ferroelectric material can be freely rotated or reversed. Therefore, they become aligned under electric field. All mentioned mechanisms can coexist in the dielectric material. However, a frequency of an applied electric field strongly influences the contribution of each effect to the total polarization. The dipole reorientation related polarization occurs only in the relatively low frequency regime, i.e. below 10^8 Hz [119]. Ionic polarization follows up to frequencies in the range from 10^{10} to 10^{12} Hz, whereas electronic polarization extends to 10^{16} Hz [119].

The polarization ***P*** can be quantified experimentally by determination of an electric displacement ***D*** through electric capacitance measurement. In the case of a capacitor, the value of electric displacement is simply equal to the electric charge, stored in the capacitor, per unit area. The relation between electric displacement and polarization is following

$$\vec{D} = \varepsilon_0 \vec{E} + \vec{P} = \varepsilon_0 \varepsilon \vec{E} \tag{1.2}$$

where ε_0 means the vacuum permittivity, ε denotes relative permittivity of the material which is a tensor in general. The polarization of a ferroelectric is nearly equal to electric displacement, since ionic displacement is significantly higher than the contribution from vacuum dielectric.

In the case of ferroelectric material the polarization versus an electric field is non-linear relation what is represented by the closed curve called the hysteresis loop (Fig. 1.4). Every ferroelectric crystal consists of the domains. These areas retain

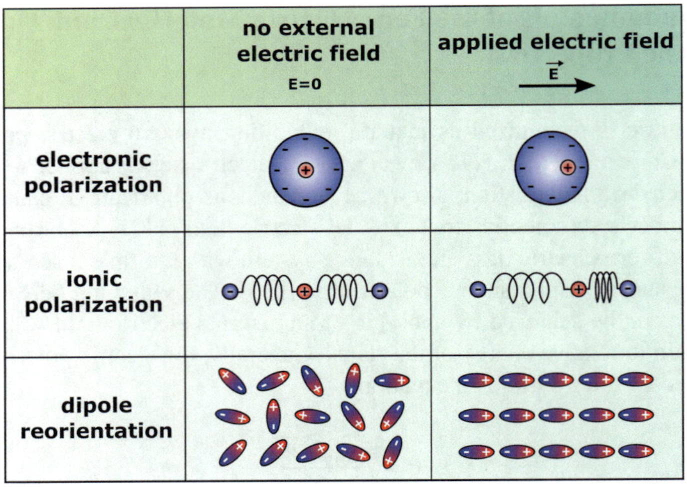

	no external electric field E=0	applied electric field $\vec{E}$
electronic polarization		
ionic polarization		
dipole reorientation		

Fig. 1.3 The schematic presenting three primary contributions to electric polarization

homogeneous polarization [90]. When no electric file is applied to the material ($E = 0$) the spontaneous polarizations (P_S) of different domains can be positive or negative. Thus, the net polarization of the material is zero ($P = 0$). If the electric field is increased starting from a virgin polydomain state with zero net macroscopic polarization, the polarization rises due to induced electronic, ionic and dipolar types of polarization. The *P*–*E* curve follows the dependence depicted by the dashed line in Fig. 1.4. Further enhancement of the electric field results in the alignment of electric dipoles in the material. Finally, a monodomain state is achieved and the polarization reaches a saturation value ($+P_S$). If then electric field is reduced, the polarization decreases as represented by the solid curve in Fig. 1.4. When electric field attains zero, the polarization has nonzero positive value what is called remanent polarization ($+P_R$). In order to set the polarization at zero value, the negative electric field should be applied to the material, known as a coercive field ($-E_C$). As the negative increase in the electric field is continued, a reverse saturation of the polarization is observed ($-P_S$). A change of electric field toward zero leads to variation of polarization which reaches nonzero negative value ($-P_R$). The ferroelectric hysteresis loop is completed if the electric field is increased from zero to sufficiently high values for which saturation of the polarization ($+P_R$) is accomplished again. As described above, the polarization of the ferroelectric crystal can be switched hysteretically between two states (positive or negative) by applying an electric field. This unique property of the ferroelectric materials is attractive for use in the nonvolatile memories [120, 121].

A sharp peak in the temperature dependence of dielectric constant of ferroelectric material occurs at Curie temperature indicating a transition from ferroelectric to non-ferroelectric (paraelectric) phase. In the paraelectric phase, i.e. above the Curie point, the temperature dependence of dielectric constant is described by the Curie–Weiss law

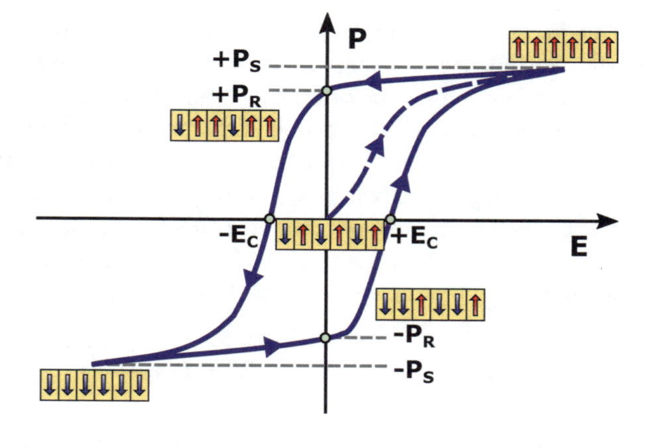

Fig. 1.4 A typical ferroelectric hysteresis loop combined with diagrams presenting the domains orientation. The detailed description is provided in the text

$$\varepsilon = \varepsilon_0 + \frac{C}{T - T_0} \quad (1.3)$$

where C means the Curie constant and T_0 is the Curie–Weiss temperature which is different from the Curie temperature ($T_0 < T_C$) for a first-order phase transition [90]. In the case of the second-order phase transition the Curie–Weiss temperature equals to the Curie temperature ($T_0 = T_C$).

1.3 Ferroelectricity at the Nanoscale

A great interest in the nanotechnology has gained the development of the new methods of ferroelectric nanomaterials fabrication [122, 123]. There approaches for synthesis of low dimensional ferroelectric nanostructures can be divided into two main groups: "bottom-up" and "top-down". The first group involves vapor phase growth, hydrothermal synthesis, solvothermal method, ultrasonic or microwave methods. The ball milling of bulk crystals and lithography the most common "top-down" technologies. The selected methods of preparation of ferroelectric chalcohalide nanomaterials will be discussed in the next chapter of this book.

The size of the material and its dimensionality are the crucial parameters that influence the ferroelectric properties at the nanoscale [124]. A ferroelectricity can be considered as a cooperative phenomenon resulting from the arrangement of charge dipoles within a crystal structure [122]. Thus, the long- and short-range alignment of electric dipoles can be significantly different comparing nanostructures and bulk crystal of the same ferroelectric compound. In Ref. [125] the Ginsburg–Devonshire theory was used to examine the size effect in thin films of $SrTiO_3$ and $Ba_xSr_{1-x}TiO_3$. It was found that the spatial correlation of the polarization and boundary conditions for the ferroelectric polarization on electrodes are two main phenomena responsible for the size effect. A dielectric constant of the thin film of $SrTiO_3$ was higher than this parameter determined for bulk single crystal of the same material. Park and coworkers investigated $PbSc_{1/2}Ta_{1/2}O_3$ nanocrystals via measurements of temperature dependence of dielectric constant, variable temperature x-ray diffraction, and differential scanning calorimetry [126]. They presented that the Curie temperature was reduced and phase transition became more diffuse in nature with a decrease of the nanoparticle size. Similarly, a drop in a phase transition was observed due to a reduction of nanoparticle size of $PbTiO_3$ [127] and $SrBi_2Ta_2O_9$ [128]. The Landau phenomenological theory of the size effect on phase transitions in ferroelectric particles was discussed in Ref. [129]. The following relation was proposed to illustrate an influence of particle size on the Curie temperature

$$T_C = T_{C0} - \frac{6D}{\delta Ad} \tag{1.4}$$

where T_C is the Curie temperature of the nanoparticle, T_{C0} denotes the Curie temperature of the bulk crystal, d is diameter of the nanoparticle, D and A are the parameters related with the correlation length, δ means the extrapolation length describing the difference between the surface and bulk [129]. The size of the nanoparticle, below which ferroelectricity disappears or it is unstable, is known as a critical size. For example, this parameter is equal to 2.6 nm, 12.6 nm, and 44 nm in the case of nanoparticles of $SrBi_2Ta_2O_9$ [128], $PbTiO_3$ [127], and $BaTiO_3$ [129], respectively. In contrast to these results, a gained ferroelectric response was reported for 10 nm ball-milled $BaTiO_3$ nanoparticles and nanocubes [130]. It was attributed to the hybrid structure of the nanoparticles which consisted of an inorganic core and an organic functional crystalline shell. Despite the many studies, described above, an explanation of the size effect in ferroelectric nanomaterials is still not clear and needs further investigation.

1.4 Engineering for Tuning of Phase Transition Temperature

The chemical composition modification and strain engineering are two main approaches that are used to tune the Curie temperature of the chalcohalide ferroelectrics. According to Ref. [131], an increase of bismuth concentration in the $Bi_xSb_{1-x}SI$ mixed crystals from 0 to 0.75 leads to a decrease of Curie temperature from 293 K to almost zero value. Similar results were presented in Ref. [81]. The fall of phase transition temperature due to substitution of antimony with bismuth seems to be comparable to the hydrostatic pressure effect reported in Ref. [132]. This suggests that a change of Sb atoms to Bi atoms in the $Bi_xSb_{1-x}SI$ crystals may act like an internal pressure which leads to a decrease of Curie temperature when value of x increases. Nitsche and others examined the ferroelectric compounds derived from SbSI by partial or total substitution of Sb by Bi or As, S by Se or O, and I by Br or Cl [79]. They observed an elevated value of the phase transition temperature (T_C = 338 K) for $SbS_{0.8}O_{0.2}I$. When SbSI is doped with chlorine, its Curie temperature is also higher in comparison to the pure SbSI [89, 133, 134]. The Curie temperature of 340 K was determined for $SbSCl_{0.2}I_{0.8}$ in Ref. [89, 134]. It was concluded in Ref. [134], that an increase in phase transition temperature originates from the lattice deformation in *a* and *b* directions. Grigas and coworkers showed that SbSI ceramic, which consisted of aligned single crystals along the ferroelectric c-axis, exhibited high value of Curie temperature $T_C = 331$ K [88]. It can be related with formation of impurities in the SbSI ceramics fabricated by the Bridgeman method. However, this effect has not been explained clearly so far.

The strain engineering of ferroelectric materials is commonly applied to enlarge the ferroelectric order, alter ferroelectric susceptibilities, and generate new modes of response [135]. The influence of strain on the Curie temperature and band structure of low-dimensional SbSI was studied in Ref. [136]. The temperature-dependent steady-state photoluminescence spectroscopy (TDPL) was used to investigate ferroelectric properties of SbSI. TDPL experiments revealed that the Curie temperature of SbSI under mechanical strain increased by over 60 K in comparison to unstrained crystal. These results were analyzed and quantified taking into consideration Landau theory and Kern formulation [136]. It should be underlined that the strain engineering is a simple and convenient method to stabilize the ferroelectric phase of SbSI above room temperature. This approach can be used in the future to construct and utilize ferroelectric devices operating at elevated temperatures.

References

1. N. Peter Papoh, Introductory chapter: chalcogen chemistry—the footprint into new materials development, in *Chalcogen Chemistry* (IntechOpen, Rijeka, 2019), p. Ch. 1
2. E.-Z.M. Ebeid, M.B. Zakaria, Thermal analysis in recycling and waste management, in *Thermal Analysis*, ed. by E.-Z.M. Ebeid, M.B. Zakaria (Elsevier, 2021), pp. 247–300
3. M.A. Busch, Halogen chemistry, in *Reference Module in Chemistry, Molecular Sciences and Chemical Engineering* (Elsevier, 2018)
4. L. Bouëssel Du Bourg, E. Furet, A. Lecomte, L. Le Pollès, S. Kohara, C.J. Benmore, E. Bychkov, D. Le Coq, Experimental and theoretical insights into the structure of tellurium chloride glasses. Inorg. Chem. **57**, 2517 (2018)
5. S. Suehara, O. Noguera, T. Aizawa, T. Sasaki, J. Lucas, Ab Initio calculation of chain structures in chalcohalide glasses. J. Non. Cryst. Solids **354**, 168 (2008)
6. W. Khan, S. Hussain, J. Minar, S. Azam, Electronic and thermoelectric properties of ternary chalcohalide semiconductors: first principles study. J. Electron. Mater. **47**, 1131 (2018)
7. A.C. Wibowo, C.D. Malliakas, H. Li, C.C. Stoumpos, D.Y. Chung, B.W. Wessels, A.J. Freeman, M.G. Kanatzidis, An unusual crystal growth method of the chalcohalide semiconductor, β-$Hg_3S_2Cl_2$: a new candidate for hard radiation detection. Cryst. Growth Des. **16**, 2678 (2016)
8. A.C. Wibowo, C.D. Malliakas, D.Y. Chung, J. Im, A.J. Freeman, M.G. Kanatzidis, Mercury bismuth chalcohalides, $Hg_3Q_2Bi_2Cl_8$ (Q = S, Se, Te): syntheses, crystal structures, band structures, and optical properties. Inorg. Chem. **52**, 2973 (2013)
9. X. Zhang, K. Liu, J.Q. He, H. Wu, Q.Z. Huang, J.H. Lin, Z.Y. Lu, F.Q. Huang, Antiperovskite chalco-halides $Ba_3(FeS_4)Cl$, $Ba_3(FeS_4)Br$, and $Ba_3(FeSe_4)Br$ with spin super-super exchange. Sci. Rep. **5**, 15910 (2015)

10. L. Wang, S.J. Hwu, A new series of chalcohalide semiconductors with composite $CdBr_2/Sb_2Se_3$ lattices: synthesis and characterization of $CdSb_2Se_3Br_2$ and indium derivatives $InSb_2S_4X$ (X = Cl and Br) and InM_2Se_4Br (M = Sb and Bi). Chem. Mater. **19**, 6212 (2007)
11. H.J. Zhao, P.F. Liu, Synthesis, crystal and electronic structure, and optical property of the pentanary chalcohalide $Ba_3KSb_4S_9Cl$. J. Solid State Chem. **232**, 37 (2015)
12. Y.J. Zheng, Y.F. Shi, C. Bin Tian, H. Lin, L.M. Wu, X.T. Wu, Q.L. Zhu, An unprecedented pentanary chalcohalide with Mn atoms in two chemical environments: unique bonding characteristics and magnetic properties. Chem. Commun. **55**, 79 (2019)
13. E.M. El-Fawal, Visible light-driven $BiOBr/Bi_2S_3$@CeMOF heterostructured hybrid with extremely efficient photocatalytic reduction performance of nitrophenols: modeling and optimization. ChemistrySelect **6**, 6904 (2021)
14. T. Li, X. Wang, Y. Yan, D.B. Mitzi, Phase stability and electronic structure of prospective Sb-based mixed sulfide and iodide 3D perovskite $(CH_3NH_3)SbSI_2$. J. Phys. Chem. Lett. **9**, 3829 (2018)
15. E. Wlaźlak et al., Heavy pnictogen chalcohalides: the synthesis, structure and properties of these rediscovered semiconductors. Chem. Commun. **54**, 12133 (2018)
16. P.I. Rentzeperis, Crystal growth and structure of chalcohalogenides and chalcogenides of the general formulae $A_m{}^{V}$ $B_n{}^{VI}$ $C_p{}^{VII}$ and A_2B_3 with A = As, Sb, Bi; B = S, Se, Te and C = Cl, Br, I. Prog. Cryst. Growth Charact. Mater. **21**, 113 (1991)
17. M. Nowak, M. Jesionek, K. Mistewicz, Fabrication techniques of group 15 ternary chalcohalide nanomaterials, in *Nanomaterials Synthesis: Design, Fabrication and Applications*, ed. by Y. Beeran Pottathara, S. Thomas, N. Kalarikkal, Y. Grohens, V. Kokol (Elsevier, 2019), pp. 337–384
18. M. Nowak, M. Jesionek, K. Mistewicz, Applications of group 15 ternary chalcohalide nanomaterials, in *Industrial Applications of Nanomaterials*, ed. by S. Thomas, Y. Grohens, N. Pottathara (Elsevier, 2019), pp. 225–282
19. K.T. Butler, S. McKechnie, P. Azarhoosh, M. Van Schilfgaarde, D.O. Scanlon, A. Walsh, Quasi-Particle Electronic Band Structure and Alignment of the V-VI-VII Semiconductors SbSI, SbSBr, and SbSeI for Solar Cells. Appl. Phys. Lett. **108**, 112103 (2016)
20. R. Nie, M. Hu, A.M. Risqi, Z. Li, S. Il Seok, Efficient and stable antimony selenoiodide solar cells, Adv. Sci. **8**, 2003172 (2021)
21. H. Shi, W. Ming, M.H. Du, Bismuth chalcohalides and oxyhalides as optoelectronic materials. Phys. Rev. B **93**, 104108 (2016)
22. S.Z.M. Murtaza, P. Vaqueiro, Rapid synthesis of chalcohalides by ball milling: preparation and characterisation of BiSI and BiSeI. J. Solid State Chem. **291**, 121625 (2020)
23. R.E. Brandt et al., Searching for "Defect-Tolerant" photovoltaic materials: combined theoretical and experimental screening. Chem. Mater. **29**, 4667 (2017)
24. R. Nie, H.S. Yun, M.J. Paik, A. Mehta, B.W. Park, Y.C. Choi, S. Il Seok, Efficient solar cells based on light-harvesting antimony sulfoiodide, Adv. Energy Mater. **8**, 1701901 (2018)
25. D.W. Davies, K.T. Butler, J.M. Skelton, C. Xie, A.R. Oganov, A. Walsh, Computer-aided design of metal chalcohalide semiconductors: from chemical composition to crystal structure. Chem. Sci. **9**, 1022 (2018)
26. B. Peng, K. Xu, H. Zhang, Z. Ning, H. Shao, G. Ni, J. Li, Y. Zhu, H. Zhu, C.M. Soukoulis, 1D SbSeI, SbSI, and SbSBr with high stability and novel properties for microelectronic, optoelectronic, and thermoelectric applications. Adv. Theory Simul. **1**, 1700005 (2018)
27. F. Palazon, Metal chalcohalides: next generation photovoltaic materials? Sol. RRL **6**, 2100829 (2022)
28. Y.C. Choi, K.W. Jung, Recent progress in fabrication of antimony/bismuth chalcohalides for lead-free solar cell applications. Nanomaterials **10**, 2284 (2020)
29. S. Farooq, T. Feeney, J.O. Mendes, V. Krishnamurthi, S. Walia, E. Della Gaspera, J. van Embden, High Gain Solution-Processed Carbon-Free BiSI Chalcohalide Thin Film Photodetectors. Adv. Funct. Mater. **31**, 2104788 (2021)

30. M.M. Frutos, M.E.P. Barthaburu, L. Fornaro, I. Aguiar, Bismuth chalcohalide-based nanocomposite for application in ionising radiation detectors. Nanotechnology **31**, 225710 (2020)
31. Y. He et al., Controlling the vapor transport crystal growth of $Hg_3Se_2I_2$ hard radiation detector using organic polymer. Cryst. Growth Des. **19**, 2074 (2019)
32. H. Xu, X. Wang, Q. Nie, Y. He, P. Zhang, T. Xu, S. Dai, X. Zhang, Glass formation and properties of Ge-Ga-Te-ZnI_2 far infrared chalcohalide glasses. J. Non. Cryst. Solids **383**, 212 (2014)
33. C. Fourmentin, X.H. Zhang, E. Lavanant, T. Pain, M. Rozé, Y. Guimond, F. Gouttefangeas, L. Calvez, IR GRIN lenses prepared by ionic exchange in chalcohalide glasses. Sci. Rep. **11**, 11081 (2021)
34. S. Simsek, H. Koc, S. Palaz, O. Oltulu, A.M. Mamedov, E. Ozbay, Band Gap and Optical Transmission in the Fibonacci Type One-Dimensional $A^5B^6C^7$ Based Photonic Crystals. Phys. Status Solidi Curr. Top. Solid State Phys. **12**, 540 (2015)
35. S. Simsek, S. Palaz, A.M. Mamedov, E. Ozbay, Fibonacci sequences quasiperiodic $A^5B^6C^7$ ferroelectric based photonic crystal: FDTD analysis. Integr. Ferroelectr. **183**, 26 (2017)
36. M. Arumugam, M.Y. Choi, Recent progress on Bismuth Oxyiodide (BiOI) photocatalyst for environmental remediation. J. Ind. Eng. Chem. **81**, 237 (2020)
37. S. Güler-KIllç, Ç. KIllç, Crystal and electronic structure of BiTeI, AuTeI, and PdTeI compounds: a dispersion-corrected density-functional study. Phys. Rev. B Condens. Matter Mater. Phys. **91**, 245204 (2015)
38. V.A. Kulbachinskii, V.G. Kytin, A.A. Kudryashov, A.N. Kuznetsov, A.V. Shevelkov, On the electronic structure and thermoelectric properties of BiTeBr and BiTeI single crystals and of BiTeI with the addition of BiI_3 and CuI. J. Solid State Chem. **193**, 154 (2012)
39. S. Li, L. Xu, X. Kong, T. Kusunose, N. Tsurumachi, Q. Feng, $Bi_{13}S_{18}X_2$-based solar cells ($X = Cl, Br, I$): photoelectric behavior and photovoltaic performance. Phys. Rev. Appl. **15**, 34040 (2021)
40. S.R. Kavanagh, C.N. Savory, D.O. Scanlon, A. Walsh, Hidden spontaneous polarisation in the chalcohalide photovoltaic absorber $Sn_2SbS_2I_3$. Mater. Horizons **8**, 2709 (2021)
41. D. Tiwari, F. Cardoso-Delgado, D. Alibhai, M. Mombrú, D.J. Fermín, Photovoltaic performance of phase-pure orthorhombic BiSI thin-films. ACS Appl. Energy Mater. **2**, 3878 (2019)
42. Y.C. Choi, E. Hwang, Controlled growth of BiSi Nanorod-based films through a two-step solution process for solar cell applications. Nanomaterials **9**, 1650 (2019)
43. Y.C. Choi, K.W. Jung, One-step solution deposition of antimony selenoiodide films via precursor engineering for lead-free solar cell applications. Nanomaterials **11**, 3206 (2021)
44. A.C. Wibowo, C.D. Malliakas, Z. Liu, J.A. Peters, M. Sebastian, D.Y. Chung, B.W. Wessels, M.G. Kanatzidis, Photoconductivity in the chalcohalide semiconductor, SbSeI: a new candidate for hard radiation detection. Inorg. Chem. **52**, 7045 (2013)
45. K. Mistewicz, M. Jesionek, M. Nowak, M. Kozioł, SbSeI Pyroelectric nanogenerator for a low temperature waste heat recovery. Nano Energy **64**, 103906 (2019)
46. S. Johnsen, Z. Liu, J.A. Peters, J.H. Song, S. Nguyen, C.D. Malliakas, H. Jin, A.J. Freeman, B.W. Wessels, M.G. Kanatzidis, Thallium chalcohalides for X-Ray and γ-Ray detection. J. Am. Chem. Soc. **133**, 10030 (2011)
47. M. Tamilselvan, A.J. Bhattacharyya, Antimony Sulphoiodide (SbSI), a narrow band-gap non-oxide ternary semiconductor with efficient photocatalytic activity. RSC Adv. **6**, 105980 (2016)
48. G. Chen, W. Li, Y. Yu, Q. Yang, Fast and low-temperature synthesis of one-dimensional (1D) single-crystalline SbSI microrod for high performance photodetector. RSC Adv. **5**, 21859 (2015)
49. K. Mistewicz et al., A simple route for manufacture of photovoltaic devices based on chalcohalide nanowires. Appl. Surf. Sci. **517**, 146138 (2020)
50. K. Mistewicz, M. Nowak, D. Stróż, A ferroelectric-photovoltaic effect in SbSI nanowires. Nanomaterials **9**, 580 (2019)
51. P. Kwolek, K. Pilarczyk, T. Tokarski, J. Mech, J. Irzmański, K. Szaciłowski, Photoelectrochemistry of N-type antimony sulfoiodide nanowires. Nanotechnology **26**, 105710 (2015)

52. Y.C. Choi, E. Hwang, D.H. Kim, Controlled growth of SbSI thin films from amorphous Sb_2S_3 for low-temperature solution processed chalcohalide solar cells. APL Mater. **6**, 121108 (2018)
53. Z. Ran, X. Wang, Y. Li, D. Yang, X.G. Zhao, K. Biswas, D.J. Singh, L. Zhang, Bismuth and antimony-based oxyhalides and chalcohalides as potential optoelectronic materials. Npj Comput. Mater. **4**, 14 (2018)
54. H. Rodot, A. Hrubý, J. Horák, Amorphous semiconducting AsSeI. Czechoslov. J. Phys. **21**, 1213 (1971)
55. C. Liu, X.J. Wang, Room temperature synthesis of $Bi_4O_5I_2$ and Bi_5O_7I Ultrathin nanosheets with a high visible light photocatalytic performance. Dalt. Trans. **45**, 7720 (2016)
56. A. Alzamly et al., Construction of BiOF/BiOI nanocomposites with tunable band gaps as efficient visible-light photocatalysts. J. Photochem. Photobiol. A Chem. **375**, 30 (2019)
57. M. Li, Y. Cui, Y. Jin, H. Li, Facile hydrolysis synthesis of $Bi_4O_5Br_2$ photocatalyst with excellent visible light photocatalytic performance for the degradation of resorcinol. RSC Adv. **6**, 47545 (2016)
58. X. Xiong, T. Zhou, X. Liu, S. Ding, J. Hu, Surfactant-mediated synthesis of single-crystalline Bi_3O_4Br nanorings with enhanced photocatalytic activity. J. Mater. Chem. A **5**, 15706 (2017)
59. B. Xu, Y. Gao, Y. Li, S. Liu, D. Lv, S. Zhao, H. Gao, G. Yang, N. Li, L. Ge, Synthesis of Bi_3O_4Cl nanosheets with oxygen vacancies: the effect of defect states on photocatalytic performance. Appl. Surf. Sci. **507**, 144806 (2020)
60. L. Xinping, H. Tao, H. Fuqiang, W. Wendeng, S. Jianlin, Photocatalytic activity of a Bi-based oxychloride Bi_3O_4Cl. J. Phys. Chem. B **110**, 24629 (2006)
61. S. Palaz, O. Oltulu, A.M. Mamedov, E. Ozbay, AVBVICVII ferroelectrics as novel materials for phononic crystals. Ferroelectrics **511**, 12 (2017)
62. M. Nowak, Photoferroelectric nanowires, in *Nanowires Science and Technology* (IntechOpen, Rijeka, 2010), p. Ch. 13
63. J. Kreisel, M. Alexe, P.A. Thomas, A photoferroelectric material is more than the sum of its parts. Nat. Mater. **11**, 260 (2012)
64. H. Li, F. Li, Z. Shen, S.T. Han, J. Chen, C. Dong, C. Chen, Y. Zhou, M. Wang, Photoferroelectric perovskite solar cells: principles advances and insights. Nano Today **37**, 101062 (2021)
65. K.T. Butler, J.M. Frost, A. Walsh, Ferroelectric materials for solar energy conversion: photoferroics revisited. Energy Environ. Sci. **8**, 838 (2015)
66. R. Kern, An electro-optical and electromechanical effect in SbSI. J. Phys. Chem. Solids **23**, 249 (1962)
67. R. Kniep, H.D. Reski, Chalcogenide iodides of arsenic. Angew. Chemie Int. Ed. English **20**, 212 (1981)
68. G.P. Voutsas, P.J. Rentzeperis, The crystal structure of the paraelectric bismuth thiochloride, BiSCl. Zeitschrift Für Krist. Cryst. Mater. **152**, 109 (1980)
69. G.P. Voutsas, P.J. Rentzeperis, Crystal structure of bismuth selenochloride, BiSeCl. Zeitschrift Für Krist. Cryst. Mater. **177**, 117 (1986)
70. E. Dönges, Über Chalkogenohalogenide Des Dreiwertigen Antimons Und Wismuts. II. Über Selenohalogenide Des Dreiwertigen Antimons Und Wismuts Und Über Antimon(III)-selenid Mit 2 Abbildungen. ZAAC J. Inorg. Gen. Chem. **263**, 280 (1950)
71. A.M. Ganose, K.T. Butler, A. Walsh, D.O. Scanlon, Relativistic electronic structure and band alignment of BiSI and BiSeI: candidate photovoltaic materials. J. Mater. Chem. A **4**, 2060 (2016)
72. A. Audzijonis, R. Sereika, L. Žigas, R. Žaltauskas, A. Kvedaravicius, Lattice dynamics of ferroelectric SbSBr crystal. Ferroelectrics **413**, 434 (2011)
73. A. Audzijonis, G. Gaigalas, L. Žigas, A. Pauliukas, B. Šalkus, R. Žaltauskas, A. Kvedaravičius, A. Čerškus, J. Narušis, Investigation of the electronic structure of the SbSeBr cluster. Cent. Eur. J. Phys. **6**, 415 (2008)
74. G.P. Voutsas, P.J. Rentzeperis, The crystal structure of antimony selenoiodide, SbSeI. Zeitschrift Für Krist. Cryst. Mater. **161**, 111 (1982)

75. A.G. Papazoglou, P.J. Rentzeperis, The crystal structure of antimony telluroiodide, SbTeI. Zeitschrift Für Krist. Cryst. Mater. **165**, 159 (1983)
76. M. Balkanski, J.Y. Prieur, A. Almeida, Ferroelectric phase transition of SbSBr. Ferroelectrics **54**, 261 (1984)
77. T. Inushima, A. Okamoto, K. Uchinokura, E. Matsuura, Observation of a phase transition in SbSBr single crystals grown by vapor transport method. J. Phys. Soc. Japan **48**, 2167 (1980)
78. E. Furman, O. Brafman, J. Makovsky, Phonons and ferroelectric phase transitions in SbSBr and SbSI and their solid solutions. Phys. Rev. B **8**, 2341 (1973)
79. R. Nitsche, H. Roetschi, P. Wild, New ferroelectric V. VI. VII compounds of the SbSI type, Appl. Phys. Lett. **4**, 210 (1964)
80. H. Akkus, A. Kazempour, H. Akbarzadeh, A.M. Mamedov, Band structure and optical properties of SbSeI: density-functional calculation. Phys. Status Solidi Basic Res. **244**, 3673 (2007)
81. K. Ishikawa, Y. Shikata, K. Toyoda, Dielectric properties of $Sb_{1-x}Bi_xSI$ crystals. Phys. Status Solidi **25**, K187 (1974)
82. S. Surthi, S. Kotru, R.K. Pandey, Preparation and electrical properties of ferroelectric SbSI films by pulsed laser deposition. J. Mater. Sci. Lett. **22**, 591 (2003)
83. K. Imai, S. Kawada, M. Ida, Anomalous pyroelectric properties of SbSi single crystals. J. Phys. Soc. Jpn **21**, 1855 (1966)
84. W.A. Smith, J.P. Doughertyt, L.E. Cross, Pyroelectricity in SbSI. Ferroelectrics **33**, 3 (1981)
85. A. Mansingh, T.S. Rao, I-V and C-V characteristics of ferroelectric SbSI(Film)-Si-metal. Ferroelectrics **50**, 263 (1983)
86. M. Yoshida, K. Yamanaka, Y. Hamakawa, Semiconducting and dielectric properties of C-Axis oriented Sbsi thin film. Jpn. J. Appl. Phys. **12**, 1699 (1973)
87. E. Fatuzzo, G. Harbeke, W.J. Merz, R. Nitsche, H. Roetschi, W. Ruppel, Ferroelectricity in SbSI. Phys. Rev. **127**, 2036 (1962)
88. J. Grigas, A. Kajokas, A. Audzijonis, L. Igas, Peculiarities and properties of SbSI Electroceramics. J. Eur. Ceram. Soc. **21**, 1337 (2001)
89. A. Audzijonis, L. Žigas, R. Sereika, R. Žaltauskas, Origin of ferroelectric phase transition in $SbSCl_xI_{1-x}$ mixed crystals. Int. J. Mod. Phys. B **28**, 1450209 (2014)
90. M. Iqbal Khan, T. Chandra Upadhyay, *General Introduction to Ferroelectrics*, in *Multifunctional Ferroelectric Materials*, ed. by D. R. Sahu (IntechOpen, Rijeka, 2021), p. Ch. 2
91. P. Szperlich, Piezoelectric $A^{15}B^{16}C^{17}$ compounds and their nanocomposites for energy harvesting and sensors: a review. Materials **14**, 6973 (2021)
92. Y. Purusothaman, N.R. Alluri, A. Chandrasekhar, S.J. Kim, Photoactive piezoelectric energy harvester driven by antimony Sulfoiodide (SbSI): A $A_VB_{VI}C_{VII}$ Class ferroelectric-semiconductor compound. Nano Energy **50**, 256 (2018)
93. B. Toroń, K. Mistewicz, M. Jesionek, M. Kozioł, D. Stróż, M. Zubko, Nanogenerator for dynamic stimuli detection and mechanical energy harvesting based on compressed SbSeI nanowires. Energy **212**, 118717 (2020)
94. K. Mistewicz, M. Jesionek, H.J. Kim, S. Hajra, M. Kozioł, Ł Chrobok, X. Wang, Nanogenerator for determination of acoustic power in ultrasonic reactors. Ultrason. Sonochem. **78**, 105718 (2021)
95. K. Mistewicz, Pyroelectric nanogenerator based on an SbSI-TiO_2 nanocomposite. Sensors **22**, 69 (2022)
96. J.Z. Xin, C.G. Fu, W.J. Shi, G.W. Li, G. Auffermann, Y.P. Qi, T.J. Zhu, X.B. Zhao, C. Felser, Synthesis and Thermoelectric properties of rashba semiconductor BiTeBr with intensive texture. Rare Met. **37**, 274 (2018)
97. H. Garot, D'un Produit Résultant de l'action Réciproque Du Sulfure d'antimoine et de l'iode. J. Pharm **10**, 511 (1824)
98. R. Nitsche, W.J. Merz, Photoconduction in ternary V-VI-VII compounds. J. Phys. Chem. Solids **13**, 154 (1960)

99. A. Kikuchi, Y. Oka, E. Sawaguchi, Crystal structure determination of SbSi. J. Phys. Soc. Japan **23**, 337 (1967)
100. S. Ueda, I. Tatsuzaki, Y. Shindo, Change in the dielectric constant of SbSI caused by illumination. Phys. Rev. Lett. **18**, 453 (1967)
101. D. Berlincourt, H. Jaffe, W.J. Merz, R. Nitsche, Piezoelectric effect in the ferroelectric range in SbSI. Appl. Phys. Lett. **4**, 61 (1964)
102. K. Hamano, T. Nakamura, Y. Ishibashi, T. Ooyane, Piezoelectric property of SbSI single crystal. J. Phys. Soc. Japan **20**, 1886 (1965)
103. J.F. Li, D. Viehland, A.S. Bhalla, L.E. Cross, Pyro-Optic studies for infrared imaging. J. Appl. Phys. **71**, 2106 (1992)
104. S. Surthi, S. Kotru, R.K. Pandey, SbSI films for ferroelectric memory applications. Integr. Ferroelectr. **48**, 263 (2002)
105. A. Mansingh, T.S. Rao, Growth and characterization of flash-evaporated ferroelectric antimony sulphoiodide thin films. J. Appl. Phys. **58**, 3530 (1985)
106. S. Narayanan, R. K. Pandey, Physical vapor deposition of antimony sulpho-iodide (SbSI) thin films and their properties, in *IEEE International Symposium on Applications of Ferroelectrics* (1994), pp. 309–311
107. C. Wang et al., SbSI nanocrystals: an excellent visible light photocatalyst with efficient generation of singlet oxygen. ACS Sustain. Chem. Eng. **6**, 12166 (2018)
108. A.V. Gomonnai, I.M. Voynarovych, A.M. Solomon, Y.M. Azhniuk, A.A. Kikineshi, V.P. Pinzenik, M. Kis-Varga, L. Daroczy, V.V. Lopushansky, X-Ray diffraction and raman scattering in SbSI nanocrystals. Mater. Res. Bull. **38**, 1767 (2003)
109. M. Tasviri, Z. Sajadi-Hezave, SbSI nanowires and CNTs encapsulated with SbSI as photocatalysts with high visible-light driven photoactivity. Mol. Catal. **436**, 174 (2017)
110. C. Wang, K. Tang, Q. Yang, B. Hai, G. Shen, C. An, W. Yu, Y. Qian, Synthesis of novel SbSI nanorods by a hydrothermal method. Inorg. Chem. Commun. **4**, 339 (2001)
111. G. Peng, H. Lu, Y. Liu, D. Fan, The construction of a single-crystalline SbSI nanorod array-WO3heterostructure photoanode for high PEC performance. Chem. Commun. **57**, 335 (2021)
112. J. Varghese, C. O'Regan, N. Deepak, R.W. Whatmore, J.D. Holmes, Surface roughness assisted growth of vertically oriented ferroelectric SbSI nanorods. Chem. Mater. **24**, 3279 (2012)
113. C. Wang et al., Nonlinear optical response of SbSI nanorods dominated with direct band gaps. J. Phys. Chem. C **125**, 15441 (2021)
114. A.K. Pathak, M.D. Prasad, S.K. Batabyal, One-dimensional SbSI crystals from Sb, S, and I mixtures in ethylene glycol for solar energy harvesting. Appl. Phys. A Mater. Sci. Process. **125**, 213 (2019)
115. S. Manoharan, D. Kesavan, P. Pazhamalai, K. Krishnamoorthy, S.J. Kim, Ultrasound irradiation mediated preparation of antimony sulfoiodide (SbSI) nanorods as a high-capacity electrode for electrochemical supercapacitors. Mater. Chem. Front. **5**, 2303 (2021)
116. O. Gladkovskaya, I. Rybina, Y.K. Gun'Ko, A. Erxleben, G.M.O'Connor, Y. Rochev, Water-based ultrasonic synthesis of SbSI nanoneedles. Mater. Lett. **160**, 113 (2015)
117. A.F. Devonshire, Theory of ferroelectrics. Adv. Phys. **3**, 85 (1954)
118. K.M. Rabe, M. Dawber, C. Lichtensteiger, C.H. Ahn, J.M. Triscone, *Modern physics of ferroelectrics: essential background, in topics in applied physics*, vol. 105 (Springer, Berlin Heidelberg, Berlin, Heidelberg, 2007), pp.1–30
119. K. Han, Q. Wang, Polymers for thin film capacitors: energy storage—Li conducting polymers, in *Polymer Science: A Comprehensive Reference*, vol. 10, ed. by K. Matyjaszewski, M. Möller (Elsevier, Amsterdam, 2012), pp.811–830
120. X. Chai, J. Jiang, Q. Zhang, X. Hou, F. Meng, J. Wang, L. Gu, D.W. Zhang, A.Q. Jiang, Nonvolatile Ferroelectric field-effect transistors. Nat. Commun. **11**, 2811 (2020)
121. P. Sharma, Q. Zhang, D. Sando, C.H. Lei, Y. Liu, J. Li, V. Nagarajan, J. Seidel, Nonvolatile ferroelectric domain wall memory. Sci. Adv. **3**, e1700512 (2017)
122. J. Varghese, R.W. Whatmore, J.D. Holmes, Ferroelectric nanoparticles, wires and tubes: synthesis, characterisation and applications. J. Mater. Chem. C **1**, 2618 (2013)

123. L. Liang, X. Kang, Y. Sang, H. Liu, One-dimensional ferroelectric nanostructures: synthesis, properties, and applications. Adv. Sci. **3**, 1500358 (2016)
124. A. Rüdiger, R. Waser, Size effects in nanoscale ferroelectrics. J. Alloys Compd. **449**, 2 (2008)
125. O.G. Vendik, S.P. Zubko, L.T. Ter-Martirosayn, Experimental evidence of the size effect in thin ferroelectric films. Appl. Phys. Lett. **73**, 37 (1998)
126. Y. Park, K.M. Knowles, K. Cho, Particle-size effect on the ferroelectric phase transition in $PbSc_{1/2}Ta_{1/2}O_3$ ceramics. J. Appl. Phys. **83**, 5702 (1998)
127. K. Ishikawa, K. Yoshikawa, N. Okada, Size effect on the ferroelectric phase transition in $PbTiO_3$ ultrafine particles. Phys. Rev. B **37**, 5852 (1988)
128. T. Yu, Z.X. Shen, W.S. Toh, J.M. Xue, J. Wang, Size effect on the ferroelectric phase transition in $SrBi_2Ta_2O_9$ nanoparticles. J. Appl. Phys. **94**, 618 (2003)
129. W.L. Zhong, Y.G. Wang, P.L. Zhang, B.D. Qu, Phenomenological study of the size effect on phase transitions in ferroelectric particles. Phys. Rev. B **50**, 698 (1994)
130. Y.A. Barnakov, I.U. Idehenre, S.A. Basun, T.A. Tyson, D.R. Evans, Uncovering the mystery of ferroelectricity in zero dimensional nanoparticles. Nanoscale Adv. **1**, 664 (2019)
131. M.K. Teng, M. Massot, M. Balkanski, S. Ziolkiewicz, Atomic substitution and ferroelectric phase transition in $Bi_xSb_{1-x}SI$. Phys. Rev. B **17**, 3695 (1978)
132. P.S. Peercy, Raman scattering near the tricritical point in SbSI. Phys. Rev. Lett. **35**, 1581 (1975)
133. B. Garbarz-Glos, Dielectric properties of SbSI-modifed in phase transition region. Ferroelectrics **292**, 137 (2003)
134. R. Sereika, R. Zaltauskas, V. Lapeika, S. Stanionytė, R. Juškenas, Structural changes in chlorine-substituted SbSI. J. Appl. Phys. **126**, 114101 (2019)
135. A.R. Damodaran, J.C. Agar, S. Pandya, Z. Chen, L. Dedon, R. Xu, B. Apgar, S. Saremi, L.W. Martin, New modalities of strain-control of ferroelectric thin films. J. Phys. Condens. Matter **28**, 263001 (2016)
136. Y. Wang, Y. Hu, Z. Chen, Y. Guo, D. Wang, E.A. Wertz, J. Shi, Effect of strain on the curie temperature and band structure of low-dimensional SbSI. Appl. Phys. Lett. **112**, 183104 (2018)

Chapter 2
The Methods of Fabrication of the Chalcohalide Nanostructures

2.1 Mechanical Milling of Bulk Crystals

Mechanical milling is one of the "top-down" approach in nanotechnology, which allows to fabricate different nanostructured materials. This method is based on a solid-state interaction between the fresh powder surfaces of the reactant materials at room temperature [1]. It does not require use of sophisticated apparatus and complicated procedures. Mechanical milling is suitable for preparation of compounds or alloys, which cannot be obtained via common melting and casting techniques. However, a problem with temperature control and stabilization is an important disadvantage of the mechanical milling. This issue is a great challenge in the case of high energy ball milling.

A fast synthesis of bismuth chalcohalides by ball milling was described in Ref. [2]. The powders of bismuth sulfoiodide (BiSI) and bismuth selenoiodide (BiSeI) were fabricated from bismuth triiodide (BiI_3), granular bismuth, sulfur, and selenium weighted in the stoichiometric ratio. Similar strategy to synthesize BiSI and $Bi_{19}S_{27}I_3$ semiconductors was shown in Ref. [3]. The elemental bismuth and sulfur were grinded to prepare bismuth sulfide (Bi_2S_3), which reacted with BiI_3 to form final products. The X-ray diffraction (XRD) of the prepared materials proved that BiSI and $Bi_{19}S_{27}I_3$ were highly crystalline and their crystallites had average sizes of 17.26 nm and 17.65 nm, respectively. The ball milling was also used to fabricate antimony sulfoiodide (SbSI) nanorods [4]. SbSI single crystals, grown by chemical transport, were inserted into a stainless steel vial equipped with a hardened steel ball. The material preparation was completed after 50 h. The average diameter of nanorods was 70 nm, what was determined XRD studies and electron microscopy. It should be concluded that the mechanical milling is a simple method of nanomaterials preparation. Therefore, it seems to be promising for future synthesis of other chalcohalide nanocrystals.

K. Mistewicz, *Low-Dimensional Chalcohalide Nanomaterials*, NanoScience and Technology, https://doi.org/10.1007/978-3-031-25136-8_2

2.2 Liquid-Phase Exfoliation

A liquid-phase exfoliation (LPE) is frequently used for fabrication of two-dimensional nanomaterials, like graphene [5], MoS_2 nanosheets [6], or Bi_2Se_3 [7]. In this method, the nanostructures of certain material are peeled off from its bulk counterpart under ultrasonic irradiation. The liquid-phase exfoliation of bismuth oxychloride (BiOCl) nanosheets was described in Ref. [8]. This process was performed for 4 h using BiOCl bulk powder dispersed in the N-methyl-2-pyrrolidone (NMP). A two-step liquid cascade centrifugation was applied to separate the BiOCl nanosheets from a sediment containing un-unexfoliated material. The average length of nanosheets was equal to 380 nm, whereas mean thickness was not determined. The BiOCl nanosheets were also produced via a gas-phase exfoliation [9]. Yu and coworkers presented a large-scale liquid-phase exfoliation of monolayered bismuth oxybromide (BiOBr) nanosheets with a thickness of approximately 0.85 nm [10]. The bulk BiOBr, obtained by a hydrolysis method, was added to the formamide and subjected to the ultrasounds for 6 h. The nanosheets of BiOBr were found to be suitable for application as efficient photocatalysts due to optimized band structure, high specific surface area, abundant active sites, and enhanced charge separation [10]. According to Ref. [11], two-dimensional nanoplates of bismuth tellurochloride (BiTeCl) can be mechanically exfoliated from bulk source material. The nanoplates thickness of 7.4 nm was confirmed by applying atomic force microscopy (AFM). The liquid exfoliation of the SbSI bulk crystals into the nanorods was reported in Ref. [12]. After 10 h of sonication, the processed suspension was centrifuged in order to isolate SbSI nanocrystals from un-unexfoliated residuals. Nanorods diameter of about 200 nm and length of a few micrometers were determined using a scanning electron microscopy (SEM). The atomic ratio of 1.3:1:1 for antimony, sulfur, and iodine was derived from an energy-dispersive spectrometry [12], what is close to the theoretical chemical composition of SbSI.

2.3 Vapor Phase Growth

Vapor phase synthesis belongs to the "bottom-up" methods of fabrication of various nanomaterials [13]. This method was applied by Peng and coworkers to prepare BiOCl nanostructures on different substrates [14]. The elemental bismuth together with moisturized and hydrolyzed $AuCl_3$ were inserted into a horizontal furnace equipped with quartz tube and heated at the temperature of 523 K for 1 h. A pure nitrogen was utilized as a carrier gas in this experiment. A change in evaporation source and temperature of the substrate resulted in modification of the morphology of the final product. The nanobelts, nanowires, nanoflakes of BiOCl were grown on glass slide, bare Si substrate, and Si substrate coated with Bi/Au film, respectively. The conversion of bismuth telluride (Bi_2Te_3) into ultrathin BiTeCl and BiTeBr at high temperatures (463–473 K) in the presence of $BiCl_3$/$BiBr_3$ vapors was proposed in Ref. [15]. Another approach in the preparation of BiTeBr is a synthesis from elemental bismuth, tellurium, and bismuth tribromide [16]. In a typical fabrication procedure, the aforementioned reagents were heated at 873 K for 10 h. Then, temperature was reduced to 673 K and maintained for 7 days.

Thin films of SbSI with different morphologies can be prepared by the vacuum thermal evaporation of SbSI single crystals [17], the conversion of Sb_2S_3 using SbI_3 vapor [18], and vapor phase deposition on anodic aluminum oxide (AAO) substrates [19]. When SbSI single crystal was evaporated through resistance heating of a tungsten filament in a vacuum, the continuous film of this compound was achieved [17]. However, the further annealing of the SbSI layers was performed at temperatures from 373 to 473 K to improve crystallinity of the material. XRD survey showed that the obtained films were predominantly oriented along *c*-axis. Similarly, the SbSI nanorods aligned parallel to the *c*-axis were grown via vapor phase deposition of Sb_2S_3 and SbI_3 mixture on AAO/Ti/Si substrate [19]. It was performed using a quartz tube which was inserted into the two-zone tube furnace. The temperature of the source and substrate zone were elevated accordingly up to 673 K and 523 K. The piezoresponse force microscopy (PFM) of an individual SbSI nanorod revealed the existence of the oppositely oriented ferroelectric domains what confirmed a switchable ferroelectricity and piezoelectricity in the prepared material. The thin film of SbSI microneedles was formed through reaction between the SbI_3 vapor and amorphous Sb_2S_3 [18]

$$Sb_2S_3 + SbI_3 \rightarrow 3SbSI. \quad (2.1)$$

The process, described by Eq. (2.1), was accomplished in an inert atmosphere to avoid an oxidation of the final product. The target, coated with the antimony triiodide, was heated at 523 K for 5 min. In result, the SbI_3 sublimed and reacted with the Sb_2S_3. The synthesized material was additionally heated for 10 min to enhance its crystallinity. Afterwards, it was cooled down and washed with ethanol in order to obtain pure antimony sulfoiodide without residual SbI_3. SEM investigations of the prepared material proved needle like morphology of SbSI microcrystals which widths and lengths varied in the ranges from 10 to 100 nm and from 1 μm to 5 μm, respectively. The films of SbSI microcrystals were recognized as promising for application in the self-powered photodetectors with high signal-to-noise ratio and a short response time [18].

2.4 Hydrothermal and Solvothermal Methods

Hydrothermal and solvothermal processes are performed in aqueous or organic solutions at elevated temperatures over 373 K and pressures higher than 1 atm [20]. The hydrothermal method refers to synthesis in which water is used as a solvent, whereas solvothermal process is based on the application of non-aqueous solution. The hydro/solvothermal methods were frequently applied to fabricate low-dimensional chalcohalides of bismuth [21–23], antimony [24–26], and their composites [27–29]. The first reports on synthesis of the bulk monocrystals of pnictogen chalcohalides were published in the early 70s of the twentieth century [30]. It should be noted that certain bismuth chalcohalides can be naturally produced in the environment. For instance, BiSI, known also as demicheleite-(I) mineral, was found in La Fossa Crater, Vulcano in Italy [31].

The fabrication method of pure bismuth oxyiodide (BiOI) as well as a composite of BiOI and activated carbon (BiOI/C) was shown in Ref. [32]. A standard procedure of synthesis of the BiOI/C composite is depicted in Fig. 2.1. At first stage, the suspension of the activated carbon, ethylene glycol (EG), and bismuth nitrate pentahydrate ($Bi(NO_3)_3{\cdot}5H_2O$) was prepared and stirred at temperature of 303 K for 15 min. Then, the solution of potassium iodide (KI) and glycol was added to the vessel with the reagents and agitated for 30 min to homogenize the mixture. Afterwards, the reagents were inserted into the Teflon-lined stainless steel autoclave and processed for 12 h at temperature of 413 K. After the reaction was completed, the product was centrifuged, washed, and dried at 333 K. The fabricated material exhibited a heterojunction hierarchical structure, since BiOI nanosheets tightly grew on the carbon surface, as presented in Fig. 2.1. The BiOI/C composite was recognized as an efficient photocatalyst for rhodamine B degradation due to its large specific surface area, high quantum efficiency, short diffusion pathways of the pollutant, and sufficient interfacial interaction sites [32].

Pancielejko and associates developed novel two-step fabrication of $BiOBr/Bi_2WO_6$ heterojunction which involved an anodic oxidation of tungsten foil and hydrothermal processing of as-anodized oxide in ionic liquid containing bismuth precursor [33]. The N-butylpyridinium bromide [BPy][Br] and bismuth nitrate pentahydrate were used as ionic liquid and bismuth precursor, accordingly. Different mixtures of reagents were prepared in order to examine an influence of molar ratio of ionic liquid to bismuth precursor (*r*) on the properties of the $BiOBr/Bi_2WO_6$ heterojunction. The nanostructures with various morphologies and chemical compositions were obtained (Fig. 2.2). The determined crystallite sizes were 27.9 nm and 9.4 nm for pristine BiOBr (Fig. 2.2a) and Bi_2WO_6 (Fig. 2.2b), respectively. In the case of $BiOBr/Bi_2WO_6$ heterojunctions, the BiOBr nanoplates

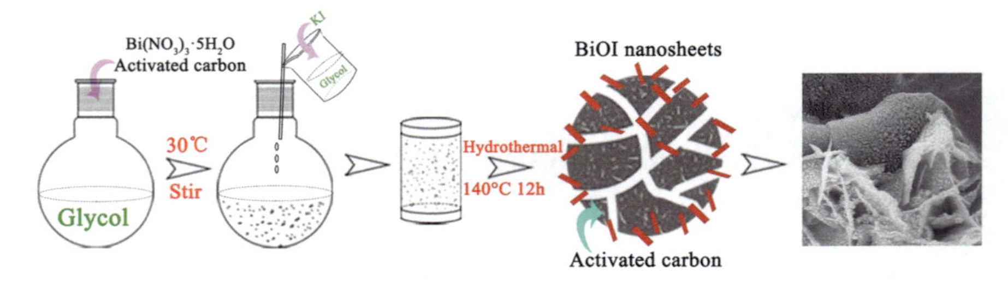

Fig. 2.1 A scheme presenting synthesis of BiOI/activated carbon micro/nanocomposites. Detailed description is provided in the text. Reprinted from Hou et al. [32] under the terms of the Creative Commons Attribution 4.0 International License (CC BY 4.0). Copyright (2017) Springer Nature

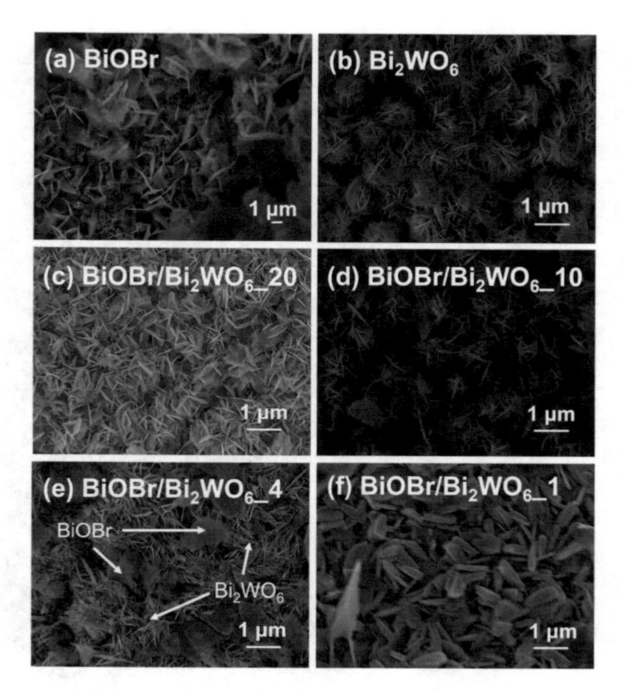

Fig. 2.2 The SEM micrographs of pristine **a** BiOBr, **b** Bi_2WO_6, and their composites: **c** $BiOBr/Bi_2WO_6$ (r = 1:20), **d** $BiOBr/Bi_2WO_6$ (r = 1:10), **e** $BiOBr/Bi_2WO_6$ (r = 1:4), **f** $BiOBr/Bi_2WO_6$ (r = 1:1), where r refers to the molar ratio of the used ionic liquid to bismuth precursor ($Bi(NO_3)_3{\cdot}5H_2O$). Reprinted from Pancielejko et al. [33] under the terms of the Creative Commons Attribution 4.0 International License (CC BY 4.0). Copyright (2021) Elsevier

were grown on the surface of flower-like Bi_2WO_6, as presented in Figs. 2.2c–f. Average crystallite sizes of the Bi_2WO_6 and BiOBr were dependent on the molar ratio of ionic liquid to Bi precursor and varied in the ranges from 8.9 nm to 32.4 nm and from 6.6 nm to 28.9 nm, respectively.

The pristine Bi_2WO_6 nanoparticles (Fig. 2.3a–c), pure BiOCl nanosheets (Fig. 2.3d–f), and the heterojunction of these materials (Fig. 2.3g, h) were fabricated using the hydrothermal and solvothermal methods [34]. In order to construct the Bi_2WO_6-BiOCl composite, the aqueous solution of $Bi(NO_3)_3{\cdot}5H_2O$, KCl, NaOH, and Bi_2WO_6 was prepared, stirred, and transferred into the Teflon-lined stainless steel autoclave, where the reagent were heated at 433 K for 24 h. The fabricated materials were examined using different techniques, including a high resolution transmission electron microscopy (HRTEM) and selected-area electron diffraction (SAED). The dimensions of pristine Bi_2WO_6 nanoparticles were in the range from 5 nm to10 nm, whereas BiOCl nanosheets sizes were between 0.5 μm and 1 μm. TEM and HRTEM studies revealed that Bi_2WO_6 nanoparticles were grown on the surface of BiOCl nanosheets. This confirmed successful formation of Bi_2WO_6-BiOCl heterojunctions which played a key role for favoring the charge separation to achieve strong photocatalytic activity.

Zhao and others fabricated microspheres of pristine BiOCl and Fe^{3+}-grafted BiOCl (Fe/BiOCl) using the hydrothermal method [35]. In the case of BiOCl preparation, the mixture of glycerol, $Bi(NO_3)_3{\cdot}5H_2O$, KCl, and deionized water was homogenized and inserted into the Teflon-lined stainless-steel autoclave. In the next step, the

Fig. 2.3 The TEM and HRTEM images of (**a**, **c**) Bi_2WO_6 nanoparticles, (**d**, **f**) BiOCl nanosheets, (**g**, **h**) Bi_2WO_6–BiOCl nanocomposites, the SAED patterns of (**b**) Bi_2WO_6 and (**e**) BiOCl. Reprinted from Guo et al. [34] under the terms of the Creative Commons Attribution 4.0 International License (CC BY 4.0). Copyright (2020) Springer Nature

reagents were heated at 383 K for 8 h. The obtained material was further calcined at 673 K. The hybrid Fe/BiOCl was produced according the same procedure, excepting an addition of $Fe(NO_3)_3{\cdot}9H_2O$ to the mixture of reagents. It was found in Ref. [35] that BiOCl exhibited a microsphere-like structure (Fig. 2.4a) with a diameter reaching up to 2 μm. The microspheres consisted of the nanoplates (Fig. 2.4b) with average width and thickness of approximately 70 nm and 20 nm, respectively. The diameter of Fe/BiOCl microspheres (Fig. 2.4c) was reduced to 1 μm due to Fe^{3+} grafting. Similarly, a thickness of nanosheets, assembled into the Fe/BiOCl microspheres (Fig. 2.4d), was decreased to 15 nm.

The morphology and physical properties of a material, fabricated via hydro/solvothermal synthesis, depend strongly on the temperature of the process and used reagents. Usually, antimony sulfoiodide is grown hydrothermally from antimony trichloride ($SbCl_3$), thiourea ($(NH_2)_2CS$), and iodine (I_2) or ammonium iodide (NH_4I). A formation of SbSI can be described by the following equation [26]

Fig. 2.4 The SEM micrographs of (**a**, **b**) BiOCl microspheres and (**c**, **d**) Fe/BiOCl microstructures. Reprinted from Zhao et al. [35] under the terms of the Creative Commons Attribution 4.0 International License (CC BY 4.0). Copyright (2018) Springer Nature

$$SbCl_3 + (NH_2)_2CS + \frac{1}{2}I_2 + \frac{5}{2}H_2O \rightarrow SbSI + \frac{1}{4}O_2 + CO_2 + 2NH_4Cl + HCl. \tag{2.2}$$

When the stoichiometric amounts of the reagents were utilized, the SbSI microrods were grown [24, 25]. An excessive iodine may support formation of SbSI nanorods [26]. A detailed comparison of hydro/solvothermal routes of preparation of various chalcohalide nanomaterials is presented in Table 2.1.

2.5 Synthesis Under Ultrasonic Irradiation

A sonochemistry is based on the application of the ultrasound to support the chemical reactions without the need of high temperature or pressure treatment. The chemical and physical effects of ultrasound result from an acoustic cavitation which involves a creation, growth, and implosive collapse of gas bubbles in the liquid. The cycles of compression and rarefaction pressure are created in the liquid medium, when an ultrasonic wave propagates through it. If negative pressure is applied to a liquid, the

Table 2.1 The comparison of different chalcohalide nanomaterials prepared using hydrothermal and solvothermal methods (used symbols and abbreviations: g-C_3N_4—graphitic carbon nitride; rGO—reduced graphene oxide; t_s—time of the hydro/solvothermal treatment; T_s—temperature of the hydro/solvothermal synthesis)

Material	Method	Morphology of the material	T_s, K	t_s, h	References
BiOBr	Solvothermal	Microspheres composed of interlaced nanosheets	413	8	[29]
	Solvothermal	Microflowers	433	12	[36]
	Hydrothermal	Nano-roundels	443	12	[37]
	Solvothermal	Nanosheets	454	5	[38]
	Hydrothermal	Microspheres containing self-assembled nanosheets	453	10	[22]
BiOCl	Hydrothermal	Nanoplates assembled into the microspheres	383	8	[35]
	Hydrothermal	Hierarchical microstructure composed from nanoplates	393	3	[39]
	Hydrothermal	Single-cystalline nanosheets	433		[23]
	Solvothermal	Nano-micro-spheres	433	12	[40]
BiOI	Hydrothermal	Hierarchical structure constructed by nanoplates	433	12	[41]
	Solvothermal	Flower-like microspheres composed of nanosheets	413	24	[42]
BiSI	Hydrothermal	Nanoparticles	453	10	[27]
	Solvothermal	Nanorods	453	20	[43]
	Solvothermal	Nanorods	453	20	[44]
	Hydrothermal	Nanorods	473	24	[45]

(continued)

Table 2.1 (continued)

Material	Method	Morphology of the material	T_s, K	t_s, h	References
$Bi_{13}S_{18}I_2$	Solvothermal	Nanorods		12	[21]
BiTeI	Hydrothermal	Submicrometer hollow spheres	463–473	10	[46]
SbSI	Hydrothermal	Microrods	433	4	[24]
	Hydrothermal	Microrods	473	24	[25]
	Hydrothermal	Nanorods	453–463	8–10	[26]
BiOBr/$BiFeWO_6$ heterojunction	Hydrothermal	BiOBr nanosheets decorated with $BiFeWO_6$ nanoparticles	453	12	[47]
BiOBr/Bi_2WO_6 heterojunction	Hydrothermal	BiOBr nanoplates on the surface of flower-like Bi_2WO_6	433	24	[33]
BiOBr/Bi_2S_3 heterostructure	Solvothermal	BiOBr nanosheets and Bi_2S_3 nanorods gathered into urchin-like spheres	393	12	[48]
$Bi_2O_2CO_3$-BiOI heterostructure	Hydrothermal	BiOI nanosheets grown on $Bi_2O_2CO_3$ rods			[49]
BiOI/C heterostructure	Solvothermal	Nanosheets organized into the micro-hierarchical structure	413	12	[32]
BiSI-rGO composite	Hydrothermal	Nanoflower structure composed of nanoneedles	433	15	[50]
Bi_2WO_6-BiOCl heterostructure	Hydrothermal and solvothermal	Heterojunction of Bi_2WO_6 nanoparticles and BiOCl nanosheets	433	24	[34]
Bi_2WO_6-BiSI heterojunction	Hydrothermal	Nanoparticles aggregated in flakes	453	10	[27]
$BiVO_4$/BiOF heterojunction	Solvothermal	Nanoparticles grown on the surface of larger particles	453	12	[51]

(continued)

Table 2.1 (continued)

Material	Method	Morphology of the material	T_s, K	t_s, h	References
carbon spheres – $BiOI/BiOIO_3$ heterojunction	Hydrothermal	Nanoplates and nanoparticles	423	6	[52]
g-C_3N_4/BiOBr composite	Hydrothermal	Flower-like structure self-assembled from nanosheets	433	12	[53]
S-doped BiOBr	Solvothermal	Microspheres composed of nanosheets	433	10	[54]

Table 2.2 A summary of different one-dimensional chalcohalide nanomaterials prepared using sonochemical method (used symbols: d_A—the average diameter of the one-dimensional nanostructure; t_s—duration time of the ultrasonic treatment; T_s—temperature of the sonochemical synthesis)

Material	Material morphology	Range of diameters, nm	d_A, nm	Solvent	T_s, K	t_s, min	References
SbSI	Nanowires	20–50		Water	317		[58]
	Nanowires		30	Isopropanol	333	360	[59]
	Nanowires	10–50		Methanol	338	80	[60]
	Nanowires	10–50		Ethanol			[61]
	Nanowires	40–100		Ethanol	323	120	[62]
	Nanowires	20–180	69	Ethanol	323	120	[63]
	Nanorods	500–1000		Ethanol		60	[64]
	Nanowires			Ethanol	323	120	This work
	Nanorods	200–500		Ethylene glycol			[65]
	Nanoneedles			Water	323	45	[66]
SbSeI	Nanowires			Ethanol	323	120	[67]
	Nanowires	20–50		Ethanol	323	120	[68]
	Nanowires	10–50	41	Ethanol	320	120	[69]
$SbS_{1-x}Se_xI$	Nanowires	10–50		Ethanol	323	120	[70]
SbSI-TiO_2 composite	Nanowires	10–220	68	Ethanol	323	240	[71]

gas micro-bubbles are formed. They collapse when reach the critical sizes. According to Ref. [55], the large temperature of approximately 5000 K, pressure of order of 1000 atm, and extremely huge heating or cooling rates of 10^{10} K/s can be generated due to bubble collapse in the acoustic cavitation. These exceptional conditions favor growth of nanomaterials [56, 57], including chalcohalide nanostructures (Table 2.2).

As presented in Table 2.2 the sonochemical method is commonly used for fabrication of SbSI, its derivatives, and SbSeI which is isostructural to the SbSI. These compounds can be synthesized under ultrasonic irradiation at the mild conditions, i.e. a relatively low temperature (317–338 K) and a normal pressure (1 atm). It is a significant advantage in comparison to the hydrothermal or vapor-phase growth which require heating the reagents at a high temperature (e.g.: 473 K [25], 673 K [19]). In addition, the time of ultrasonic processing (45–360 min) is much shorter than the duration of hydrothermal synthesis (24 h [25]) or ball milling of the bulk crystals (50 h [4]). Furthermore, the sonochemical method is a facile, cheap, and allows to fabricate substantial amount of the material in one step.

A change of the color and consistency of the material during a sonochemical synthesis of the SbSI nanowires is presented in Fig. 2.5a–d. A growth of the SbSI nanowires led to a change of the diffuse reflectance spectrum (Fig. 2.5e). The absorption edge became more sharp with an increase of time of ultrasonic irradiation. The

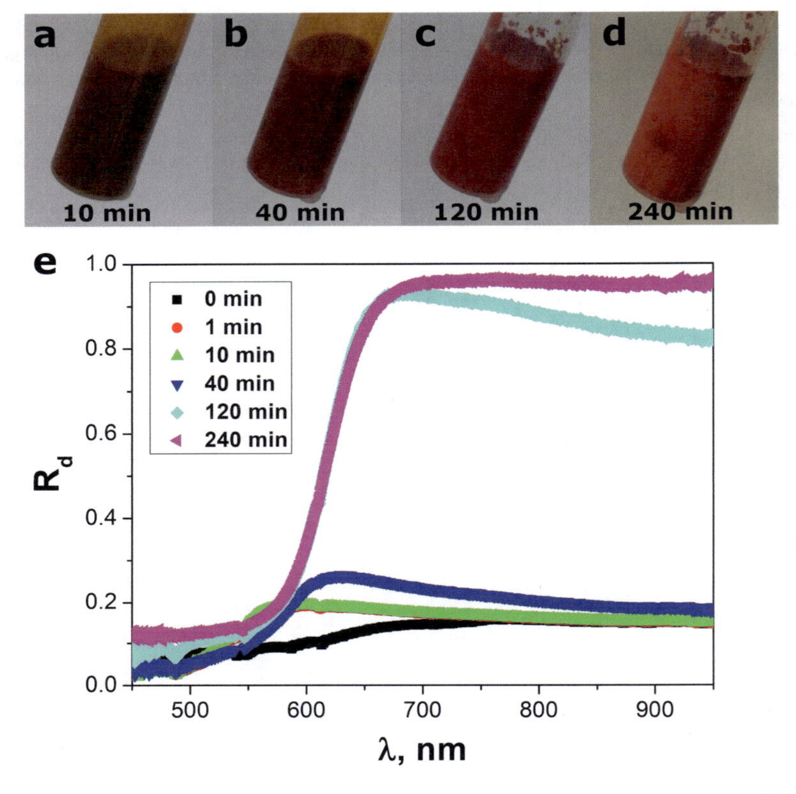

Fig. 2.5 Photographs of synthesis of the SbSI gel after **a** 10 min, **b** 40 min, **c** 120 min, **d** 240 min of ultrasonic irradiation and **e** diffuse reflectance spectra recorded during material preparation

positions of the absorption edges observed after 120 min and 240 min were almost the same suggesting that the sonochemical synthesis was completed within 120 min.

Antimony sulfoiodide grows into the one-dimensional nanostructures, such as nanorods, nanoneedles, and nanowires (Fig. 2.6). If the elements (Sb, S, and I_2) are used as the reagents, a mechanism of ultrasonic synthesis of SbSI in ethanol can be described as follows. At first, antimony triiodide is formed when antimony reacts with iodine dissolved in ethanol.

$$2Sb + 3I_2 \rightarrow 2SbI_3. \tag{2.3}$$

The ethanol is decomposed within the collapsing bubbles [72]

$$C_2H_5OH \leftrightarrow C_2H_4 + H_2O. \tag{2.4}$$

Ultrasonic irradiation of water results in formation of highly reactive H^* and OH^* radicals [56, 57]

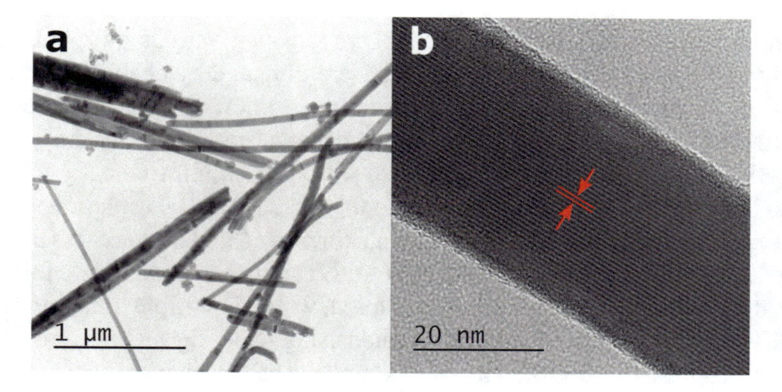

Fig. 2.6 The typical **a** TEM micrograph of multiple agglomerated SbSI nanowires fabricated using sonochemical method and **b** HRTEM image of an individual SbSI nanowire. The arrows in (**b**) indicate the interplanar spacing of 0.6527(18) nm for (110) planes in SbSI. Reprinted from Mistewicz et al. [62] under the terms of the Creative Commons Attribution 4.0 International License (CC BY 4.0). Copyright (2020) MDPI

$$H_2O \rightarrow H^* + OH^*. \tag{2.5}$$

In the next step, sulfur reacts with H^* radicals leading to creation of hydrogen sulfide

$$S + 2H^* \rightarrow H_2S. \tag{2.6}$$

The SbSI is formed as a product of reaction between the hydrogen sulfide and antimony triiodide

$$H_2S + SbI_3 \rightarrow SbSI + H_2 + I_2. \tag{2.7}$$

The SbSI molecules connect with each other and self-assemble to form double $[(SbSI)_\infty]_2$ chains that are connected with weak Van der Waals interactions. Finally, the SbSI grows along the *c*-axis into the one-dimensional nanowire, as shown in Fig. 2.6.

2.6 Microwave Synthesis

Until now, the microwaves have been employed for synthesis of a few bismuth chalcohalides: BiOCl [73], BiOBr [73], $Bi_{19}S_{27}Br_3$ [74], and BiOBr-reduced graphene oxide (rGO) composite [75]. It was presented in Ref. [73] that the microwave irradiation can be used for growth of BiOX nanostructures in mannitol solution, where X refers to different elements: chlorine, bromine, or iodine. The BiOCl, BiOBr, and BiOI were fabricated when NaCl, KBr, and KI were added to the mannitol solution as halogen precursors, respectively. Moreover, various morphologies and sizes of the prepared BiOX nanostructures were obtained by changing the concentrations of the mannitol and halides). The simple microwave-assisted method of fabrication of BiOBr/rGO nanocomposite is depicted in Fig. 2.7. Firstly, the solution of $Bi(NO_3)_3{\cdot}5H_2O$ and ethylene glycol was agitated ultrasonically for 15 min. The graphene oxide and KBr were added to the $Bi(NO_3)_3{\cdot}5H_2O$ solution which was subsequently stirred mechanically for 30 min. Such prepared mixture was transferred into the Teflon-lined vessel and subjected to the microwaves at 363 K for 15 min. Finally, the product of the synthesis was filtered, washed multiple times, and dried at 353 K. The prepared material was investigated using scanning and transmission electron microscopies. The BiOBr/rGO nanocomposite had a spherical structure formed by the assembly of numerous nanosheets. It was found in Ref. [75] that incorporation of reduced graphene oxide into bismuth oxybromide prevented the agglomeration of the nanoparticles. The BiOBr/rGO nanocomposite exhibited high photocatalytic activity toward removal of methylene blue from the aqueous solutions.

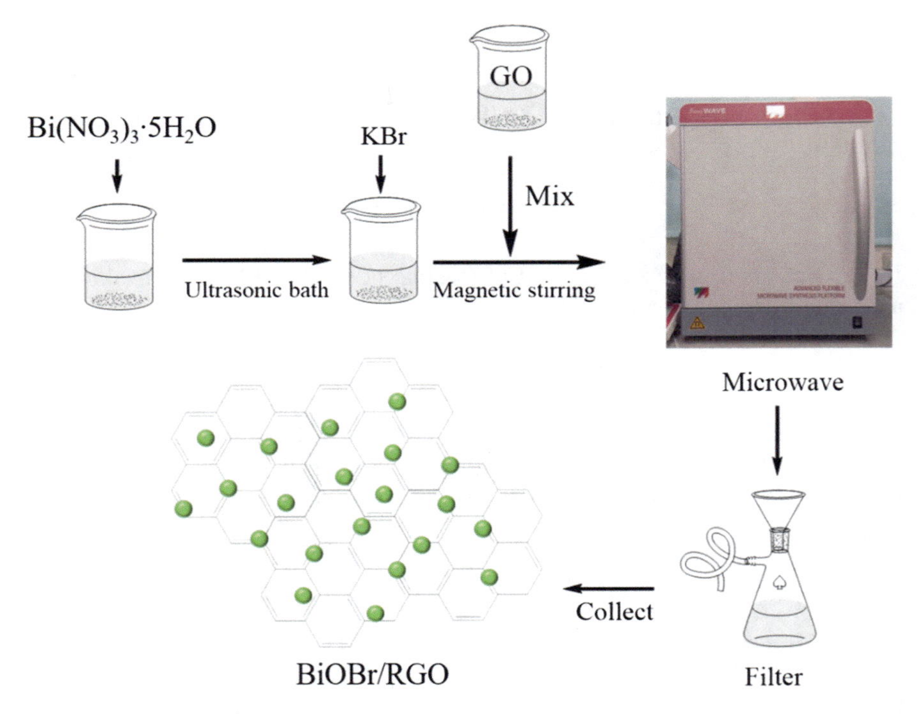

Fig. 2.7 A scheme presenting microwave-assisted preparation of BiOBr/rGO nanocomposite. Detailed description is provided in the text. Reprinted from Shih et al. [75] under the terms of the Creative Commons Attribution 4.0 International License (CC BY 4.0). Copyright (2021) MDPI

2.7 Laser/Heat-Induced Crystallization

The formation of crystalline inclusions in the halcogenide glasses can be achieved under heat [76, 77] and laser treatment [78–81]. The ferroelectric nanostructures of SbSI were fabricated by vacuum melting of SbSI and As_2S_3 which were homogenized at temperature of 800–870 K for 24 h [77]. The heat treatment conditions influenced the size of the crystalline inclusions in the $(As_2S_3)_{100-x}(SbSI)_x$ glass matrix, where x varied from 50 to 100. The size effect was found to be responsible for determining the dielectric properties of prepared glasses. The vacuum melting method was also used for fabrication of the $(As_2S_3)_{100-x}(SbSI)_x$ and $(As_2Se_3)_{100-x}(SbSI)_x$ glassy alloys, for which x was in the range from 0 to 40 [76].

Savytskii and coworkers presented a modification of SbSI glass using 488 nm radiation emitted by the argon laser [78]. When the SbSI glass was subjected to the laser radiation with the power density lower than 200 W/mm^2, laser-induced expansion of the material occurred. The evaporation of SbI_3 and strong surface erosion were observed for the power density of the laser radiation higher than 400 W/mm^2. The optimal value of the laser beam power density for surface crystallization of the SbSI was in the range from 250 W/mm^2 to 320 W/mm^2. The crystallization of SbSI from the amorphous $(As_2S_3)_{0.45}(SbSI)_{0.55}$ layers was accomplished using 647.1 nm krypton laser [80]. This effect was noticed when the power density of the laser radiation was higher than the threshold value which was dependent on the thermal history of the $(As_2S_3)_{0.45}(SbSI)_{0.55}$ film.

2.8 Electrospinning

The electrospinning method has been applied so far for preparation of the nanofibers containing different bismuth chalcohalide nanomaterials (Table 2.3). The bismuth oxyiodide nanofibers were fabricated by using polyacrylonitrile (PAN) as the supporting polymer [82]. Further annealing of this material at 773 K for 5 h resulted in a decomposition of PAN and a conversion of the nanofibers into the plate-like nano/microstructures. The two-step process was applied to produce TiO_2/BiOI heterostructured nanofibers [83]. This method involved electrospinning of the TiO_2 nanofibers and solvothermal synthesis of the TiO_2/BiOI heterostructures carried out in the Teflon-lined autoclave at 433 K for 18 h. The BiOI nanosheets with an average lateral dimension of about 300 nm were grown on the surfaces of the TiO_2 nanofibers which diameters and lengths were in the ranges from 400 to 550 nm and 15 to 45 μm, respectively. Similarly, two-step method [84] was developed to prepare hierarchical nanostructures of BiOBr/AgBr on carbon nanofibers (CNFs). At first, the nanofibers of BiOBr/AgBr composite were fabricated via electrospinning from the solution composed of $Bi(NO_3)_3 \cdot 5H_2O$, KBr, $AgNO_3$, and PAN dissolved in N, N′-dimethylformamide (DMF). In the second stage of material preparation, the composite was carbonized at high temperatures of 423 K, 773 K, and 1073 K for 1 h,

Table 2.3 The bismuth chalcohalide nanomaterials fabricated via electrospinning (used symbols and abbreviations: CNF—carbon nanofiber; PAN—polyacrylonitrile; PVP—polyvinylpyrrolidone; *U*—the applied voltage between the collector and a tip of the needle)

Material	Material morphology	Used polymer	*U*, kV	Flow rate, mm/h	References
BiOI	nanofibers mat	PAN	20		[82]
	array of nanofibers incorporated with the nanoflakes	chitosan	15		[85]
TiO_2/BiOI heterostructure	nanofibers	PVP	25	5.4	[83]
BiOBr/AgBr/CNF composite	BiOBr/AgBr hierarchical nanostructures on CNFs	PAN	15	0.12	[84]

2 h, and 4 h, accordingly. The BiOBr/AgBr/CNF composite was further examined as a promising photocatalyst for rhodamine-B and p-nitrophenol decomposition under the visible-light irradiation.

2.9 Successive Ionic Layer Adsorption and Reaction

The successive ionic layer adsorption and reaction (SILAR) is recognized as a method suitable for preparation of homogeneous and large area thin films. It is based on immersion of the substrate into separately placed cations and anions [86]. The main advantages of this technique are simplicity and repeatability. The SILAR was utilized to fabricate the nanostructured films of bismuth chalcohalides (BiOI [87–90], $BiOBr_{1-x}I_x$ [91]) and the Fe_2O_3/BiOI composite [92].

A deposition of bismuth oxyiodide layers on the fluorine doped tin oxide (FTO) substrates was accomplished through SILAR method [88]. Two various precursors were used as the cation and anion sources. The deposition process was performed for following concentrations of the $Bi(NO_3)_3{\cdot}5H_2O$ and KI: 2 mM, 5 mM, 6 mM, 7 mM, 8 mM, and 10 mM. The withdrawal speed was adjusted to 0.2 mm/s. The number of the cycles was 30. Fabricated BiOI films were at 373 K for 1 h. According to [88], the precursors concentration influenced strongly the morphology of obtained material (Fig. 2.8). The BiOI nanoparticles were grown at low concentration of precursors (5 mM). The wider BiOI flakes were observed for increased precursors concentration (6 mM). When the precursors concentration was higher than 7 mM, the rod-like structures were formed. They consisted of numerous BiOI flakes, as shown in Fig. 2.8). It was concluded in Ref. [88] that increase in the precursor concentration results in rise of the reaction probability between anion and cation. It leads to more

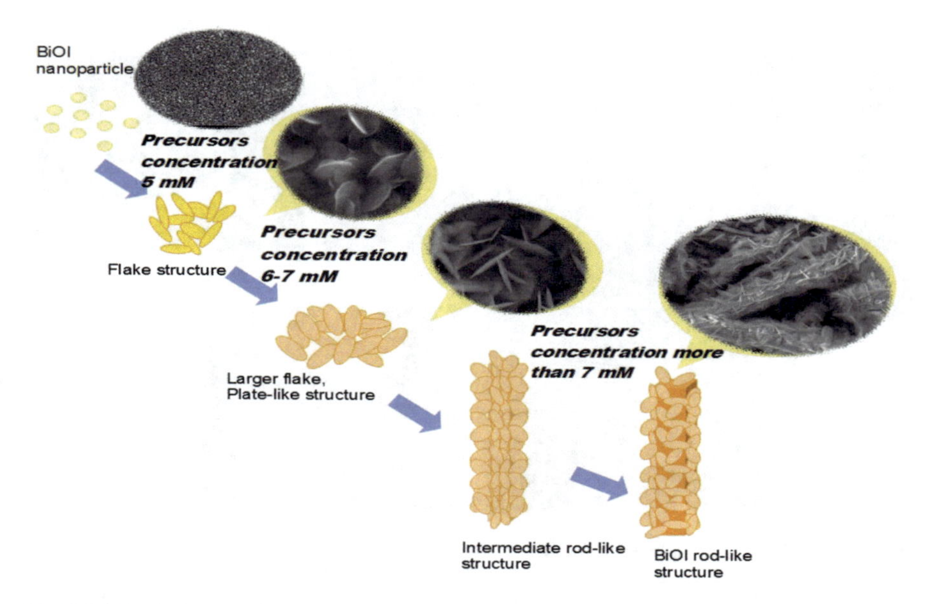

Fig. 2.8 The influence of precursors concentration on the morphology of BiOI fabricated using SILAR method. Reprinted from Putri et al. [88] under the terms of the Creative Commons Attribution 4.0 International License (CC BY 4.0). Copyright (2019) Elsevier

collisions between reactants in the concentrated solution, what supports the formation of BiOI flakes with higher sizes.

Abdul-Manaf and coworkers [90] presented similar approach in BiOI films preparation to this described above. They applied the SILAR technique under the same molar ratio of $Bi(NO_3)_3 \cdot 5H_2O$ and KI. In the first stage of the deposition procedure, the glass or FTO substrate was immersed into the $Bi(NO_3)_3 \cdot 5H_2O$ aqueous solution for 10 s. Then, it was inserted into the KI solution for 20 s. Afterwards, it was put into the deionized water, rinsed, and dried. This cycle was repeated 30 times. In the last step, the deposited BiOI films were annealed at different temperatures. The as-deposited BiOI film was composed of BiOI flakes agglomerated in the flower-like structures (Fig. 2.9a). The size of the BiOI flakes was enhanced from approximately 0.8 μm (Fig. 2.9b) to about 3.2 μm due to annealing at 623 K (Fig. 2.9c). The thermal treatment of the material at higher temperatures (Fig. 2.9d, e) resulted in many disadvantageous effects, such as loss of material through sublimation and formation of a polycrystalline structure with mixed phases [90]. The annealing of the BiOI films at the temperature of 623 K was found to be optimal for their future applications in the lead-free solar cells.

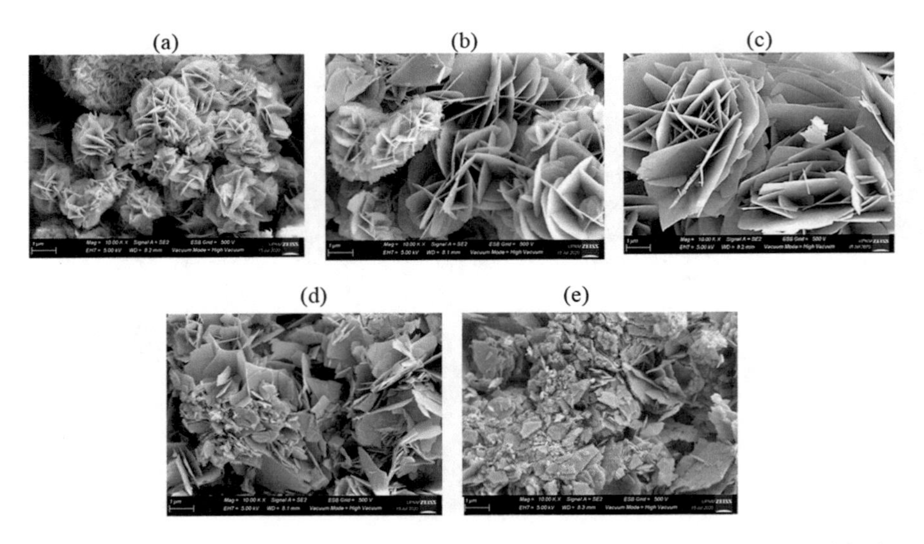

Fig. 2.9 SEM micrographs of BiOI thin films as-deposited via SILAR method **a** and further annealed at the temperatures of **b** 523 K, **c** 623 K, **d** 723 K, and **e** 823 K. Reprinted from Abdul-Manaf et al. [90] under the terms of the Creative Commons Attribution 4.0 International License (CC BY 4.0). Copyright (2021) IOP Publishing

References

1. M. Sherif El-Eskandarany, A. Al-Hazza, L.A. Al-Hajji, N. Ali, A.A. Al-Duweesh, M. Banyan, F. Al-Ajmi, Mechanical milling: a superior nanotechnological tool for fabrication of nanocrystalline and nanocomposite materials. Nanomaterials
2. S.Z.M. Murtaza, P. Vaqueiro, Rapid synthesis of chalcohalides by ball milling: preparation and characterisation of BiSI and BiSeI. J. Solid State Chem. **291**, 121625 (2020)
3. Z. Li, Q. Zhang, L. Wu, W. Gu, Y. Liu, Mechanochemical synthesis of BiSI and $Bi_{19}S_{27}I_3$ semiconductor materials. Adv. Powder Technol. **30**, 1985 (2019)
4. A.V. Gomonnai, I.M. Voynarovych, A.M. Solomon, Y.M. Azhniuk, A.A. Kikineshi, V.P. Pinzenik, M. Kis-Varga, L. Daroczy, V.V. Lopushansky, X-Ray diffraction and raman scattering in SbSI nanocrystals. Mater. Res. Bull. **38**, 1767 (2003)
5. Z. Li et al., Mechanisms of liquid-phase exfoliation for the production of graphene. ACS Nano **14**, 10976 (2020)
6. D. Sahoo, B. Kumar, J. Sinha, S. Ghosh, S.S. Roy, B. Kaviraj, Cost effective liquid phase exfoliation of MoS_2 nanosheets and photocatalytic activity for wastewater treatment enforced by visible light. Sci. Rep. **10**, 10759 (2020)
7. J. Shen et al., Liquid phase exfoliation of two-dimensional materials by directly probing and matching surface tension components. Nano Lett. **15**, 5449 (2015)
8. J.N. Coleman, Y. Nalawade, J. Pepper, A. Harvey, A. Griffin, D. Caffrey, A.G. Kelly, All-printed dielectric capacitors from high-permittivity, liquid-exfoliated BiOCl nanosheets. ACS Appl. Electron. Mater. **2**, 3233 (2020)
9. Y. Shi, J. Li, C. Mao, S. Liu, X. Wang, X. Liu, S. Zhao, X. Liu, Y. Huang, L. Zhang, Van der waals gap-rich BiOCl atomic layers realizing efficient, pure-water CO_2-to-CO photocatalysis. Nat. Commun. **12**, 5923 (2021)
10. H. Yu, H. Huang, K. Xu, W. Hao, Y. Guo, S. Wang, X. Shen, S. Pan, Y. Zhang, Liquid-phase exfoliation into monolayered BiOBr nanosheets for photocatalytic oxidation and reduction. ACS Sustain. Chem. Eng. **5**, 10499 (2017)

11. Y. Liu, J. Yin, Z. Tan, M. Wang, J. Wu, Z. Liu, H. Peng, Electrical and photoresponse properties of inversion asymmetric topological insulator BiTeCl nanoplates. ChemNanoMat **3**, 406 (2017)
12. C. Wang et al., Nonlinear optical response of SbSI nanorods dominated with direct band gaps. J. Phys. Chem. C **125**, 15441 (2021)
13. M. Malekzadeh, M.T. Swihart, Vapor-phase production of nanomaterials. Chem. Soc. Rev. **50**, 7132 (2021)
14. H. Peng, C.K. Chan, S. Meister, X.F. Zhang, Y. Cui, Shape evolution of layer-structured bismuth oxychloride nanostructures via low-temperature chemical vapor transport. Chem. Mater. **21**, 247 (2009)
15. D. Hajra, R. Sailus, M. Blei, K. Yumigeta, Y. Shen, S. Tongay, Epitaxial synthesis of highly oriented 2D janus rashba semiconductor BiTeCl and BiTeBr layers. ACS Nano **14**, 15626 (2020)
16. J.Z. Xin, C.G. Fu, W.J. Shi, G.W. Li, G. Auffermann, Y.P. Qi, T.J. Zhu, X.B. Zhao, C. Felser, Synthesis and thermoelectric properties of rashba semiconductor BiTeBr with intensive texture. Rare Met. **37**, 274 (2018)
17. P.K. Ghosh, A.S. Bhalla, L.E. Cross, Preparation and electrical properties of thin films of antimony sulphur iodide (SbSI). Ferroelectrics **51**, 29 (1983)
18. K.C. Gödel, U. Steiner, Thin film synthesis of SbSI micro-crystals for self-powered photodetectors with rapid time response. Nanoscale **8**, 15920 (2016)
19. J. Varghese, C. O'Regan, N. Deepak, R.W. Whatmore, J.D. Holmes, Surface roughness assisted growth of vertically oriented ferroelectric SbSI nanorods. Chem. Mater. **24**, 3279 (2012)
20. O. Icten, Functional nanocomposites: promising candidates for cancer diagnosis and treatment, in *Synthetic Inorganic Chemistry: New Perspectives*, ed. by J.M. Hamilton (Elsevier, 2021), pp. 279–340
21. S. Li, L. Xu, X. Kong, T. Kusunose, N. Tsurumachi, Q. Feng, Bismuth chalcogenide iodides $Bi_{13}S_{18}I_2$ and BiSI: solvothermal synthesis, photoelectric behavior, and photovoltaic performance. J. Mater. Chem. C **8**, 3821 (2020)
22. X. Wang, F. Zhang, Y. Yang, Y. Zhang, L. Liu, W. Lei, Controllable synthesis and photocatalytic activity of nano-BiOBr photocatalyst. J. Nanomater. **2020**, 1013075 (2020)
23. J. Jiang, K. Zhao, X. Xiao, L. Zhang, Synthesis and facet-dependent photoreactivity of BiOCl single-crystalline nanosheets. J. Am. Chem. Soc. **134**, 4473 (2012)
24. G. Chen, W. Li, Y. Yu, Q. Yang, Fast and low-temperature synthesis of one-dimensional (1D) single-crystalline SbSI microrod for high performance photodetector. RSC Adv. **5**, 21859 (2015)
25. I. Cho, B.K. Min, S.W. Joo, Y. Sohn, One-dimensional single crystalline antimony sulfur iodide, SbSI. Mater. Lett. **86**, 132 (2012)
26. C. Wang, K. Tang, Q. Yang, B. Hai, G. Shen, C. An, W. Yu, Y. Qian, Synthesis of novel SbSI nanorods by a hydrothermal method. Inorg. Chem. Commun. **4**, 339 (2001)
27. R. Zhang, K. Zeng, T. Zhang, Enhanced visible-light-driven photocatalytic activity of Bi_2WO_6-BiSI Z-scheme heterojunction photocatalysts for tetracycline degradation. Int. J. Environ. Anal. Chem. 1 (2020)
28. N. Sharma, Z. Pap, S. Garg, K. Hernádi, Hydrothermal synthesis of BiOBr and BiOBr/CNT composites, their photocatalytic activity and the importance of early $Bi_6O_6(OH)_3(NO_3)_3{\cdot}1.5H_2O$ formation. Appl. Surf. Sci. **495**, 143536 (2019)
29. B. Cui, W. An, L. Liu, J. Hu, Y. Liang, Synthesis of CdS/BiOBr composite and its enhanced photocatalyticdegradation for rhodamine. Appl. Surf. Sci. **319**, 298 (2014)
30. V.I. Popolitov, B.N. Litvin, A.N. Lobachev, Hydrothermal crystallization of semiconducting compounds of group AV BVI CVII (AV : Sb, Bi; BVI : S, Se, Te; CVII; I, Br, Cl). Phys. Status Solidi **3**, K1 (1970)
31. F. Demartin, C.M. Gramaccioli, I. Campostrini, Demicheleite-(I), BiSI, a new mineral from La Fossa Crater, Vulcano, Aeolian Islands, Italy. Mineral. Mag. **74**, 141 (2010)
32. J. Hou, K. Jiang, M. Shen, R. Wei, X. Wu, F. Idrees, C. Cao, Micro and nano hierachical structures of BiOI/activated carbon for efficient visible-light-photocatalytic reactions. Sci. Rep. **7**, 11665 (2017)

33. A. Pancielejko, J. Łuczak, W. Lisowski, A. Zaleska-Medynska, P. Mazierski, Novel two-step synthesis method of thin film heterojunction of BiOBr/Bi_2WO_6 with improved visible-light-driven photocatalytic activity. Appl. Surf. Sci. **569**, 151082 (2021)
34. M. Guo, Z. Zhou, S. Yan, P. Zhou, F. Miao, S. Liang, J. Wang, X. Cui, Bi_2WO_6–BiOCl heterostructure with enhanced photocatalytic activity for efficient degradation of oxytetracycline. Sci. Rep. **10**, 18401 (2020)
35. Q. Zhao, Y. Xing, Z. Liu, J. Ouyang, C. Du, Synthesis and characterization of modified BiOCl and their application in adsorption of low-concentration dyes from aqueous solution. Nanoscale Res. Lett. **13**, 69 (2018)
36. T. Senasu, T. Chankhanittha, K. Hemavibool, S. Nanan, Solvothermal synthesis of BiOBr photocatalyst with an assistant of PVP for visible-light-driven photocatalytic degradation of fluoroquinolone antibiotics. Catal. Today **384–386**, 209 (2022)
37. Y. Zhang, P. Cao, X. Zhu, B. Li, Y. He, P. Song, R. Wang, Facile construction of BiOBr ultra-thin nano-roundels for dramatically enhancing photocatalytic activity. J. Environ. Manage. **299**, 113636 (2021)
38. Y. Mi, H. Li, Y. Zhang, N. Du, W. Hou, Synthesis and photocatalytic activity of BiOBr nanosheets with tunable crystal facets and sizes. Catal. Sci. Technol. **8**, 2588 (2018)
39. E. Bárdos, V.A. Márta, S. Fodor, E.Z. Kedves, K. Hernadi, Z. Pap, Hydrothermal crystallization of bismuth oxychlorides (BiOCl) using different shape control reagents. Materials **14**, 2261 (2021)
40. T. Senasu, T. Narenuch, K. Wannakam, T. Chankhanittha, S. Nanan, Solvothermally grown BiOCl catalyst for photodegradation of cationic dye and fluoroquinolone-based antibiotics. J. Mater. Sci. Mater. Electron. **31**, 9685 (2020)
41. Y. Li, J. Wang, H. Yao, L. Dang, Z. Li, Efficient decomposition of organic compounds and reaction mechanism with BiOI photocatalyst under visible light irradiation. J. Mol. Catal. A Chem. **334**, 116 (2011)
42. J. Xia, S. Yin, H. Li, H. Xu, L. Xu, Q. Zhang, Enhanced photocatalytic activity of Bismuth Oxyiodine (BiOI) porous microspheres synthesized via reactable ionic liquid-assisted solvothermal method. Colloids Surf A Physicochem. Eng. Asp. **387**, 23 (2011)
43. M.M. Frutos, M.E.P. Barthaburu, L. Fornaro, I. Aguiar, Bismuth chalcohalide-based nanocomposite for application in ionising radiation detectors. Nanotechnology **31**, 225710 (2020)
44. I. Aguiar, M. Mombrú, M.P. Barthaburu, H.B. Pereira, L. Fornaro, Influence of solvothermal synthesis conditions in BiSI nanostructures for application in ionizing radiation detectors. Mater. Res. Express **3**, 25012 (2016)
45. J. Lee, B.K. Min, I. Cho, Y. Sohn, Synthesis and characterization of 1-D BiSI and 2-D BiOI nanostructures. Bull. Korean Chem. Soc. **34**, 773 (2013)
46. C. Wang, K. Tang, Q. Yang, J. Hu, Y. Qian, Fabrication of BiTeI submicrometer hollow spheres. J. Mater. Chem. **12**, 2426 (2002)
47. C. Lu, W. Wu, H. Zhou, In Situ fabrication of BiOBr/$BiFeWO_6$ heterojunction with excellent photodegradation activity under visible light. J. Solid State Chem. **303**, 122465 (2021)
48. E.M. El-Fawal, Visible light-driven BiOBr/Bi_2S_3@CeMOF heterostructured hybrid with extremely efficient photocatalytic reduction performance of nitrophenols: modeling and optimization. ChemistrySelect **6**, 6904 (2021)
49. Y. Peng, H. Qian, N. Zhao, Y. Li, Synthesis of a novel 1D/2D $Bi_2O_2CO_3$–BiOI heterostructure and its enhanced photocatalytic activity. Catalysts **11**, 1284 (2021)
50. H. Sun, X. Xiao, V. Celorrio, Z. Guo, Y. Hu, C. Kirk, N. Robertson, A Novel method to synthesize BiSI uniformly coated with RGO by chemical bonding and its application as a supercapacitor electrode material. J. Mater. Chem. A **9**, 15452 (2021)
51. H. Razavi-Khosroshahi, S. Mohammadzadeh, M. Hojamberdiev, S. Kitano, M. Yamauchi, M. Fuji, $BiVO_4$/BiOX (X = F, Cl, Br, I) heterojunctions for degrading organic dye under visible light. Adv. Powder Technol. **30**, 1290 (2019)
52. J. Wu et al., Hydrothermal synthesis of carbon spheres—BiOI/$BiOIO_3$ heterojunctions for photocatalytic removal of gaseous Hg^0 under visible light. Chem. Eng. J. **304**, 533 (2016)

53. J. Wu, Y. Xie, Y. Ling, Y. Dong, J. Li, S. Li, J. Zhao, Synthesis of flower-like g-C_3N_4/BiOBr and enhancement of the activity for the degradation of bisphenol a under visible light irradiation. Front. Chem. **7**, 649 (2019)
54. Y. Liu, Z. Hu, J.C. Yu, Photocatalytic degradation of ibuprofen on S-Doped BiOBr. Chemosphere **278**, 130376 (2021)
55. K.S. Suslick, W.B. McNamara, Y. Didenko, Hot spot conditions during multi-bubble cavitation, in *Sonochemistry and sonoluminescence*, ed. by L.A. Crum, T.J. Mason, J.L. Reisse, K.S. Suslick (Springer, Netherlands, Dordrecht, 1999), pp. 191–204
56. X. Hangxun, B.W. Zeiger, K.S. Suslick, Sonochemical synthesis of nanomaterials. Chem. Soc. Rev. **42**, 2555 (2013)
57. M.A. Dheyab, A.A. Aziz, M.S. Jameel, Recent advances in inorganic nanomaterials synthesis using sonochemistry: a comprehensive review on iron oxide, gold and iron oxide coated gold nanoparticles. Molecules
58. M. Nowak et al., Fabrication and characterization of SbSI gel for humidity sensors. Sens. Actuators A Phys. **210**, 119 (2014)
59. P. Kwolek, K. Pilarczyk, T. Tokarski, J. Mech, J. Irzmański, K. Szaciłowski, Photo-electrochemistry of N-Type antimony sulfoiodide nanowires. Nanotechnology **26**, 105710 (2015)
60. M. Tasviri, Z. Sajadi-Hezave, SbSI nanowires and CNTs encapsulated with SbSI as photocatalysts with high visible-light driven photoactivity. Mol. Catal. **436**, 174 (2017)
61. A. Starczewska, M. Nowak, P. Szperlich, B. Toroń, K. Mistewicz, D. Stróz, J. Szala, Influence of humidity on impedance of SbSI gel. Sens. Actuators A Phys. **183**, 34 (2012)
62. K. Mistewicz, M. Kępińska, M. Nowak, A. Sasiela, M. Zubko, D. Stróż, Fast and efficient piezo/photocatalytic removal of methyl orange using SbSI nanowires. Materials **13**, 4803 (2020)
63. K. Mistewicz et al., A simple route for manufacture of photovoltaic devices based on chalcohalide nanowires. Appl. Surf. Sci. **517**, 146138 (2020)
64. S. Manoharan, D. Kesavan, P. Pazhamalai, K. Krishnamoorthy, S.J. Kim, Ultrasound irradiation mediated preparation of antimony Sulfoiodide (SbSI) nanorods as a high-capacity electrode for electrochemical supercapacitors. Mater. Chem. Front. **5**, 2303 (2021)
65. A.K. Pathak, M.D. Prasad, S.K. Batabyal, One-dimensional SbSI crystals from Sb, S, and I mixtures in ethylene glycol for solar energy harvesting. Appl. Phys. A Mater. Sci. Process. **125**, 213 (2019)
66. O. Gladkovskaya, I. Rybina, Y.K. Gun'Ko, A. Erxleben, G.M. O'Connor, Y. Rochev, Water-based ultrasonic synthesis of SbSI nanoneedles, Mater. Lett. **160**, 113 (2015)
67. K. Mistewicz, M. Jesionek, M. Nowak, M. Kozioł, SbSeI pyroelectric nanogenerator for a low temperature waste heat recovery. Nano Energy **64**, 103906 (2019)
68. K. Mistewicz, A. Starczewska, M. Jesionek, M. Nowak, M. Kozioł, D. Stróż, Humidity dependent impedance characteristics of SbSeI nanowires. Appl. Surf. Sci. **513**, 145859 (2020)
69. B. Toroń, K. Mistewicz, M. Jesionek, M. Kozioł, D. Stróż, M. Zubko, Nanogenerator for dynamic stimuli detection and mechanical energy harvesting based on compressed SbSeI nanowires. Energy **212**, 118717 (2020)
70. M. Nowak, B. Kauch, P. Szperlich, D. Stróz, J. Szala, T. Rzychoń, Bober, B. Toroń, A. Nowrot, Sonochemical preparation of $SbS_{1-x}Se_xI$ nanowires. Ultrason. Sonochem. **17**, 487 (2010)
71. K. Mistewicz, Pyroelectric nanogenerator based on an SbSI-TiO_2 nanocomposite. Sensors **22**, 69 (2022)
72. S.J. Reese, D.H. Hurley, H.W. Rollins, Effect of surface acoustic waves on the catalytic decomposition of ethanol employing a comb transducer for ultrasonic generation. Ultrason. Sonochem. **13**, 283 (2006)
73. G. Li, F. Qin, R. Wang, S. Xiao, H. Sun, R. Chen, BiOX (X=Cl, Br, I) nanostructures: mannitol-mediated microwave synthesis, visible light photocatalytic performance, and Cr(VI) removal capacity. J. Colloid Interface Sci. **409**, 43 (2013)
74. C. Deng, H. Guan, X. Tian, Novel $Bi_{19}S_{27}Br_3$ superstructures: facile microwave-assisted aqueous synthesis and their visible light photocatalytic performance. Mater. Lett. **108**, 17 (2013)

75. K.Y. Shih, Y.L. Kuan, E.R. Wang, One-step microwave-assisted synthesis and visible-light photocatalytic activity enhancement of BiOBr/RGO nanocomposites for degradation of methylene blue. Materials
76. M. Barj, O.A. Mykaylo, D.I. Kaynts, O.V. Gorina, O.G. Guranich, V.M. Rubish, Formation and structure of crystalline inclusions in As_2S_3-SbSI and As_2Se_3-SbSI systems glass matrices. J. Non. Cryst. Solids **357**, 2232 (2011)
77. D.I. Kaynts, A.P. Shpak, V.M. Rubish, O.A. Mykaylo, O.G. Guranich, P.P. Shtets, P.P. Guranich, Formation of ferroelectric nanostructures in $(As_2S_3)_{100-x}(SbSI)_x$ glassy matrix. Ferroelectrics **371**, 28 (2008)
78. D. Savytskii, B. Knorr, V. Dierolf, H. Jain, Challenges of CW laser-induced crystallization in a chalcogenide glass. Opt. Mater. Express **3**, 1026 (2013)
79. Y.M. Azhniuk, P. Bhandiwad, V.M. Rubish, P.P. Guranich, O.G. Guranich, A.V. Gomonnai, D.R.T. Zahn, Photoinduced changes in the structure of As_2S_3-based SbSI nanocrystal-containing composites studied by raman spectroscopy. Ferroelectrics **416**, 113 (2011)
80. Y.M. Azhniuk, A. Villabona, A.V. Gomonnai, V.M. Rubish, V.M. Marjan, O.O. Gomonnai, D.R.T. Zahn, Raman and AFM studies of $(As_2S_3)_{0.45}(SbSI)_{0.55}$ thin films and bulk glass. J. Non. Cryst. Solids **396–397**, 36 (2014)
81. Y.M. Azhniuk, V. Stoyka, I. Petryshynets, V.M. Rubish, O.G. Guranich, A.V. Gomonnai, D.R.T. Zahn, SbSI nanocrystal formation in As-Sb-S-I glass under laser beam. Mater. Res. Bull. **47**, 1520 (2012)
82. V.J. Babu, R.S.R. Bhavatharini, S. Ramakrishna, Electrospun BiOI nano/microtectonic plate-like structure synthesis and UV-light assisted photodegradation of ARS dye. RSC Adv. **4**, 19251 (2014)
83. Y. Zhang, S. Liu, Z. Xiu, Q. Lu, H. Sun, G. Liu, TiO_2/BiOI heterostructured nanofibers: electrospinning-solvothermal two-step synthesis and visible-light photocatalytic performance investigation. J. Nanoparticle Res. **16**, 2375 (2014)
84. Q. Huang, G. Jiang, H. Chen, L. Li, Y. Liu, Z. Tong, W. Chen, Hierarchical nanostructures of BiOBr/AgBr on electrospun carbon nanofibers with enhanced photocatalytic activity. MRS Commun. **6**, 61 (2016)
85. X. Yang, X. Li, L. Zhang, J. Gong, Electrospun template directed molecularly imprinted nanofibers incorporated with BiOI nanoflake arrays as photoactive electrode for photoelectrochemical detection of triphenyl phosphate. Biosens. Bioelectron. **92**, 61 (2017)
86. S.S. Kale, R.S. Mane, H. Chung, M.Y. Yoon, C.D. Lokhande, S.H. Han, Use of successive ionic layer adsorption and reaction (SILAR) method for amorphous titanium dioxide thin films growth. Appl. Surf. Sci. **253**, 421 (2006)
87. A.A. Putri, S. Kato, N. Kishi, T. Soga, Angle dependence of synthesized BiOI prepared by dip coating and its effect on the photovoltaic performance. Jpn. J. Appl. Phys. **58**, SAAD09 (2019)
88. A.A. Putri, S. Kato, N. Kishi, T. Soga, Relevance of precursor molarity in the prepared bismuth oxyiodide films by successive ionic layer adsorption and reaction for solar cell application. J. Sci. Adv. Mater. Devices **4**, 116 (2019)
89. A.A. Abuelwafa, R.M. Matiur, A.A. Putri, T. Soga, Synthesis, structure, and optical properties of the nanocrystalline Bismuth Oxyiodide (BiOI) for optoelectronic application, Opt. Mater. (Amst). **109**, 110413 (2020)
90. N.A. Abdul-Manaf, A.H. Azmi, F. Fauzi, N.S. Mohamed, The effects of micro and macro structure on electronic properties of bismuth oxyiodide thin films. Mater. Res. Express **8**, 96401 (2021)
91. H. Jia, Y. Li, Y. Mao, D. Yu, W. He, Z. Zheng, Room temperature synthesis of $BiOBr_{1-x}I_x$ thin films with tunable structure and conductivity type for enhanced photoelectric performance. RSC Adv. **10**, 41755 (2020)
92. M.J. Chang, H. Wang, H.L. Li, J. Liu, H.L. Du, Facile preparation of novel Fe_2O_3/BiOI hybrid nanostructures for efficient visible light photocatalysis. J. Mater. Sci. **53**, 3682 (2018)

Chapter 3
Strategies for Incorporation of Chalcohalide Nanomaterials into the Functional Devices

3.1 Solution Processing for Thin Films Preparation

A deposition of the thin film via solution processing involves the application of a liquid precursor on a substrate which is then converted to the desired solid layer in a subsequent post-treatment step [1]. In this method, the used solvent and processing route are crucial for the selection of an appropriate precursor [2]. The solution processing technique is suitable for scalability and commercialization of thin film devices [3]. Moreover, solution-based coating can be performed at an ambient atmosphere. It is in contrast to the physical and chemical vapor deposition methods that require well controlled atmosphere and are usually carried out in vacuum, using expensive equipment and energy-intensive processes [3]. However, a formation of high quality film through solution processing is a complex procedure affected by the many effects, including wetting phenomena, fluid mechanics, and evaporation.

Until now, the solution processing have been applied for a construction of the solar cells and photodetectors based on the chalcohalide materials, such as SbSI [4–7], SbSeI [5, 8, 9], BiSI [10–14], and $Sb_{0.67}Bi_{0.33}SI$ [15]. Usually, the preparation of the chalcohalide layer via solution processing is divided into the several steps (Fig. 3.1). In the first stage of SbSI film fabrication, the solution of $SbCl_3$, thiourea (TU), and N,N-dimethylformamide (DMF) was spin-coated on the TiO_2 blocking layer (TiO_2-BL)/F-doped SnO_2 (FTO) substrate [4]. Then, the sample was annealed at 423 K to form a layer of the amorphous Sb_2S_3 (am-Sb_2S_3). In the second step, the solution of the SbI_3 and N-methyl-2-pyrrolidone (NMP) was deposited on the sample via spin-coating and annealed for conversion into the SbSI film. The structure and morphology of the SbSI thin films were controlled by the tuning the input ratio of the $SbCl_3$ to thiourea. When this ratio was equal to 1:1.5 the SbSI nanorods were aggregated on the surface of the TiO_2-BL/FTO substrate. As this parameter was enhanced, the prepared films became more dense. The continuous layer with the thickness of 300 nm was formed for $SbCl_3$ to TU ratio of 1:3. The irregular films were obtained with the further rise of the $SbCl_3$ to TU ratio.

K. Mistewicz, *Low-Dimensional Chalcohalide Nanomaterials*, NanoScience and Technology, https://doi.org/10.1007/978-3-031-25136-8_3

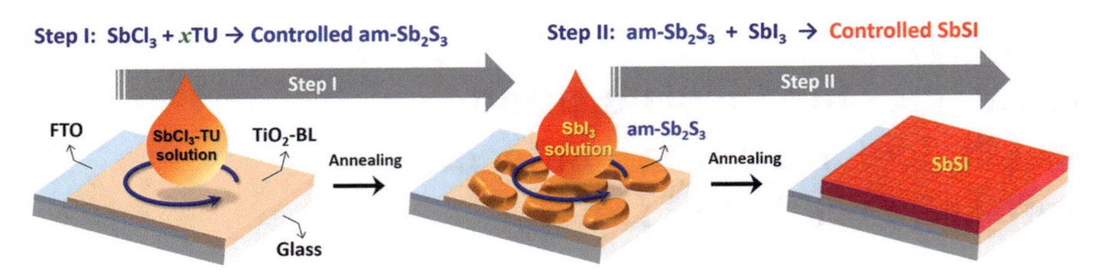

Fig. 3.1 The formation of SbSI thin film using two steps solution processing. Detailed description is provided in the text. Reprinted from Choi et al. [4] under the terms of the Creative Commons Attribution 4.0 International License (CC BY 4.0). Copyright (2018) American Institute of Physics (AIP)

The deposition of the antimony selenoiodide film via solution processing method was reported in Ref. [8] as a technology suitable for construction of the efficient and stable SbSeI solar cells. At first, a selenium single-source precursor (Se-SSP) was prepared as the solution of DMF and $[(SbL_2Cl_2)Cl]_2 \cdot (CH_3)_2CO]$, where L refers to N,N-dimethyl selenourea. The Se-SSP was spin-coated on FTO/BL/mp-TiO_2 substrate and thermally decomposed in argon gas at 423 K. The procedure of spin-coating followed by the thermal decomposition of the material was accomplished multiple times to obtain the expected amount of antimony selenide. In order to crystallize Sb_2Se_3, the sample was annealed in argon at temperature of 573 K. In the second step of the SbSeI film preparation, the solution of antimony triiodide was deposited on the FTO/BL/mp-TiO_2/Sb_2Se_3 sample and annealed in Ar gas at 423 K. Afterwards, the spin-coating cycles and thermal decomposition of the SbI_3 solution were performed similarly to those carried out in the first stage of sample preparation. At the end, the sample was washed in the DMF to remove the residual SbI_3. Fabricated SbSeI film was studied using scanning electron microscopy (SEM). The top and side views of the SbSeI layer are presented in Fig. 3.2a, b, respectively. The uniform distribution of the chemical elements (Sb, Se, and I) in the SbSeI film was proved by the energy-dispersive X-ray (EDX) linear scan (Fig. 3.2c) and elemental mapping (Fig. 3.2d). The secondary electron cut-off regions of the He I UPS spectrum and the X-ray photoelectron spectroscopy (XPS) valance level spectrum of SbSeI is shown in Fig. 3.2e. The energy levels of the materials used to prepare the FTO/BL/mp-TiO_2/SbSeI/HTM(L)/Au solar cell were determined (Fig. 3.2f). The n-type of the electrical conductivity of SbSeI was confirmed.

The two step solution processing of BiSI film preparation was described in Ref. [10]. The solution of bismuth oxide (Bi_2O_3) and thiourea was spin-coated on the TiO_2-BL/FTO substrate and heated at temperature of 473 K in the first step of sample fabrication (Fig. 3.3a). A formation of the Bi_2S_3 was confirmed by the X-ray diffraction (XRD) pattern and absorption spectrum represented by the blue curves in Figs. 3.3b, c, respectively. The optical band gap (E_g) of 1.57 eV was determined for Bi_2S_3 film. In the second stage of BiSI film preparation (Fig. 3.3a), the BiI_3 solution was dropped onto the Bi_2S_3/TiO_2-BL/FTO sample. Then, it was annealed at 473 K to convert the Bi_2S_3 to BiSI. The sample was cooled down and washed with the

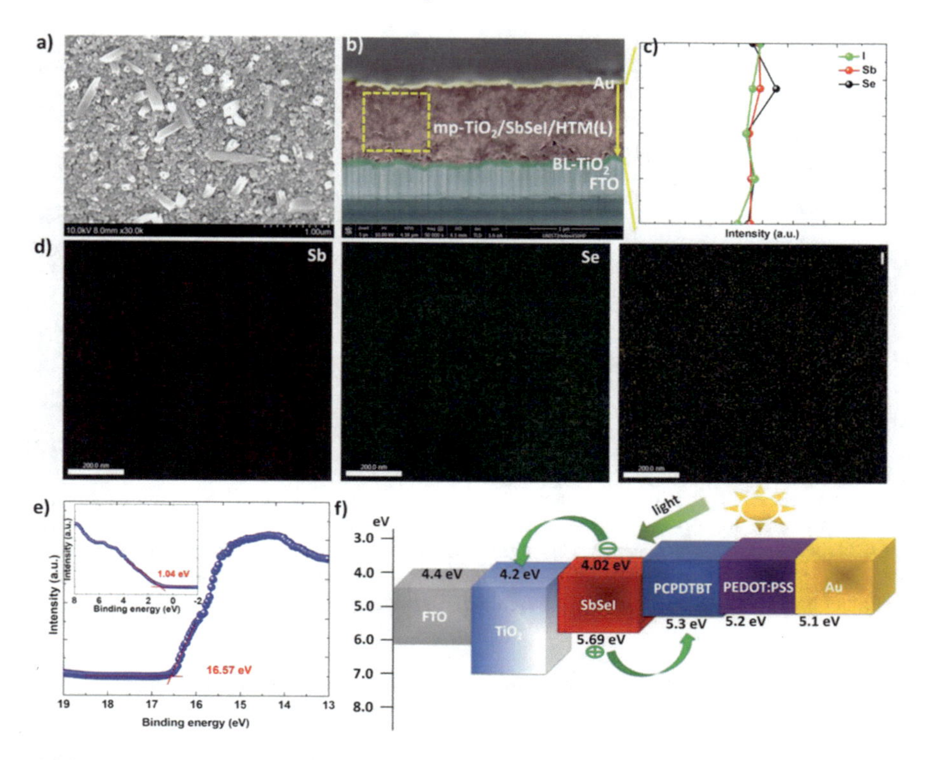

Fig. 3.2 The SEM top view **a** of the SbSeI film deposited on the glass/mp-TiO_2 substrate. The cross-sectional SEM micrograph (**b**) of FTO/BL/mp-TiO_2/SbSeI/HTM(L)/Au solar cell. The EDX data (**c**) measured from the top to bottom in mp-TiO_2/SbSeI/HTM(L) layer indicated by yellow arrow in (**b**). The EDX elemental mapping (**d**) of the area marked by the yellow rectangle in (**b**). Secondary electron cut-off region of He I UPS spectra and XPS valence level spectrum (**e**) of FTO/mp-TiO_2/SbSeI structure. The energy levels **f** of the components used to construct FTO/BL/mp-TiO_2/SbSeI/HTL/Au solar cell. Reprinted from Nie et al. [8] under the terms of the Creative Commons Attribution 4.0 International License (CC BY 4.0). Copyright (2021) John Wiley and Sons

NMP solvent to remove the remaining BiI_3. The XRD studies confirmed growth of orthorhombic BiSI (red curve in Fig. 3.3b). The BiSI film consisted of large number of nanorods with diameters in the range from 60 to 100 nm. The band gap of BiSI film of 1.61 eV was calculated from the absorption spectrum (red line in Fig. 3.3c). This value is close to the data reported in the literature for BiSI ($E_g = 1.6$ eV [16], $E_g = 1.63$ eV [17]). It was reported in Ref. [10] that when ratio of bismuth to sulfur in the Bi_2O_3-thiourea solution was 1:3, the best quality of BiSI layer was achieved. It is similar to the conclusion presented in Ref. [4] where the antimony to sulfur ratio of 1:3 was found as the optimum value for SbSI formation. The prepared BiSI/TiO_2-BL/FTO layered structure was recognized as promising for future application in the low-cost and environment-friendly solar cells.

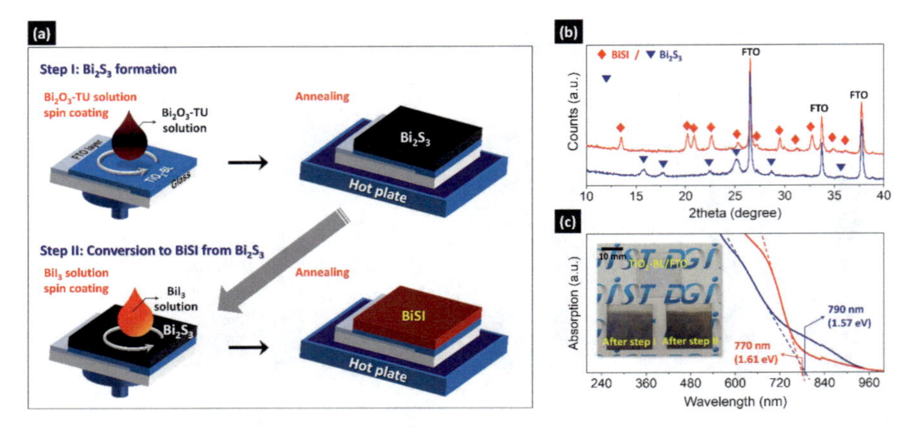

Fig. 3.3 A scheme **a** of the preparation of BiSI film on the TiO_2-BL/FTO substrate via the two-step solution process. X-ray diffraction patterns (**b**) and absorption spectra (**c**) of the samples fabricated after step I (blue curves) and step II (red curves). Inset in (**c**) shows a photograph of the TiO_2-BL/FTO substrate before material deposition, after step I and step II of the sample fabrication. Reprinted from Choi and Hwang [10] under the terms of the Creative Commons Attribution 4.0 International License (CC BY 4.0). Copyright (2019) MDPI

3.2 The Spin-Coating Deposition of Polymer Composites

Purusothaman and coworkers [18] presented fabrication of photoactive piezoelectric energy harvester based on the composite of SbSI and polymethyl methacrylate (PMMA). The solution of PMMA in toluene (3 w/v %) was prepared and stirred at temperature of 333 K. Then, 1 wt% of SbSI was added to the PMMA-toluene solution. The composite solution was stirred, spin-coated on an aluminum electrode and dried at 343 K. Deposition process was repeated multiple times to achieve full coverage of SbSI-PMMA composite on the Al electrode. The second Al electrode was attached on the top of the SbSI-PMMA film. Afterwards, the antistatic tape was bounded to the electrodes and Al/SbSI-PMMA/Al structure was encapsulated into the polydimethylsiloxane (PDMS). The prepared device was applied as mechanical energy harvester that generated the voltage response of 5 V and current output of 150 nA under a linear mechanical force of 2 N [18].

A fabrication of SbSI-polyacrylonitrile (PAN) nanocomposite and its spin-coating deposition for a facile construction of the photovoltaic devices was for the first time proposed in Ref. [19]. This method was summarized in Fig. 3.4. The SbSI nanowires were synthesized using sonochemical method. The suspension of SbSI nanowires in DMF was prepared (Fig. 3.4a) and agitated ultrasonically for 30 min to break up the agglomerates of the nanowires (Fig. 3.4b). In the next step, the PAN polymer was added to the SbSI-DMF suspension (Fig. 3.4c) which was homogenized using a magnetic stirrer. Small amount of the SbSI-PAN-DMF mixture was dropped onto the indium tin oxide (ITO) substrate with TiO_2 layer (Fig. 3.4d) and

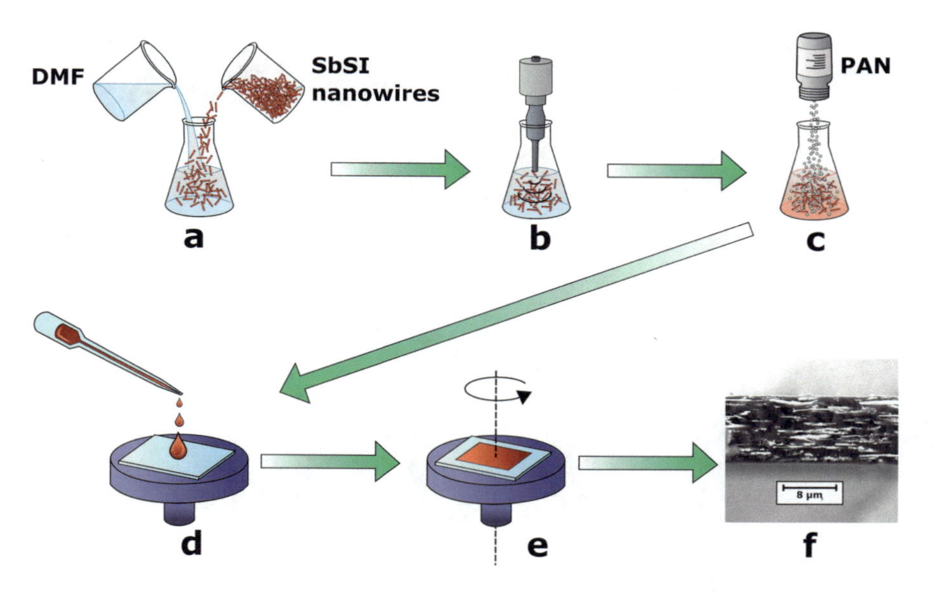

Fig. 3.4 The procedure of SbSI-PAN composite film fabrication: **a** the mixing of SbSI nanowires with DMF, **b** homogenization of the SbSI suspension in DMF under ultrasonic irradiation, **c** an addition of PAN polymer into the mixture and its mechanical stirring, **d** dropping the SbSI-PAN-DMF mixture on the TiO_2/ITO substrate, **e** the spin-coating of the material, **f** SEM micrograph of prepared SbSI-PAN film

spin-coated (Fig. 3.4e). An exemplary SEM micrograph of the fabricated SbSI-PAN nanocomposite is presented in Fig. 3.4f. The SbSI nanowires were found to be oriented horizontally. It can be explained due to the centrifugal forces originating from the spin-coating of the material. The proposed method of preparation of chalcohalide nanocomposite film was universal, fast, simple, and did not involve any thermal processing of the material. However, the application of the polymer matrix resulted in a limited photovoltaic performance of the solar cell based on the SbSI-PAN nanocomposite [19]. The increase of the SbSI nanowires in the nanocomposite led to the enhancement of the open circuit voltage and the short circuit current density of the prepared photovoltaic device.

3.3 Solution Casting

The solution casting is a simple method of nanocomposite films preparation [20–23]. In this method, the polymer phase is dissolved in volatile solvent and mixed with nanosized reinforcements [24]. Then, the solution is casted on a flat surface to form a film and solvent is evaporated. Usually, the solution casting can be carried out at relatively low temperature.

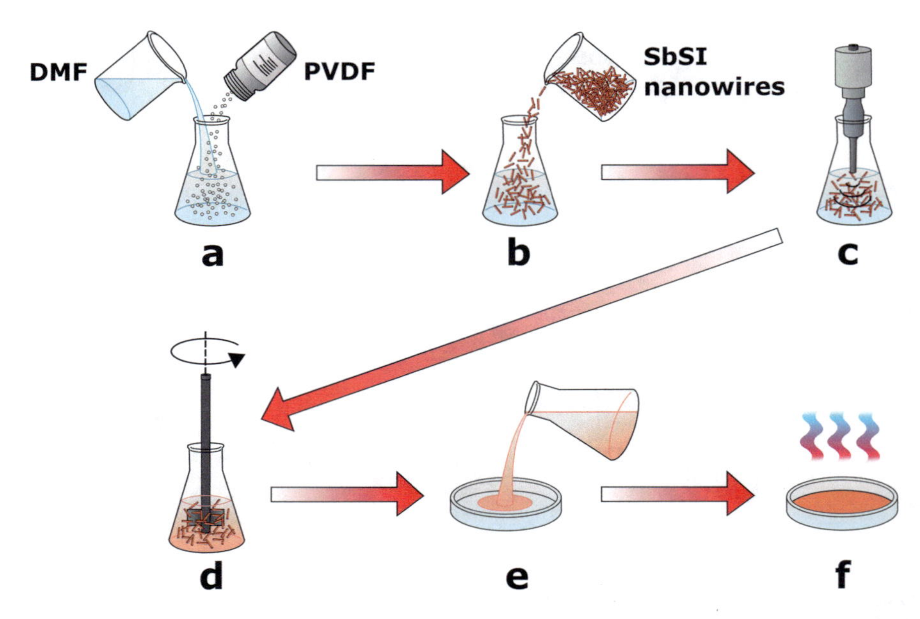

Fig. 3.5 Fabrication of SbSI-PVDF nanocomposite film via solution casting: **a** the mixing of PVDF powder with DMF, **b** an addition of the SbSI nanowires into the PVDF-DMF solution, **c** ultrasonic agitation, **d** mechanical stirring, **e** pouring the SbSI-PVDF-DMF suspension into the Petri dish, **f** material heating under vacuum in order to DMF evaporation

A fabrication of SbSI-polyvinylidene fluoride (PVDF) nanocomposite film is presented herein for the first time. At first stage, the PVDF powder was inserted into DMF in order to obtain 5% wt. solution (Fig. 3.5a). Afterwards, the xerogel of SbSI nanowires was added to the PVDF-DMF solution (Fig. 3.5b). The suspension of SbSI nanowires in PVDF-DMF was agitated ultrasonically for 15 min at temperature of 343 K (Fig. 3.5c). In the next step, the suspension was stirred mechanically with velocity of 600 rpm at 343 K for 2 h (Fig. 3.5d). The SbSI-PVDF-DMF suspension was outgassed and poured into the Petri dish (Fig. 3.5e). Finally, the prepared film was heated at 343 K for 2 h under a pressure of 50 kPa in order to DMF evaporation (Fig. 3.5f). After DMF removal, the mass of the SbSI constituted 50% of the total mass of the SbSI-PVDF nanocomposite. The SbSI-PVDF nanocomposite was further used as piezoelectric nanogenerator (Sect. 4.1.2 in Chap. 4 of this book).

3.4 The Films Deposition Via Drop-Casting

A novel preparation method of the SbSI-TiO_2 nanocomposite film has been recently developed [25]. In this approach, TiO_2 nanoparticles played the role of the binders in the synthesized nanocomposite which allowed to fabricate continuous dense films of the material. This goal could not be achieved using only SbSI nanowires without TiO_2

nanoparticles as the fillers. The drop-casting of pure SbSI nanowires led to formation of irregular and discontinuous films. Furthermore, the method of the SbSI-TiO_2 film preparation eliminated the need of an application of a binding polymer matrix [25]. In result, the electrical properties of the SbSI-TiO_2 nanocomposite originated only from the SbSI nanowires and TiO_2 nanoparticles. It was a significant advantage in comparison to the SbSI-polymer composites which electrical properties can be worse in comparison to the electrical properties of the pristine SbSI nanowires [19]. The SbSI-TiO_2 nanocomposite was synthesized from stoichiometric amounts of antimony, sulfur, and iodine mixed with TiO_2 nanoparticles which constituted about 30% of the total nanocomposite mass (Fig. 3.6a). The reagents were immersed in the ethanol and subjected to the ultrasounds for 4 h at the temperature of 323 K (Fig. 3.6b). In the next step, the sol of SbSI-TiO_2 nanocomposite and ethanol was drop-casted onto polyethylene terephthalate (PET) substrate coated with thin layer of ITO (Fig. 3.6c). This process was repeated several times to obtain dense film of SbSI-TiO_2 nanocomposite. The sample was dried (Fig. 3.6d). The Au electrode was sputtered on the top of the sample (Fig. 3.6e). The wires were attached to the device in order to investigate the electrical and pyroelectric properties of the SbSI-TiO_2 nanocomposite (Fig. 3.6f).

Figure 3.7a presents SbSI-TiO_2 nanocomposite film sandwiched between Au and ITO electrodes on the PET substrate. The SEM investigations of the SbSI-TiO_2

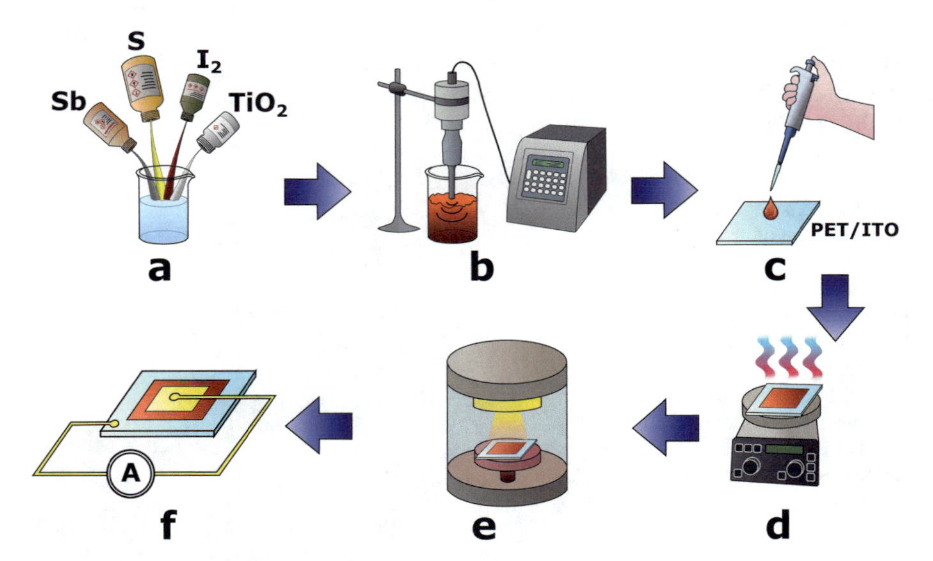

Fig. 3.6 Fabrication of the pyroelectric nanogenerator based on the SbSI-TiO_2 nanocomposite film: **a** an inserting the reagents into the ethanol, **b** sonochemical synthesis of the SbSI-TiO_2 nanocomposite, **c** drop-casting of the material on the PET/ITO substrate, **d** drying of the sample, **e** the sputtering of the electrodes, **f** connecting the device to the external measurement circuit. Reprinted from Mistewicz [25] under the terms of the Creative Commons Attribution 4.0 International License (CC BY 4.0). Copyright (2022) MDPI

nanocomposite revealed that the material was composed of the TiO_2 nanoparticles distributed around the chaotically oriented SbSI nanocrystals (Fig. 3.7b). The chemical composition of the material was determined using an energy-dispersive X-ray spectroscopy (EDS). It was confirmed that the TiO_2 constituted 29.9% of the nanocomposite total mass. This value was close to the initial amount of titanium dioxide (30%) used for fabrication of the SbSI-TiO_2 nanocomposite. The elemental atomic ratio of 0.37:0.36:0.27 for antimony, sulfur, and iodine was averaged over the sample volume. The excess concentration of antimony is a characteristic feature that was observed for SbSI with different morphologies, such as thin films [26], nanocrystals [27], nanowires [28, 29], and nanorods [30]. In the case of SbSI nanowires, this effect originates from the presence of a fuzzy shell on the crystalline core of the SbSI nanowire. The EDS analysis of the individual SbSI nanowire proved the formation of stoichiometric nanocrystal of SbSI (spot 1 in Fig. 3.7c). The agglomerated TiO_2 nanoparticles and small residuals of antimony were identified in the close vicinity of the SbSI nanowire (spot 2 in Fig. 3.7c). The TiO_2 nanoparticles were resistant to ultrasonic irradiation due to the remarkable chemical stability of this material [31].

The diameters and lengths of SbSI nanowires in the SbSI-TiO_2 nanocomposite followed a log-normal distribution and reached up to 220 nm and 8.2 μm, respectively. The average length of SbSI nanowires was 2.52(7) μm, whereas their mean diameter of 68(2) nm was calculated. This value was in good agreement with the average diameter of the SbSI nanowires of 69(3) nm determined in Ref. [19]. The SbSI-TiO_2 nanocomposite sandwiched between Au and ITO electrodes was investigated further as a pyroelectric nanogenerator. It exhibited a high pyroelectric coefficient of 264(7) nC/(cm^2·K) as well as peak and average surface power densities of 8.39(2) $\mu W/m^2$ and 2.57(2) $\mu W/m^2$, respectively.

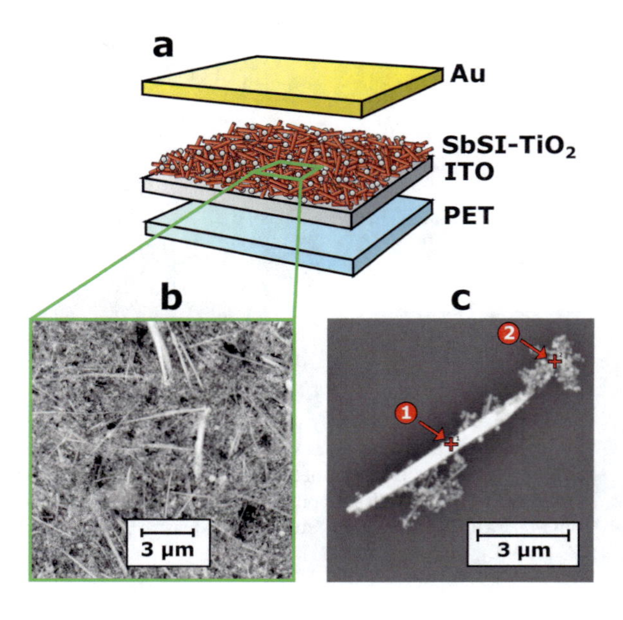

Fig. 3.7 A schematic representation of the pyroelectric nanogenerator based on the SbSI-TiO_2 nanocomposite film (**a**), its SEM micrograph (**b**), and SEM image of an individual nanowire (**c**). Reprinted from Mistewicz [25] under the terms of the Creative Commons Attribution 4.0 International License (CC BY 4.0). Copyright (2022) MDPI

3.5 High Pressure Compression of Nanowires into the Bulk Samples

A room temperature compression of the chalcohalide nanowires into the macroscopic samples was presented for the first time in Ref. [32]. The nanowires of SbSeI were grown using sonochemical method and dried to obtain the xerogel of this material. The small amount of the SbSeI xerogel was inserted in the steel cylinder which served as a mold in this process. The mold was closed with a steel piston. The strain of 100 MPa was applied to the mold. The piston was pressed against the mold with the speed of 5 mm/min. The bulk sample of compressed SbSeI nanowires was equipped with gold electrodes and encapsulated in the silicone rubber. Such prepared device was tested as a pyroelectric nanogenerator for a low temperature waste heat recovery. It was subjected to the periodic cycles of the temperature changes with the amplitude of 10 K. SbSeI device generated the pyroelectric current of 11 nA and the output power density of 0.59(4) $\mu W/m^2$. A room temperature compression of the SbSeI nanowires was applied further to construct other devices, i.e. nanosensor of humidity [33], piezoelectric nanogenerators [34, 35], and hybrid piezo/triboelectric nanogenerator [36]. The achievements in the field of room temperature compression of the chalcohalide nanowires are summarized in Table 3.1.

Scanning electron microscopy studies of the compressed chalcohalide nanowires showed that they were randomly distributed in the investigated samples (Figs. 3.8, 3.9a, b, and 3.10a–c). Moreover, a presence of the small voids between the nanowires bundles was recognized as a typical feature of the pellets prepared via high pressure compression of the chalcohalide nanowires [32, 34, 37]. The nanowires constituted from 41 to 50% of the total volumes of the pellets. It should be underlined that these values are significantly higher than packing factor of 4.7% reported for the uncompressed SbSI xerogel [38]. The rise in the compression pressure resulted in

Table 3.1 A summary of the bulk chalcohalide structures fabrication using high pressure compression of the nanowires at room temperature

Material	Compression pressure, MPa	Packing factor, %	Application of the material	References
SbSeI nanowires	100	41	Pyroelectric nanogenerator	[32]
	100	44	Humidity nanosensor	[33]
	120	50	Piezoelectric nanogenerator	[34]
	160	50	Piezoelectric nanogenerator	[35]
	120		Piezo/triboelectric nanogenerator	[36]
SbSI nanowires	160	50		[37]

the slight increase of the packing factor (Table 3.1). The porous microstructure of the compressed SbSeI nanowires system was beneficial for its application in the humidity nanosensor [33]. The incomplete compactness of this material led to the formation of the large surface area which supported ability of the gas molecules adsorption and enhanced sensitivity of humidity detection. The SEM examination of the compressed SbSeI nanowires proved that the majority of the nanowires were not damaged despite an application of the high pressure of 120 MPa [34]. The powder-like regions were observed on the top and bottom surfaces of the SbSeI pellet. This effect was attributed to the action of the piston that caused cracking of some nanowires within these areas of the sample [34]. The average surface roughness of the SbSeI pellet was 1.78(13) μm indicating that compressed SbSeI nanowires are promising for use in high-performance triboelectric nanogenerators for mechanical energy harvesting.

The high pressure compression of the chalcohalide nanowires did not influence any change of the chemical composition of the processed materials (Figs. 3.9b–e and 3.10d). It was confirmed by the numerous EDS surveys [32, 33, 37]. No other chemical elements than these expected (antimony, selenium, sulfur, and iodine) were detected in the examined samples. The concentrations of the chemical elements in

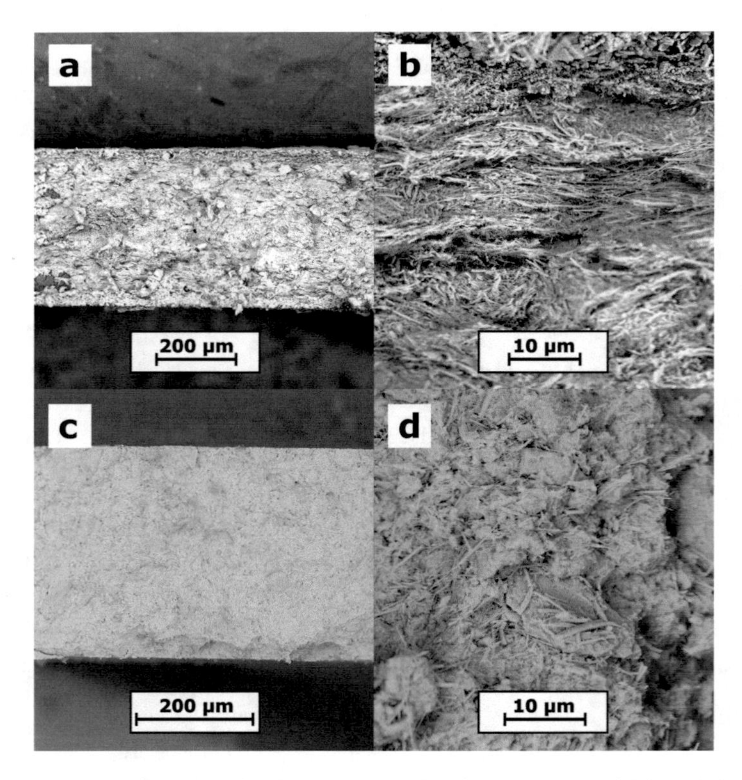

Fig. 3.8 Typical SEM images of **a**, **b** SbSI and **c**, **d** SbSeI bulk pellets fabricated through the compression of the nanowires under pressure of 160 MPa at room temperature

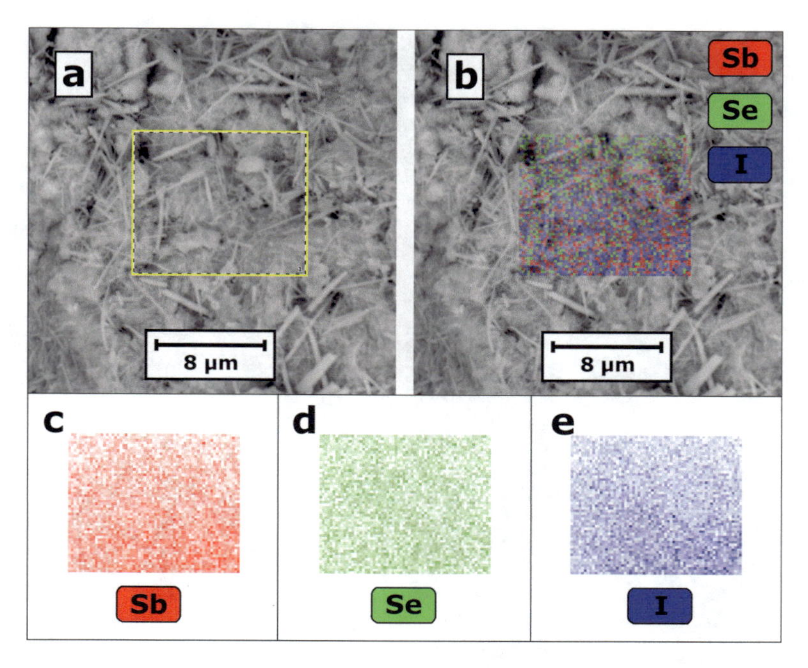

Fig. 3.9 SEM micrograph (**a**) of the compressed SbSeI nanowires and combined EDS elemental mapping (**b**) of the sample area indicated by the yellow rectangle in (**a**). The spatial distribution of the antimony (**c**), selenium (**d**), and iodine (**e**). Reprinted from Mistewicz et al. [35] under the terms of the Creative Commons Attribution 4.0 International License (CC BY 4.0). Copyright (2021) Elsevier

the investigated pellets were close to the theoretical values for stoichiometric SbSeI or SbSI. The slight deviations from the theoretical compositions were noted for both studied materials. It was due to the a little excess of the Sb, Se or S which was accompanied by a deficiency of iodine [35, 37]. This effect was ascribed to the possible evaporation of iodine from SbSeI/SbSI surfaces during EDS measurements. Furthermore, it was also suggested in Ref.[32, 37] that the chemical composition of the SbSeI/SbSI nanowires surfaces can be different than the cores of the nanowires covered with the amorphous fuzzy shells. The distribution of the chemical elements in the SbSeI pellet was almost uniform (Fig. 3.9b–e).

The Maxwell–Wagner-Sillars interfacial polarization phenomena [39] were revealed in the compressed SbSI nanowires [37]. The dielectric response of this material was examined as a function of temperature and frequency. The Cole-Davidson and Cole–Cole functions were fitted to the experimental data recorded in the ferroelectric and paraelectric phases, respectively. The spontaneous polarization was responsible for an asymmetric broadening of relaxation functions in the ferroelectric state. It was demonstrated by the pear-like shape curves in the Cole–Cole-type plots. The symmetric broadening of relaxation functions was observed in the paraelectric phase.

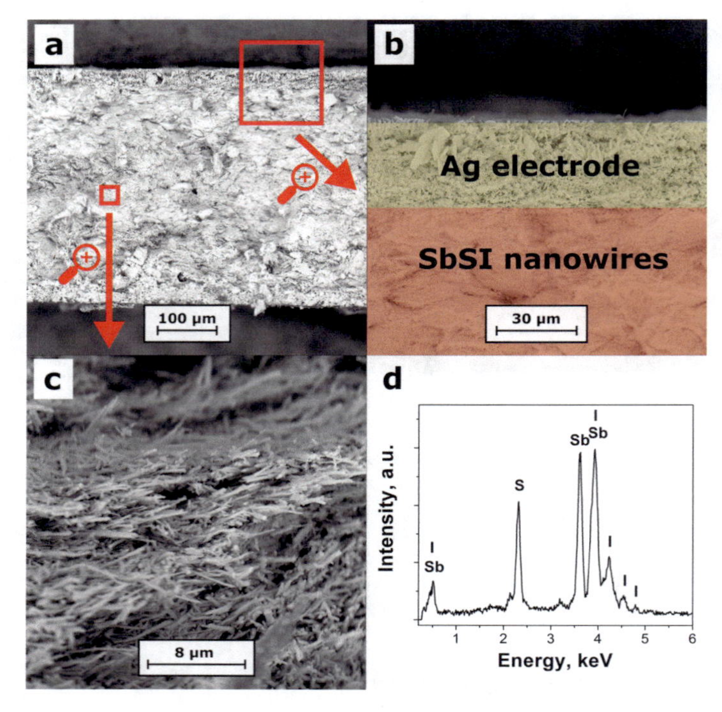

Fig. 3.10 SEM micrographs of the cross-section (**a**), edge (**b**), and interior (**c**) of the bulk sample prepared via compression of the SbSI nanowires under a pressure of 160 MPa at room temperature. The EDS spectrum (**d**) of compacted SbSI nanowires. Reprinted from Starczewska et al. [37] under the terms of the Creative Commons Attribution 4.0 International License (CC BY 4.0). Copyright (2022) MDPI

The compression of chalcohalide nanowires into the bulk samples possesses several advantages. This method is a facile and fast. It is carried out at room temperature. In addition, this technology is useful for preparation of the samples with desired dimensions and geometry that depend on the applied mold. Thus, the samples with high area of electrodes and relatively low thickness are easily fabricated. In result, an electric resistance of the sample can be reduced what is beneficial in the case of highly resistive one-dimensional ferroelectric nanomaterials. The compression of chalcohalide nanowires into the bulk samples enables to obtain porous structures with the incomplete packing factor (41–50%). They are promising for application in sensitive gas nanosensors due to large surface to volume ratio.

3.6 A Pressure Assisted Sintering

A modification of the technology, described in the section above, was presented in Ref. [16, 40]. The nanocomposite of BiSI nanorods and amorphous carbon structures was synthesized using solvothermal method [16]. The different samples were prepared using various processing conditions. The BiSI-amorphous carbon nanocomposite was compressed by applying only uniaxial pressure of about 694 MPa or 694 MPa uniaxial pressure followed by using of 5.2 GPa isostatic strain. Such prepared pellets were annealed at the temperature of 523 K for 3.5 h under the argon atmosphere. The gold electrodes were evaporated thermally on the top and bottom sides of the pellets. The samples were encapsulated with the acrylic. The pellets of the BiSI-amorphous carbon nanocomposite were tested as detectors of the ionizing radiation. The devices were exposed to the radiation, emitted by the ^{241}Am source, and corresponding electric current response was measured. It was concluded in Ref. [16] that the annealing of pellets played a crucial role in improvement of the detectors performance. The high pressure compression of BiSI-amorphous carbon nanocomposite followed by the heat treatment was found as a simpler method of radiation detectors fabrication than standard technologies based on the growth of single crystals or thin films.

Interesting approach of preparation of porous-structured photoelectrodes based on BiOI and BiSX (X = Br, I) was presented in Ref. [40]. At first, the suspension of BiSI particles in an isopropyl alcohol was drop-casted on the FTO glass substrate and subsequently dried at room temperature. This procedure was repeated three times. Afterwards, the sample was heated at 323 K for 20 min in air. Then, a quartz glass plate was pressed against the BiSI/FTO sample under a pressure of 5 MPa for 30 min. A temperature was maintained at 323 K during this experiment. Finally, the sample was cooled down to the room temperature and quartz glass plate was removed from the sample. A formation of densely-packed layer with reduced thickness was confirmed. The partial fusion of the BiSX particles at the electrode surfaces was observed. A pressure assisted treatment led to the improvement of the photoelectrochemical performance of the BiSI particles. This effect was due to a pressure-assisted sintering between the BiSI particles and enhancement of the contact area between them originating from the more densely-packed particle arrangement.

3.7 Electric Field Assisted Alignment of Nanowires

The alignment of one-dimensional nanostructures in the electric field has been demonstrated so far for different materials, including metal [41–43] and semiconductor [44–46] nanowires. When nanowire dielectric constant is higher than that of the environment, the nanowires tend to align orthogonal to the electrode edges, following the average direction of electric field [47]. In the case of ferroelectric nanowires, they possess the high electric dipole moment which should support the change of nanowire orientation under an electric field. The first report on alignment of SbSI nanowires under an electric field was provided in Ref. [48]. This technology was applied further by Mistewicz and coworkers to construct many functional devices, such as gas sensors, photodetector, and a prototype of solar cell (Table 3.2). All mentioned devices can be divided into two types. The first group refers to the SbSI nanowires bridged between microelectrodes with a gap (d) of 1 μm which was comparable to the average length of nanowires. The second type of the devices were constructed through an alignment of the SbSI nanowires between microscopic electrodes with relatively large electrode distance d = 250 μm. It was shown in Ref. [49] that the alignment of the SbSI nanoparticles in an external electric field can be also utilized for a mediated switching of nematic liquid crystals. The ferroelectric nanoparticles of SbSI reoriented and held liquid crystal molecules in a direction of nanoparticles' orientation even if the applied electric field attempted to orient a liquid crystal in the orthogonal direction [49].

The SbSI xerogel consists of the large number of chaotically oriented nanowires, as presented in Fig. 3.11a. The device based on the SbSI xerogel (Fig. 3.11b) exhibits following drawbacks. The electrical properties of the SbSI xerogel depend on many undesired effects, including the random distribution of contacts between separate nanowires, and contacts between the nanowires and the electrodes. Therefore, the reliability and repeatability of the devices based on the SbSI xerogel are strongly limited. These flaws are eliminated in the arrays of a few SbSI nanowires aligned between microelectrodes (Fig. 3.11c). Due to the lack of connections between the individual nanowires, the performance of such device (Fig. 3.11d) can be significantly improved in comparison to the SbSI xerogel. The nanowires alignment seems to be

Table 3.2 A comparison of the devices constructed through the alignment of the SbSI nanowires in an electric field (used abbreviations and symbols: d—electrode distance; IDEs—interdigitated electrodes; MEs—microelectrodes)

Used substrate	d, μm	Application of the device	References
Al_2O_3 substrate with Pt IDEs	250	Photodetector	[48]
Glass substrate with Au MEs	1	Hydrogen and oxygen sensor	[50]
Si/SiO_2 substrate with Au MEs	1	Piezoelectric generator	[51]
Glass substrate with Au MEs	1	Nitrous oxide sensor	[52]
Si/SiO_2 substrate with Au MEs	1	Humidity sensor	[53]
Glass substrate with Au MEs	1	Ammonia sensor	[29]
Si/SiO_2 substrate with Au MEs	1	Photodetector	[54]
Al_2O_3 substrate with Pt IDEs	250	Photovoltaic device	[55]

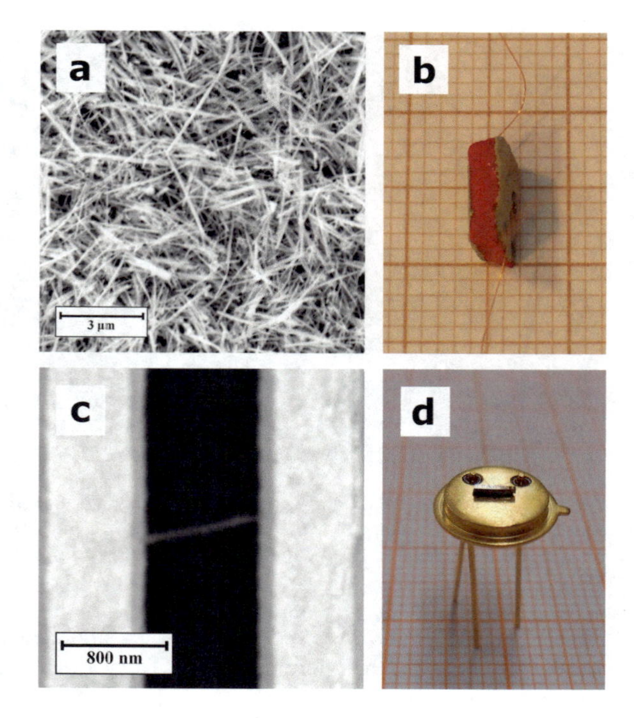

Fig. 3.11 The SEM image (**a**) of the humidity sensor (**b**) based on SbSI xerogel. The SEM image (**c**) of the SbSI nanowire aligned in electric field between gold microelectrodes on Si/SiO_2 substrate which was assembled in the standardized metal semiconductor package (**d**). Reprinted from Mistewicz et al. [53] under the terms of the Creative Commons Attribution 4.0 International License (CC BY 4.0). Copyright (2017) Springer Nature

important in the case of the piezoelectric nanogenerators [51] based on the SbSI nanowires which the best piezoelectric properties occur along their c-axis.

In a typical procedure of the SbSI nanowires alignment [29, 52, 53], the SbSI xerogel was dispersed in toluene under ultrasonic irradiation. Afterwards, a droplet of dispersed solution was dropped onto the special chip or substrate. The electric field of $5 \cdot 10^5$ V/m was applied to the electrodes during the SbSI sol deposition. Finally, the toluene was evaporated when the sample was dried at the elevated temperature. It was possible to deposit a desired amount of the SbSI nanowires on the substrate by adjusting nanowires concentration and repeating of the coating process.

Different substrates with various electrode distances were used to verify the scalability of the electric field assisted alignment of the SbSI nanowires. The SbSI nanowires were deposited onto the substrates equipped with metal electrodes separated by a gap of 1000 µm (Fig. 3.12a, e), 250 µm (Fig. 3.12b, f), 2 µm (Fig. 3.12c, g), and 1 µm (Fig. 3.12d, h). In all these cases the applied electric field was equal to $5 \cdot 10^5$ V/m. The SEM investigations (Fig. 3.12e–h) confirmed that this method is scalable and it does not depend on the electrode distance. However, one can remember that in order to align the nanowires in the large scale successfully, the high voltage must be applied.

The alignment of the SbSI nanowires can be followed by their ultrasonic welding to the microelectrodes. This technology was described in detail in Ref. [52]. The ultrasonic processing of the SbSI nanowires was performed at room temperature for 60 ms using the ultrasonic wave with frequency and amplitude of 70 kHz and

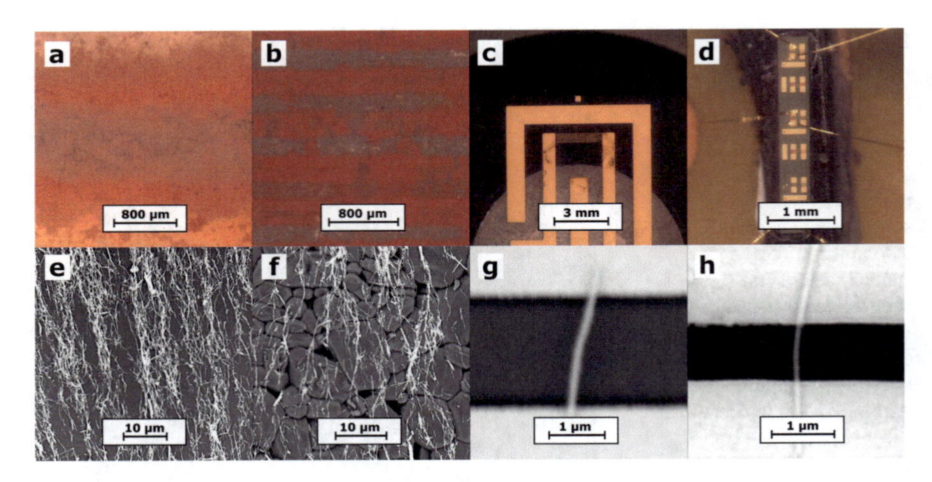

Fig. 3.12 The optical photographs of the SbSI nanowires aligned between electrodes on different substrates with various electrode distances: **a** SiO_2 (d = 1 mm), **b** Al_2O_3 (d = 250 μm), **c** SiO_2 (d = 2 μm), **d** SiO_2/Si (d = 1 μm), and corresponding SEM images (**e**–**h**)

0.04 μm, respectively. After the welding was completed, the ends of SbSI nanowires were embedded into the microelectrodes. The formation of the bonds between SbSI nanowires and microelectrodes was proved by the SEM and atomic force microscopy (AFM) investigations. It was observed in Ref. [52] that some nanowires were curved due to the ultrasonic welding. Furthermore, approximately 20% of the all SbSI nanowires were moved away from processed area of the sample. The ultrasonic welding of the SbSI nanowires resulted in increase of the electric conductance of the SbSI/Au junctions by over 400%. This effect was explained due to the enhancement of the effective interface surface between SbSI nanowires and microelectrodes.

The Arrhenius plot of electric current flowing through the aligned SbSI nanowires is presented in Fig. 3.13b. It consisted of the two linear curves with different slopes corresponding to the paraelectric and ferroelectric phases. Such change in the activation energy at the Curie temperature is typical for the ferroelectric material [56, 57]. Thus, two different theoretical dependences were least squares fitted to the experimental data by applying well known relation

$$I(T) = I_0 \cdot \exp\left(-\frac{E_A}{k_B T}\right), \tag{3.1}$$

where I_0 means the pre-exponential coefficient of the electric current, E_A refers to the activation energy, and k_B is the Boltzmann constant. The values of activation energies of 0.3083(6) eV and 0.6422(6) eV were calculated for ferroelectric and paraelectric phases, respectively. The Curie temperature of 291(2) K was determined as the intersection of the linear dependencies in the Arrhenius plot of an electric conductance (Fig. 3.13b). The literature data of Curie temperatures evaluated for

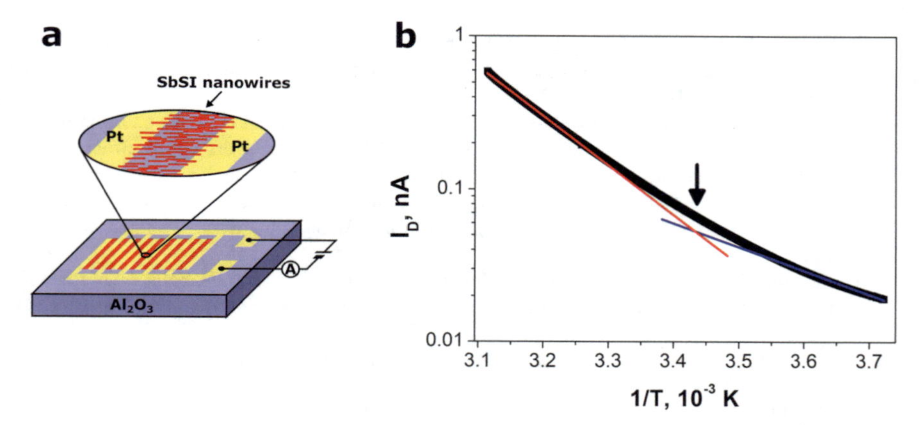

Fig. 3.13 A scheme of the SbSI nanowires aligned between Pt electrodes on the Al_2O_3 substrate. The Arrhenius plot (**b**) of electric current flowing through the aligned SbSI nanowires at the constant bias voltage. The blue and red lines in (**b**) represent the best fitted Eq. (3.1) in ferroelectric and paraelectric phases, respectively. The arrow in (**b**) indicates the Curie temperature. The values of fitted parameters are provided in the text. Reprinted from Mistewicz et al. [55] under the terms of the Creative Commons Attribution 4.0 International License (CC BY 4.0). Copyright (2019) MDPI

bulk crystals, thin films, ceramics, and nanowires of SbSI was compared in Table 1.3 in the Chap. 1 of this book.

References

1. K. Tufail Chaudhary, Thin film deposition: solution based approach, in *Thin Films* (IntechOpen, Rijeka, 2021), p. Ch. 10
2. R.M. Pasquarelli, D.S. Ginley, R. O'hayre, Solution processing of transparent conductors: from flask to film. Chem. Soc. Rev. **40**, 5406 (2011)
3. M. Eslamian, Inorganic and organic solution-processed thin film devices. Nano-Micro Lett. **9**, 3 (2017)
4. Y.C. Choi, E. Hwang, D.H. Kim, Controlled growth of SbSI thin films from amorphous Sb_2S_3 for low-temperature solution processed chalcohalide solar cells. APL Mater. **6**, 121108 (2018)
5. K.W. Jung, Y.C. Choi, Compositional engineering of antimony chalcoiodides via a two-step solution process for solar cell applications. ACS Appl. Energy Mater. (2021)
6. R. Nie, S. Il Seok, Efficient antimony-based solar cells by enhanced charge transfer. Small Methods **4**, 1900698 (2020)
7. R. Nie, H.S. Yun, M.J. Paik, A. Mehta, B.W. Park, Y.C. Choi, S. Il Seok, Efficient solar cells based on light-harvesting antimony sulfoiodide. Adv. Energy Mater. **8**, 1701901 (2018)
8. R. Nie, M. Hu, A.M. Risqi, Z. Li, S. Il Seok, Efficient and stable antimony selenoiodide solar cells. Adv. Sci. **8**, 2003172 (2021)
9. Y.C. Choi, K.W. Jung, One-step solution deposition of antimony selenoiodide films via precursor engineering for lead-free solar cell applications. Nanomaterials **11**, 3206 (2021)
10. Y.C. Choi, E. Hwang, Controlled growth of BiSI nanorod-based films through a two-step solution process for solar cell applications. Nanomaterials **9**, 1650 (2019)

11. S. Farooq, T. Feeney, J.O. Mendes, V. Krishnamurthi, S. Walia, E. Della Gaspera, J. van Embden, High gain solution-processed carbon-free BiSI chalcohalide thin film photodetectors. Adv. Funct. Mater. **31**, 2104788 (2021)
12. B. Yoo, D. Ding, J.M. Marin-Beloqui, L. Lanzetta, X. Bu, T. Rath, S.A. Haque, Improved charge separation and photovoltaic performance of BiI_3 absorber layers by use of an in situ formed BiSI interlayer. ACS Appl. Energy Mater. **2**, 7056 (2019)
13. D. Tiwari, F. Cardoso-Delgado, D. Alibhai, M. Mombrú, D.J. Fermín, Photovoltaic performance of phase-pure orthorhombic BiSI thin-films. ACS Appl. Energy Mater. **2**, 3878 (2019)
14. V. Sugathan, B. Ghosh, P.C. Harikesh, V. Kotha, P. Vashishtha, T. Salim, A. Yella, N. Mathews, Synthesis of bismuth sulphoiodide thin films from single precursor solution. Sol. Energy **230**, 714 (2021)
15. R. Nie, J. Im, S. Il Seok, Efficient solar cells employing light-harvesting $Sb_{0.67}Bi_{0.33}SI$, Adv. Mater. **31**, 1808344 (2019)
16. M.M. Frutos, M.E.P. Barthaburu, L. Fornaro, I. Aguiar, Bismuth chalcohalide-based nanocomposite for application in ionising radiation detectors. Nanotechnology **31**, 225710 (2020)
17. N.T. Hahn, A.J.E. Rettie, S.K. Beal, R.R. Fullon, C.B. Mullins, N-BiSI thin films: selenium doping and solar cell behavior. J. Phys. Chem. C **116**, 24878 (2012)
18. Y. Purusothaman, N.R. Alluri, A. Chandrasekhar, S.J. Kim, Photoactive piezoelectric energy harvester driven by antimony sulfoiodide (SbSI): A $A_VB_{VI}C_{VII}$ class ferroelectric-semiconductor compound. Nano Energy **50**, 256 (2018)
19. K. Mistewicz et al., A simple route for manufacture of photovoltaic devices based on chalcohalide nanowires. Appl. Surf. Sci. **517**, 146138 (2020)
20. T. Wu, B. Chen, Facile fabrication of porous conductive thermoplastic polyurethane nanocomposite films via solution casting. Sci. Rep. **7**, 17470 (2017)
21. M. El Achaby, F.Z. Arrakhiz, S. Vaudreuil, E.M. Essassi, A. Qaiss, Piezoelectric β-polymorph formation and properties enhancement in graphene oxide—PVDF nanocomposite films. Appl. Surf. Sci. **258**, 7668 (2012)
22. N. Jouault, D. Zhao, S.K. Kumar, Role of casting solvent on nanoparticle dispersion in polymer nanocomposites. Macromolecules **47**, 5246 (2014)
23. M. D'Arienzo et al., SiO_2/Ladder-like polysilsesquioxanes nanocomposite coatings: playing with the hybrid interface for tuning thermal properties and wettability. Coatings **10**, 913 (2020)
24. A.P. Mathew, K. Oksman, Processing of Bionanocomposites: Solution Casting, in *Handbook of Green Materials*, Vol. 5 (WORLD SCIENTIFIC, 2014), pp. 35–52
25. K. Mistewicz, Pyroelectric nanogenerator based on an SbSI-TiO_2 nanocomposite. Sensors **22**, 69 (2022)
26. S. Narayanan, R.K. Pandey, Physical vapor deposition of antimony Sulpho-Iodide (SbSI) thin films and their properties, in *IEEE International Symposium on Applications of Ferroelectrics* (1994), pp. 309–311
27. C. Wang et al., SbSI nanocrystals: an excellent visible light photocatalyst with efficient generation of singlet oxygen. ACS Sustain. Chem. Eng. **6**, 12166 (2018)
28. K. Mistewicz, M. Kępińska, M. Nowak, A. Sasiela, M. Zubko, D. Stróż, Fast and efficient piezo/photocatalytic removal of methyl orange using SbSI nanowires. Materials **13**, 4803 (2020)
29. K. Mistewicz, M. Nowak, D. Stróż, A. Guiseppi-Elie, Ferroelectric SbSI nanowires for ammonia detection at a low temperature. Talanta **189**, 225 (2018)
30. S. Manoharan, D. Kesavan, P. Pazhamalai, K. Krishnamoorthy, S.J. Kim, Ultrasound irradiation mediated preparation of antimony Sulfoiodide (SbSI) nanorods as a high-capacity electrode for electrochemical supercapacitors. Mater. Chem. Front. **5**, 2303 (2021)
31. K. Choi, J. Bang, I. kyu Moon, K. Kim, J. Oh, Enhanced photoelectrochemical efficiency and stability using nitrogen-doped TiO_2 on a GaAs photoanode. J. Alloys Compd. **843**, 155973 (2020)
32. K. Mistewicz, M. Jesionek, M. Nowak, M. Kozioł, SbSeI pyroelectric nanogenerator for a low temperature waste heat recovery. Nano Energy **64**, 103906 (2019)

33. K. Mistewicz, A. Starczewska, M. Jesionek, M. Nowak, M. Kozioł, D. Stróż, Humidity dependent impedance characteristics of SbSeI nanowires. Appl. Surf. Sci. **513**, 145859 (2020)
34. B. Toroń, K. Mistewicz, M. Jesionek, M. Kozioł, D. Stróż, M. Zubko, Nanogenerator for dynamic stimuli detection and mechanical energy harvesting based on compressed SbSeI nanowires. Energy **212**, 118717 (2020)
35. K. Mistewicz, M. Jesionek, H.J. Kim, S. Hajra, M. Kozioł, Ł Chrobok, X. Wang, Nanogenerator for determination of acoustic power in ultrasonic reactors. Ultrason. Sonochem. **78**, 105718 (2021)
36. B. Toroń, K. Mistewicz, M. Jesionek, M. Kozioł, M. Zubko, D. Stróż, A New hybrid piezo/triboelectric SbSeI nanogenerator. Energy **238**, 122048 (2022)
37. A. Starczewska, K. Mistewicz, M. Kozioł, M. Zubko, D. Stróż, J. Dec, Interfacial polarization phenomena in compressed nanowires of SbSI. Materials
38. M. Nowak, P. Szperlich, Bober, J. Szala, G. Moskal, D. Stróz, Sonochemical preparation of SbSI Gel. Ultrason. Sonochem. **15**, 709 (2008)
39. M. Samet, V. Levchenko, G. Boiteux, G. Seytre, A. Kallel, A. Serghei, Electrode polarization vs. maxwell-wagner-sillars interfacial polarization in dielectric spectra of materials: characteristic frequencies and scaling laws. J. Chem. Phys. **142**, 194703 (2015)
40. N. Nishimura, H. Suzuki, M. Higashi, R. Abe, A pressure-assisted low temperature sintering of particulate bismuth chalcohalides BiSX (X = Br, I) for fabricating efficient photoelectrodes with porous structures. J. Photochem. Photobiol. A Chem. **413**, 113264 (2021)
41. P.A. Smith, C.D. Nordquist, T.N. Jackson, T.S. Mayer, B.R. Martin, J. Mbindyo, T.E. Mallouk, Electric-field assisted assembly and alignment of metallic nanowires. Appl. Phys. Lett. **77**, 1399 (2000)
42. P. García-Sánchez, J.J. Arcenegui, H. Morgan, A. Ramos, Self-assembly of metal nanowires induced by alternating current electric fields. Appl. Phys. Lett. **106**, 23110 (2015)
43. Y. Cao, W. Liu, J. Sun, Y. Han, J. Zhang, S. Liu, H. Sun, J. Guo, A technique for controlling the alignment of silver nanowires with an electric field. Nanotechnology **17**, 2378 (2006)
44. B. Xie, H. Zhang, Q. Zhang, J. Zang, C. Yang, Q. Wang, M.Y. Li, S. Jiang, Enhanced energy density of polymer nanocomposites at a low electric field through aligned $BaTiO_3$ nanowires. J. Mater. Chem. A **5**, 6070 (2017)
45. D.V. Talapin, C.T. Black, C.R. Kagan, E.V. Shevchenko, A. Afzali, C.B. Murray, Alignment, electronic properties, doping, and on-chip growth of colloidal PbSe nanowires. J. Phys. Chem. C **111**, 13244 (2007)
46. S.Q. Li, Y.X. Liang, T.L. Guo, Z.X. Lin, T.H. Wang, Synthesis of vertically electric-field-aligned In_2O_3 nanowires. Mater. Lett. **60**, 1492 (2006)
47. H.E. Ruda, A. Shik, Principles of nanowire alignment in an electric field. J. Appl. Phys. **109**, 64305 (2011)
48. M. Nowak, Bober, B. Borkowski, M. Kępińska, P. Szperlich, D. Stróz, M. Sozańska, Quantum efficiency coefficient for photogeneration of carriers in SbSI nanowires, Opt. Mater. (Amst). **35**, 2208 (2013)
49. Y. Garbovskiy, A.V. Emelyanenko, A. Glushchenko, Inverse "Guest-Host" effect: ferroelectric nanoparticles mediated switching of nematic liquid crystals. Nanoscale **12**, 16438 (2020)
50. K. Mistewicz, M. Nowak, A. Starczewska, M. Jesionek, T. Rzychoń, R. Wrzalik, A. Guiseppi-Elie, Determination of electrical conductivity type of SbSI nanowires. Mater. Lett. **182**, 78 (2016)
51. K. Mistewicz, M. Nowak, D. Stróz, R. Paszkiewicz, SbSI nanowires for ferroelectric generators operating under shock pressure. Mater. Lett. **180**, 15 (2016)
52. K. Mistewicz, M. Nowak, R. Wrzalik, J. Śleziona, J. Wieczorek, A. Guiseppi-Elie, Ultrasonic processing of SbSI nanowires for their application to gas sensors. Ultrasonics **69**, 67 (2016)
53. K. Mistewicz, M. Nowak, R. Paszkiewicz, A. Guiseppi-Elie, SbSI nanosensors: from gel to single nanowire devices. Nanoscale Res. Lett. **12**, 97 (2017)
54. K. Mistewicz, Recent advances in ferroelectric nanosensors: toward sensitive detection of gas, mechanothermal signals, and radiation. J. Nanomater. **2018**, 2651056 (2018)

55. K. Mistewicz, M. Nowak, D. Stróż, A ferroelectric-photovoltaic effect in SbSI nanowires, Nanomaterials **9**, 580 (2019)
56. Dhananjay, J. Nagaraju, S.B. Krupanidhi, Off-centered polarization and ferroelectric phase transition in Li-Doped ZnO thin films grown by pulsed-laser ablation. J. Appl. Phys. **101**, 104104 (2007)
57. A. Mansingh, K.N. Srivastava, B. Singh, Effect of surface capacitance on the dielectric behavior of ferroelectric lead germanate. J. Appl. Phys. **50**, 4319 (1979)

Chapter 4
Devices for Energy Harvesting and Storage

4.1 Piezoelectric Nanogenerators

4.1.1 An Introduction to the Piezoelectric Nanogenerators

A global energy consumption is still increasing fulfill the growing human population needs. The production of the electric energy from the non-renewable resources like natural gas, petroleum, and coal, results in a releasing of the huge amount of the pollutants and waste to the natural environment. The use of the fossil fuels leads to the emission of greenhouse gases responsible for the climate change. Thus, there has risen a strong need for the devices and ecofriendly technologies that can be applied for the effective energy harvesting. The mechanical vibrations constitute the most common and accessible energy source in the human surrounding. On the other hand, a monitoring of the vibrations is critical for the prevention or early detection of the mechanical system failure. It can be accomplished with the piezoelectric transducers that offers better response linearity, long-term and temperature stability in comparison to the piezoresistive or capacitive sensors [1].

The piezoelectric effect can be divided into the two types: direct and converse one. The direct piezoelectric effect refers to a change in electrical polarization of the material and generation of the electrical charge under application of an external mechanical stress. In the converse piezoelectric effect, a mechanical deformation is observed when an electric field is applied to the material. The direct piezoelectric effect is commonly utilized for vibration or deformation detection and mechanical energy harvesting, whereas the converse piezoelectric is used in the piezoelectric actuators. The relation between the electric displacement (D_i) and the electric field (E_j) for the dielectric materials is described by the relation given below [2]

$$D_i = \sum_{j=1}^{3} \varepsilon_{ij} E_j, \tag{4.1}$$

K. Mistewicz, *Low-Dimensional Chalcohalide Nanomaterials*, NanoScience and Technology, https://doi.org/10.1007/978-3-031-25136-8_4

where ε_{ij} denotes the component of the a second-rank electric permittivity tensor of the dielectric material which is represented by the 3 × 3 matrix. Since this matrix is symmetric ($\varepsilon_{ij} = \varepsilon_{ji}$), it contains only six independent permittivity coefficients. In general, the tensorial representation of the strain and the electric displacement can be described in the reduced notation by the following equations [3, 4]

$$D_i = d_{ikl} T_{kl} + \varepsilon_{ik}^{T} E_k, \tag{4.2}$$

$$S_{ij} = s_{ijkl}^{E} T_{kl} + d_{kij} E_k, \tag{4.3}$$

where T_{kl} represents the stress component, d_{ikl} is the piezoelectric strain/charge coefficient, S_{ij} denotes strain component, s_{ijkl} is the elastic compliance of the material, the superscripts T and E indicate that the coefficients are given at constant stress and electric field, respectively. The piezoelectric constant, named as the piezoelectric charge coefficient or piezoelectric strain coefficient, is defined as

$$d_{ij} = \left(\frac{\partial D_i}{\partial T_j}\right)_E = \left(\frac{\partial S_j}{\partial E_i}\right)_T. \tag{4.4}$$

Furthermore, there are known other piezoelectric constants, such as the piezoelectric voltage coefficient

$$g_{ij} = -\left(\frac{\partial E_i}{\partial T_j}\right)_D = \left(\frac{\partial S_j}{\partial D_i}\right)_T, \tag{4.5}$$

piezoelectric stress coefficient

$$e_{ij} = -\left(\frac{\partial T_j}{\partial E_i}\right)_S = \left(\frac{\partial D_i}{\partial S_j}\right)_E, \tag{4.6}$$

and piezoelectric stiffness coefficient

$$h_{ij} = -\left(\frac{\partial E_i}{\partial S_j}\right)_D = -\left(\frac{\partial T_j}{\partial D_i}\right)_S. \tag{4.7}$$

The digits in the subscript of the piezoelectric coefficient describe direction of the polarization generation and the direction of the applied stress [4]. The Maxwell's displacement current ($\boldsymbol{J}_D$) is represented by the equation

$$\overrightarrow{J_D} = \frac{\partial \vec{D}}{\partial t} = \varepsilon_0 \frac{\partial \vec{E}}{\partial t} + \frac{\partial \vec{P}}{\partial t}, \tag{4.8}$$

where $\boldsymbol{P}$ is the electric polarization, ε_0 denotes the electric permittivity in vacuum. A definition of the electric polarization was provided in the Sect. 1.2 in the Chap. 1 of this book. The electric polarization can be expressed using formula [5]

$$P_i = e_{ijk} S_{jk}. \tag{4.9}$$

Assuming the constant electric field, the Eqs. (4.8) and (4.9) can be combined in the following relation

$$J_{Di} = \frac{\partial P_i}{\partial t} = e_{ijk} \left(\frac{\partial S}{\partial t} \right)_{jk}. \tag{4.10}$$

According to Eq. (4.10), the output current density of the piezoelectric nanogenerator is a linear function of the rate of the strain change. When the external electric field is zero and the polarization occurs only along z-axis the displacement current can be described as follows [5]

$$J_{Dz} = \frac{\partial P_z}{\partial t} = \frac{\partial \sigma_p(z)}{\partial t}, \tag{4.11}$$

where σ_p means the piezoelectric polarization charges on the surface. The open circuit voltage (U_{oc}) of the piezoelectric nanogenerator is represented by the equation

$$U_{oc} = \frac{z \sigma_p(z)}{\varepsilon}. \tag{4.12}$$

The working principle of the piezoelectric nanogenerator can be explained briefly as follows. Usually, the device consists of the piezoelectric material sandwiched between electrodes (Fig. 4.1a). When the vertical mechanical deformation is applied to the device, the piezoelectric polarization charges with density of σ_p are generated at the top and bottom sides of the piezoelectric film, as illustrated in Fig. 4.1b. The corresponding density σ describes the distribution of the free carriers on the device electrodes. The magnitude of the piezoelectric effect can be evaluated by measuring the open circuit voltage (Eq. 4.12) or short circuit current flowing through an external load (Eq. 4.11). The enhancement of the force, applied to the nanogenerator, leads to the increase of the polarization charge density (Fig. 4.1c). The piezoelectric phenomenon can be used for the strain/stress detection or energy harvesting through conversion of the mechanical energy into the electricity.

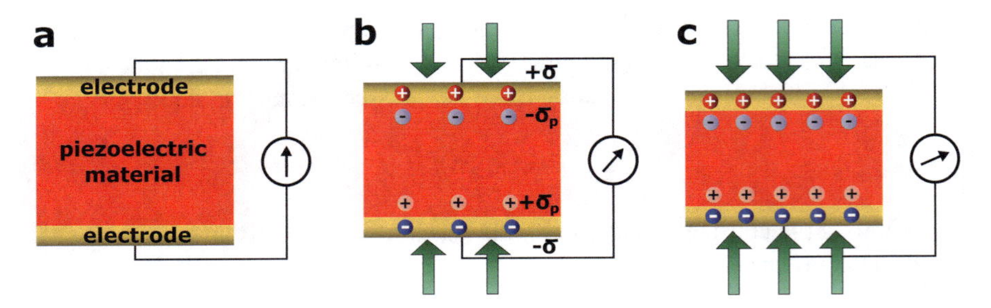

Fig. 4.1 A scheme presenting the working principle of the piezoelectric nanogenerator under unstrained condition (**a**), with applied stress (**b**), and under increased stress condition (**c**). The detailed description is provided in the text

4.1.2 Devices for Detection of Low Frequency Vibrations and Mechanical Energy Harvesting

Antimony sulfoiodide (SbSI) is known as the member of the chalcohalide family of materials with outstanding piezoelectric properties. The high piezoelectric strain coefficient of bulk SbSI crystals was studied as a function of temperature in Ref. [6]. This parameter attained large peak value $d_{33} = 2000$ pC/N at a temperature close to the Curie temperature. It was shown in Ref. [7] that the electromechanical coupling factor of SbSI single crystal reached a huge value of $k_{33} = 0.9$ just below Curie temperature. Moreover, k_{33} coefficient exhibited almost constant value (0.75–0.85) in the broad temperature range from approximately 120 K to 273 K. Bhalla and coworkers fabricated composites of SbSI crystals and Spurrs epoxy and investigated their piezoelectric properties [8]. The piezoelectric coefficients $d_{31} = 10$ pC/N and $d_{33} = 1500$ pC/N were determined for the sample that contained the highest concentration of SbSI (65%) in the composite volume. It was suggested in Ref. [9] that SbSI exhibit unique piezoelectric behavior similar to the ferroelectric hydrogen-bonded system for which phase transition occurs near room temperature. Antimony selenoiodide (SbSeI) is isostructural to SbSI. It possesses also piezoelectric properties which were confirmed theoretically [10, 11] as well as experimentally [12, 13]. An existence of large vertical piezoelectricity in the two-dimensional hexagonal chromium chalcohalides was predicted on the basis of density functional theory [14]. However, the piezoelectric properties of the numerous chalcohalide materials have not been studied so far.

The first report on application of SbSI nanowires in ferroelectric/piezoelectric nanogenerator was published by Mistewicz and coworkers in 2016 [15]. The SbSI nanowires were grown using sonochemical method and aligned in an electric filed between gold microelectrodes. The detailed descriptions of the ultrasonic synthesis of SbSI nanowires and their electric field assisted alignment are provided in the Chap. 2 (Sect. 2.5) and Chap. 3 (Sect. 3.7), respectively. The nanogenerator consisted of an array of a few SbSI nanowires welded ultrasonically to the microelectrodes. The air gun was pointed at the device and used to generate the shock waves with velocity of

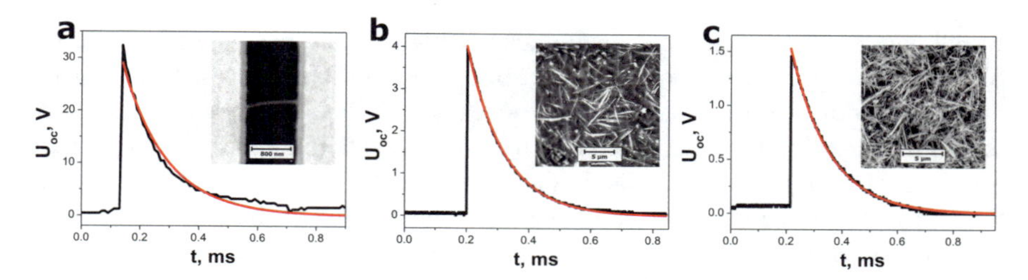

Fig. 4.2 The transient characteristics of the voltage pulse generated in **a** array of a few SbSI nanowires aligned between microelectrodes, **b** SbSI/silicone rubber nanocomposite, and **c** SbSI xerogel under a shock pressure of 5.9 MPa. The red curves represent the theoretical formula (4.13) best fitted to the experimental data. Values of the fitted parameters are presented in Table 4.1. The insets show the scanning electron microscopy (SEM) images of the selected areas of examined devices

130 m/s and pressure of 5.9 MPa. The transient characteristic of the voltage pulse was measured (Fig. 4.2a). The amplitude of the voltage response attained 29.0(7) V, what corresponded to the huge electric field of about $E = 3{\cdot}10^7$ V/m. A generation of the electric field in the SbSI nanowires under a shock pressure can originate from the two possible effects: force-electric and piezoelectric phenomena. The force-electric effect occurs, when the applied stress/strain is sufficient to reorientate the polarization. In such case, the bound charges are released and the depolarization of the ferroelectric material is observed [16–18]. This process leads to the remnant polarization loss and a decrease or elimination of the piezoelectric effect [19]. In the piezoelectric effect, remnant polarization remains unchanged under the shock compression. Herein, additional experiments were conducted to compare the response of the array of a few aligned SbSI nanowires with the voltage outputs of the SbSI/silicone rubber nanocomposite (Fig. 4.2b), and SbSI xerogel (Fig. 4.2c) under a shock pressure of 5.9 MPa.

According to Ref. [20], the voltage response is decreased exponentially after a shock compression of the ferroelectric material

$$U_{oc}(t) = U_{max}\exp\left(-\frac{t - t_0}{\tau}\right) \quad \text{for } t > t_0, \tag{4.13}$$

where: U_{max} is the peak value of the open circuit voltage generated in the ferroelectric material under shock compression, t_0 means the time when the device was subjected to a shock pressure, and τ denotes a time constant. The experimental results, presented in Fig. 4.2, were best fitted with theoretical relation (4.13). The values of the fitted parameters are given in Table 4.1. The determined time constant τ was found to be an order of magnitude larger in comparison to a response time which was less than 10 μs. It was attributed to the long lifetime of depolarizing potential and screening of the charges. The electric field, generated in the an array of a few aligned SbSI nanowires, was much higher than this parameter reported in the literature for other ferroelectric materials (Table 4.2). The values of U_{max} and E_{max} measured for aligned SbSI nanowires were significantly higher than those registered for the SbSI/silicone rubber nanocomposite and SbSI xerogel (Table 4.1) which contained a large number of chaotically oriented nanowires (Fig. 4.1b, c). The higher response of the SbSI/silicone rubber nanocomposite in comparison to this for SbSI xerogel may result from better flexibility. The additional experiments, presented herein for the first time, proved that the nanowires alignment and elimination of the contacts between separate nanowires seem to be crucial factors for achieving the large amplitude of the ferroelectric/piezoelectric nanogenerator response.

The mechanical energy harvesting using piezoelectric nanowires can be more effective in comparison to application of the conventional thin film based piezoelectric transducers due to the existence of the flexoelectric effect, large sensitivity to the small forces, and superior mechanical properties enabling longer operational lifetime, higher critical strain and flexibility [25]. It should be underlined, that SbSI possesses exceptional electromechanical properties. Moreover, the high surface to volume ratio makes them very sensitive to pressure impact. The alignment of SbSI nanowires allows to measure their ferroelectric or piezoelectric properties along *c*-axis, which are much higher than these properties existing in the perpendicular

Table 4.1 The values of the parameters of Eq. (4.13) fitted to the experimental data shown in Fig. 4.2

Nanogenerator	U_{max}, V	E_{max}, V/m	τ, μs
Array of a few aligned SbSI nanowires	29.0(7)	$2.9 \cdot 10^7$	150(5)
SbSI/silicone rubber nanocomposite	4.011(9)	$3.3 \cdot 10^3$	119.3(4)
SbSI xerogel	1.528(4)	$4.8 \cdot 10^2$	138.6(6)

Table 4.2 A comparison of the electric field (E) generated in the different ferroelectric materials under a shock pressure (p) at room temperature (used abbreviations: NBT—$Na_{0.5}Bi_{0.5}TiO_3$; PZST—$Pb_{0.99}Nb_{0.02}[(Zr_{0.90}Sn_{0.10})_{0.96}Ti_{0.04}]_{0.98}O_3$; PZT 52/48—$Pb(Zr_{0.52}Ti_{0.48})O_3$; PZT 95/5—$Pb_{0.99}Nb_{0.02}(Zr_{0.95}Ti_{0.05})_{0.98}O_3$)

Material	Morphology of the material	p, MPa	E, 10^6 V/m	Reference
PZT 52/48	Bulk ceramic	3800	3.4	[21]
	Bulk ceramic	1500	3.4(5)	[22]
PZST	Bulk ceramic	2470	3.4	[23]
PZT 95/5	Bulk ceramic	4600	3.7	[24]
	Bulk ceramic	2400	7.3	[20]
NBT	Bulk ceramic	4930	8.58	[19]
SbSI	Array of a few aligned nanowires	5.9	29.0(7)	[15]
SbSI	Xerogel (bulk sample)	5.9	$4.8 \cdot 10^{-4}$	This work
SbSI/silicone rubber	Nanocomposite (bulk sample)	5.9	$3.3 \cdot 10^{-3}$	This work

directions. It is also expected that SbSI nanowires should provide a versatile platform for self-powered sensing of different stimuli, such as strain, stress, thermal signals or radiation [26].

In 2018, Purusothaman and coworkers demonstrated photoactive piezoelectric energy harvester based on SbSI [27]. The SbSI was synthesized via a solid-state reaction from the elements (Sb, S, and I_2) at two different temperatures of 523 K and 623 K for 1 h with a ramping rate of 1 min/K. The polymethyl methacrylate (PMMA) was used as a polymer matrix interface (Fig. 4.3a). In a typical procedure, the PMMA was stirred in toluene at elevated temperature (333 K) for 1 h. Then, the SbSI (1 wt%) was added to the solution of the PMMA and toluene. After the mixture was agitated for 1 h, it was spin-coated on aluminum electrode. The spin-coating was performed several times in order to obtain full coverage of the electrode. The Al electrode was placed on the top of the SbSI-PMMA composite. The antistatic tape was attached to the electrodes. Finally, the sample was encapsulated with the polydimethylsiloxane (PDMS). The device generated the piezoelectric peak-to-peak voltage of about 5 V (Fig. 4.3b) and electric current of 0.15 μA (Fig. 4.3c) under the mechanical excitation with force and frequency of 2 N and 1.27 Hz, respectively [27]. The maximum power density of 4.6 $\mu W/m^2$ was obtained for optimum load matching resistance of 10 MΩ (Fig. 4.3d). The piezoelectric response of the SbSI-PMMA composite was studied as a function of the light illumination. When the material was illuminated with the visible radiation (630 nm), the charge carriers were photogenerated in the photoactive semiconducting SbSI. They recombined with the charge carriers arising from the piezoelectric potential generated in the SbSI-PMMA composite under mechanical strain. Therefore, an increase of the light illumination resulted in an evident decrease of the piezoelectric voltage. The SbSI-PMMA piezoelectric nanogenerator was successfully applied to power a liquid crystal display (LCD) and green light emitting diodes (LEDs), as presented in Fig. 4.3e. The existence of the piezo-phototronic effect in the SbSI confirmed the suitability of this material for its further applications in hybrid self-powered devices for the multisource energy harvesting.

The composite of SbSI nanowires and cellulose was demonstrated as a piezoelectric nanogenerator for a detection of sound and shock pressure [28]. The preparation of the material can be described as follows. The cellulose fibers with diameters in the range from 10 μm to 25 μm and the length reaching up to a few millimeters were chosen as the polymer matrix. The cellulose was dispersed in the water under ultrasonic treatment. Afterwards, the SbSI xerogel was mixed ultrasonically with the suspension of the cellulose in water for 2 h. When the homogenization of the material was completed, it was deposited on a blotting paper. Then, the SbSI-cellulose nanocomposite was compressed to obtain a sheet with desired thickness of 50 μm. The small sample was cut from this sheet. The gold electrodes were sputtered on the opposite sides of the sample. The metal wires were attached to the electrodes. The sample was encapsulated in the silicone rubber in order to eliminate an influence of humidity on the electrical properties of the SbSI-cellulose nanocomposite. The open circuit voltages of 2.5 V and 24 mV were generated by the device under shock pressure of 30 bar and sound excitation with frequency of 175 Hz, respectively [28].

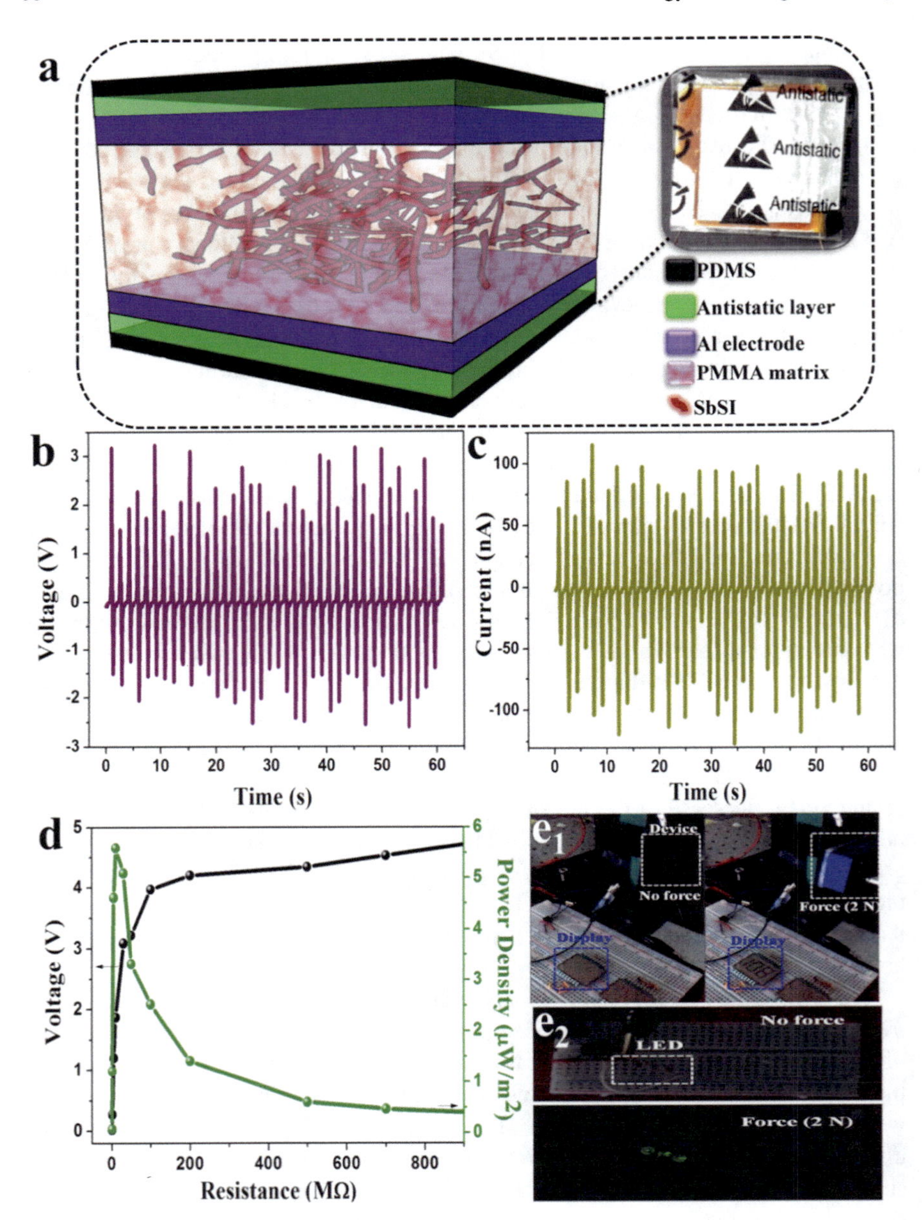

Fig. 4.3 A scheme and photograph (**a**) of piezoelectric nanogenerator based on based on the SbSI-PMMA composite, its voltage (**b**) and (**c**) current responses to the applied force of 2 N, influence of load resistance on voltage and power density (**d**), application of nanogenerator for powering liquid power display (**e1**) and light emitting diodes (**e2**). Reprinted from Purusothaman et al. [27] with permission from Elsevier. Copyright (2018) Elsevier

The fabrication of the piezoelectric fibers based on the composite of SbSI nanowires and polyvinylidene fluoride (PVDF) was presented in Ref. [29]. The suspension of the SbSI nanowires in the solution of PVDF and N,N-dimethylformamide (DMF) was prepared and homogenized. The fibers of the SbSI-PVDF nanocomposite were extruded from the melt at the temperature of 443 K and under a pressure of 10 bar using the extruder equipped with a single-hole nozzle. The diameter of the fibers was dependent on the extrusion velocity. About 300 fibers of the SbSI-PVDF nanocomposite were merged together with a silicone rubber and cut into the slab with the thickness of 1.18 mm. The top and bottom surfaces of the sample were coated with a silver paint and connected to the metal wires. Such prepared device was tested as a piezoelectric nanogenerator. The peak-to-peak voltage of approximately 2 V was measured as a response to the hitting of the sample with the frequency of 50 Hz and force of 17.8 N.

An application of the SbSeI nanowires in a piezoelectric nanogenerator was reported in Ref. [13]. The nanowires of SbSeI were grown from the chemical elements (Sb, Se, and I_2) in ethanol under ultrasonic irradiation. The SbSeI xerogel was compressed under high pressure of 120 MPa to obtain a bulk pellet of this material. This technology was discussed in detail in Chap. 3 (Sect. 3.5) of this book. The sample of compressed SbSeI nanowires was sandwiched between gold electrodes and covered with silicone rubber. The prepared device was subjected to the periodic mechanical excitation with force of 17.8 N. The frequency of the impact was adjusted in the range from 5 to 200 Hz. The maximum value of the open circuit voltage reached 0.385 V at the resonant frequency of 70 Hz. It corresponded to the maximum surface and volume power densities of 14.1 nW/cm^2 and 0.38 mW/cm^3, accordingly. The piezoelectric response of the SbSeI nanogenerator to a finger tapping was also examined. In such case, the maximum peak-to-peak value of the voltage attained 3.41 V, whereas its median was equal to 0.96 V [13].

The nanogenerator was fabricated from SbSeI xerogel via its compression under a pressure of 160 MPa at room temperature. Its piezoelectric output voltage was measured under mechanical excitation in the form of human finger tapping with two various frequencies. The average peak-to-peak voltages of 0.66(2) V (Fig. 4.4a) and 0.48(2) V (Fig. 4.4b) were determined for the excitation frequencies of 2.51 Hz and 6.41 Hz, respectively. The transient magnitude of the voltage response was dependent on the impact force applied to the SbSeI nanogenerator.

The film of SbSI/PVDF nanocomposite was prepared using a simple solution casting method. This technique was described in detail in Chap. 3 (Sect. 3.3) of this book. The gold electrodes with area of 1 cm^2 were sputtered on the opposite sides of the SbSI/PVDF layer. The piezoelectric response of the SbSI/PVDF nanocomposite film is herein presented for the first time. The piezoelectric output of the SbSI/PVDF nanogenerator was measured before and after ferroelectric poling of the device. The poling was accomplished by inserting the SbSI/PVDF nanocomposite film in silicone oil bath and applying the external electric field of 5 MV/m. The value of the applied electric field was much higher than coercive field reported in the literature for SbSI [6, 30]. The ferroelectric poling was done in order to align the ferroelectric

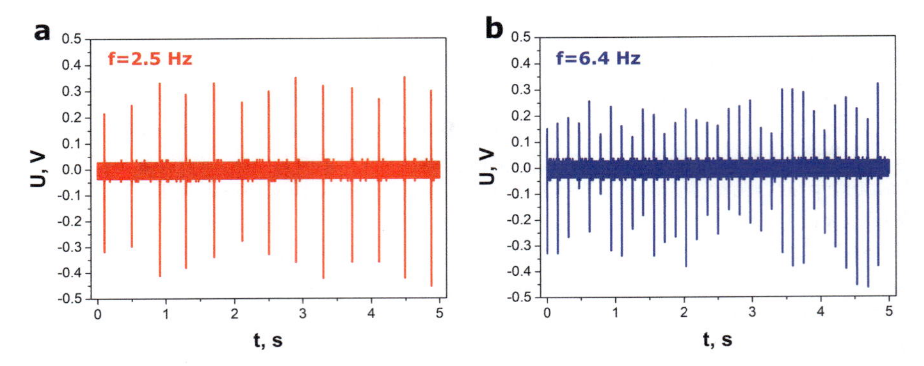

Fig. 4.4 The open circuit voltage response of the piezoelectric SbSeI nanogenerator subjected to the human finger tapping with two different excitation frequencies of **a** 2.51 Hz and **b** 6.41 Hz

dipoles in the nanocomposite and enhance its piezoelectric properties. The negligible response of the SbSI/PVDF nanogenerator was observed before ferroelectric poling. The piezoelectric output of the SbSI/PVDF nanogenerator measured after ferroelectric poling is shown in Fig. 4.5. The average peak-to-peak voltages attained the high values of 7.54(6) V, 9.03(8) V, and 6.04(2) V when SbSI/PVDF nanogenerator was hit with the frequencies of 1 Hz, 5 Hz, and 10 Hz, respectively. The average peak-to-peak currents of 7.5(1) μA, 10.7(1) μA, and 7.91(7) μA were determined for excitation frequencies of 1 Hz, 5 Hz, and 10 Hz, accordingly.

The piezoelectric performance of different chalcohalide nanogenerators is summarized in Table 4.3. The antimony sulfoiodide and antimony selenoiodide seem to be attractive for a detection of dynamic pressure changes and vibrations without a

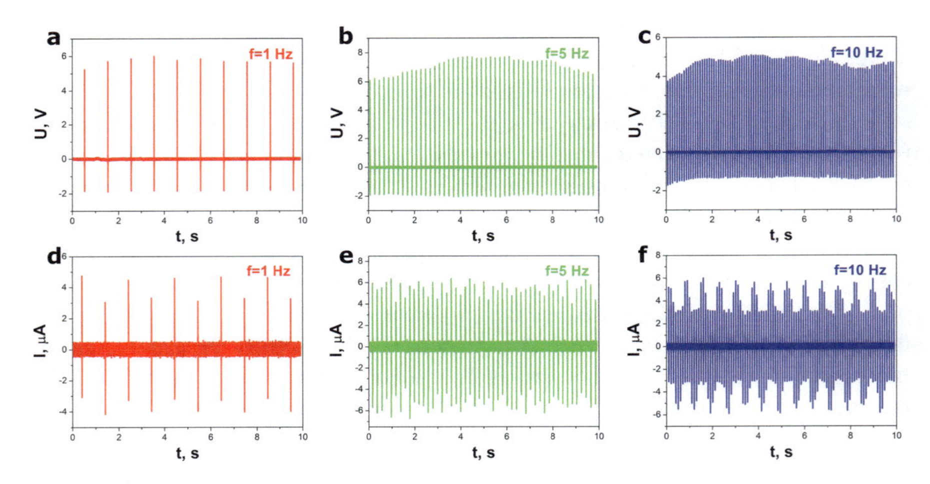

Fig. 4.5 The voltages (**a–c**) and currents (**d–f**) generated by the SbSI/PVDF nanocomposite subjected to hitting impact ($F = 50$ N) with different excitations frequencies: **a**, **d** 1 Hz, **b**, **e** 5 Hz, and **c**, **f** 10 Hz. The device was poled at external electric field of 5 MV/m

need of application of an external power source. However, the piezoelectric devices, based on chalcohalide nanomaterials, still suffer from a relatively low efficiency of energy harvesting. In the case of the devices fabricated from compressed SbSeI nanowires, they possess insufficient flexibility and low mechanical hardness [13]. Moreover, the compressed chalcohalide nanowires as well as their polymer composites have large electrical impedance that does not match with the most common electrical loads. A selection of the suitable polymer matrix may result in enhancement of the charge carriers transport and rise of the short circuit current density. The possible increase of the nanogenerator output power can be also achieved through nanowire alignment. An influence of the temperature on the piezoelectric properties of chalcohalide nanogenerators remains unknown. This effect should be studied in near future.

Table 4.3 A comparison of the piezoelectric nanogenerators based on the chalcohalide nanomaterials (used abbreviations: f—frequency of the mechanical excitation; F—force of the impact; I_{sc}—short circuit current; PAN—polyacrylonitrile; PMMA—polymethyl methacrylate; P_S—surface power density; PVDF—polyvinylidene fluoride; U_{oc}—open circuit voltage). The superscripts "avr", "max", and "pp" refer to average surface power density, maximum surface power density, and peak-to-peak open circuit voltage, respectively

Material	Excitation type	f, Hz	F, N	U_{oc}, V	P_S, μW/cm^2	References
Array of a few SbSI nanowires	Shock pressure			29		[15]
SbSI/PAN	Bending	1		0.2		[31]
SbSI/PMMA composite	Hitting	1.27	2	5 $^{(pp)}$	$4.6 \cdot 10^{-4\,(max)}$	[27]
SbSI/cellulose composite	Sonic wave	175		0.024	$2.1 \cdot 10^{-4\,(max)}$	[28]
	Shock pressure			2.5		
SbSI/PVDF composite	Vibrations	50	17.8	2.5 $^{(pp)}$	408.8	[29]
Compressed SbSeI nanowires	Finger tapping	0.75		3.41 $^{(pp)}$		[13]
	Striking	70	17.8	0.385 $^{(pp)}$	$14 \cdot 10^{-3\,(max)}$ $0.53 \cdot 10^{-3\,(avr)}$	
compressed SbSeI nanowires	Finger tapping	2.03		0.7 $^{(pp)}$		[12]
		5.56		0.55 $^{(pp)}$		
	Ultrasonic wave	$2 \cdot 10^4$		~0.13 $^{(pp)}$		
Compressed SbSeI nanowires	Finger tapping	2.51		0.66 $^{(pp)}$		This work
		6.41		0.48 $^{(pp)}$		
SbSI xerogel	Shock pressure			1.528		This work
SbSI/silicone rubber composite	Shock pressure			4.011		This work
SbSI/PVDF composite	Hitting	1	50	7.54 $^{(pp)}$	56.6 $^{(max)}$	This work
		5	50	9.03 $^{(pp)}$	96.6 $^{(max)}$	
		10	50	6.04 $^{(pp)}$	47.8 $^{(max)}$	

4.1.3 Ultrasonic Sensors

A detection of the ultrasonic waves using the self-powered piezoelectric SbSeI nanosensor was presented for the first time in Ref. [12]. The SbSeI nanowires were prepared sonochemically and compressed into a bulk pellet by applying a pressure of 160 MPa at room temperature. The gold electrodes were sputtered on the opposite sides of the SbSeI pellet. The sample was covered with silicone rubber in order to avoid the influence of water on the electric properties of SbSeI nanowires. A scheme and photograph of the fabricated device are depicted in Fig. 4.6a, b, accordingly. The nanosensor was calibrated using the VCX-750 processor that emitted the ultrasonic waves with frequency of 20 kHz and maximum acoustic power of 750 W (Fig. 4.6c). In the next step, the SbSeI nanosensor was applied to determine an acoustic power in the Sonic-6 reactor (Fig. 4.6d).

The calibration of the SbSeI sensor was carried out in two different manners. The first one was based on the fitting of the theoretical dependence to the experimental data, as shown in Fig. 4.7. It was found in Ref. [12] that the sum of two sines is the best function that describes the output voltage generated by the SbSeI sensor

$$U(t) = U_1 \sin[2\pi f_1(t - t_{01})] + U_2 \sin[2\pi f_2(t - t_{02})], \tag{4.14}$$

where U_1 and U_2 mean the amplitudes of the voltage, f_1 and f_2 are frequencies, t_{01} and t_{02} represent the time constants. The relation (4.14) was best fitted to the voltage responses of the SbSeI sensor registered under ultrasonic excitation. Its acoustic power was adjusted in the range from 150 to 750 W. The fitted parameters of the Eq. (4.14) are given in Table 4.4.

The increase of the acoustic power resulted in the rise of voltage amplitudes (U_1 and U_2), whereas the frequencies (f_1 and f_2) remained unchanged within the experimental uncertainty. The average frequencies of f_1 = 19.87(31) kHz and f_2 = 59.6(14) kHz corresponded to the fundamental and third harmonic frequencies of the VCX-750 reactor, respectively. The open-circuit voltage responses of SbSeI nanosensor to ultrasound excitation were analyzed using the Fast Fourier Transform (FFT). The FFT spectra (Fig. 4.8) consisted of the narrow sharp peaks at the fundamental (20 kHz) and third (60 kHz) frequencies. In addition, the small peaks at the other harmonics were also observed. They were attributed to the non-linear bubble oscillations [32]. The fitting of the theoretical dependence (4.14) to the experimental data (Fig. 4.7) and FFT analysis (Fig. 4.8) proved that the fundamental and third harmonic frequencies contributed mainly to the measure voltage signal. This effect is a typical for the ultrasonic transducers [32–34].

It was shown in Ref. [12] that the piezoelectric response of the SbSeI nanosensor can be expressed using following formula

$$U_{RMS}^2 = \frac{1}{2}\sum_{i=1}^{n} U_i^2 = A \cdot P_a, \tag{4.15}$$

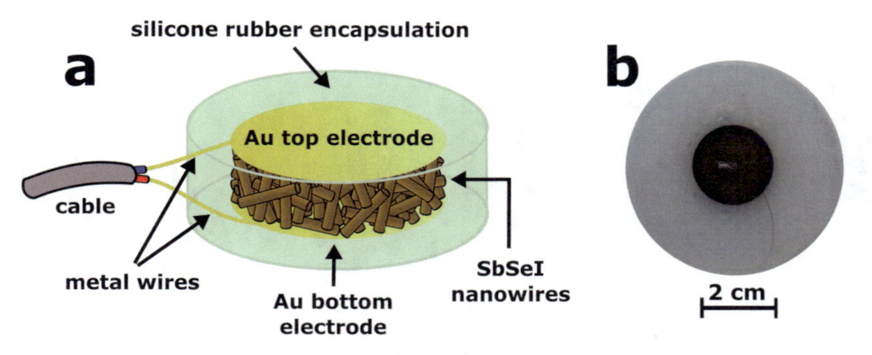

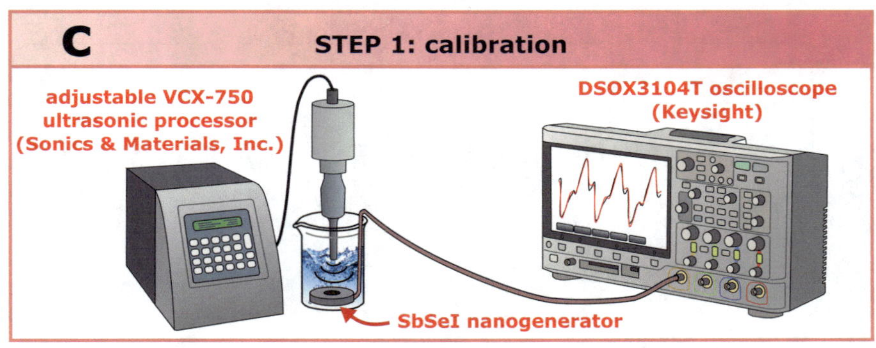

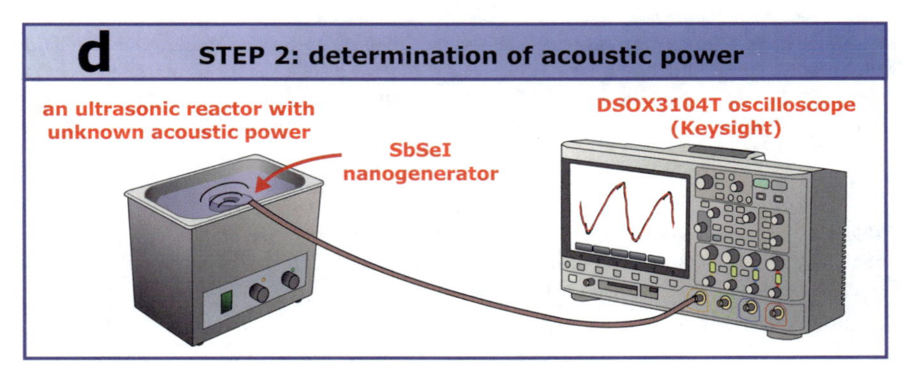

Fig. 4.6 A scheme (**a**) and photograph (**b**) of the self-powered piezoelectric nanosensor based on the compressed nanowires of SbSeI. Schematic diagrams of the measurement apparatus applied for calibration of the nanosensor (**c**) and determination of ultrasonic power in the ultrasonic reactor (**d**). Reprinted from Mistewicz et al. [12] under the terms of the Creative Commons Attribution 4.0 International License (CC BY 4.0). Copyright (2021) Elsevier

where U_{RMS} means the total root mean square (RMS) voltage generated by the SbSeI nanosensor, U_i denotes the voltage amplitude for i harmonic, n represents the total number of sinusoidal waveforms in the nanosensor response, A is a calibration coefficient determining the device sensitivity, P_a is an acoustic power of the ultrasound. The squared total RMS voltage was found to be a linear function of the acoustic

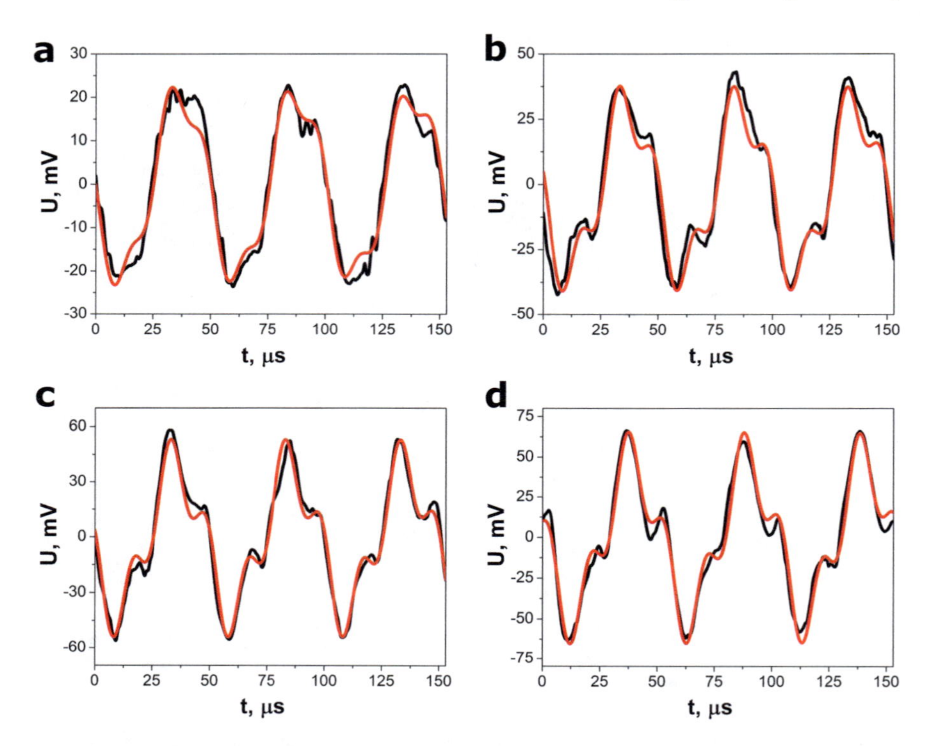

Fig. 4.7 The waveforms of voltage generated by the SbSeI nanosensor under ultrasonic excitation with a relative power of **a** 20% **b** 50%, **c** 80%, and **d** 100%. The black and red curves represent the experimental data and theoretical dependence (4.14), respectively. Reprinted from Mistewicz et al. [12] under the terms of the Creative Commons Attribution 4.0 International License (CC BY 4.0). Copyright (2021) Elsevier

Table 4.4 The parameters of Eq. (4.14) fitted to the waveforms of the voltage generated by the SbSeI nanosensor. The data was taken from Mistewicz et al. [12] under the terms of the Creative Commons Attribution 4.0 International License (CC BY 4.0). Copyright (2021) Elsevier

Relative acoustic power, %	U_1, mV	U_2, mV	f_1, kHz	F_2, kHz
20	21.07(5)	4.27(6)	19.72(1)	60.18(5)
50	32.08(6)	10.44(6)	20.070(9)	60.38(3)
80	39.13(6)	15.84(7)	19.99(1)	60.08(2)
100	47.54(7)	18.78(7)	19.701(7)	59.49(2)

power, what was in agreement with Eq. (4.15). The calibration coefficients (A) was determined by fitting the theoretical dependence (4.15) to the experimental data. Value of this parameter was used to calculate the acoustic power and conversion efficiency of the Sonic-6 reactor. The values of the acoustic power of 222(7) W and 255(8) W were determined by applying FFT analysis (Fig. 4.8) and fitting the theoretical dependence to the measured voltage waveforms (Fig. 4.7), respectively. These

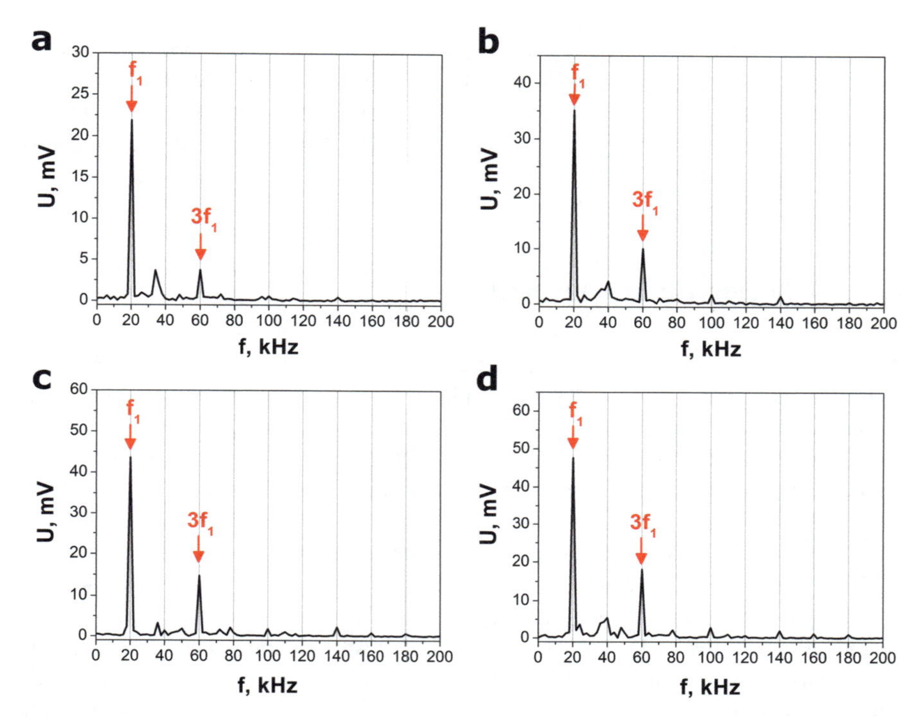

Fig. 4.8 FFT spectra of the transient characteristics of the voltage generated by the SbSeI nanosensor under ultrasonic excitation with a relative power of **a** 20%, **b** 50%, **c** 80%, and **d** 100%. Reprinted from Mistewicz et al. [12] under the terms of the Creative Commons Attribution 4.0 International License (CC BY 4.0). Copyright (2021) Elsevier

two independent methods of acoustic power determination provided similar results. Furthermore, the uncertainties of the acoustic power did not differ significantly. The method based on the Fast Fourier Transform was recognized as more sensitive, what was confirmed by slightly higher value of the calibration coefficient A. Simultaneously, this approach was found to be more demanding, since it requires much more complex computing [12].

Developed SbSeI nanosensor has a lot of advantages that are beneficial for its future application in sonochemistry and underwater acoustic measurements. First of all, it is a self-powered device. An external power supply is not needed to generate the voltage signal. Secondly, the SbSeI nanosensor operates without an amplifier. Thus, only an oscilloscope is necessary to acquire the voltage waveform generated by the SbSeI nanosensor. Thirdly, this device is portable, small, and flat. Therefore, it can be readily mounted into the ultrasonic reactor. It is in contrast to the standard needle-type hydrophones equipped with long horns [35, 36]. When the SbSeI nanosensor is calibrated for detection of acoustic waves in the certain liquid, no other characterization of the liquid is required. It is a great predominance over the calorimetry [37]. In this method, the heat capacity and mass of the liquid have to be known or determined. The last but not least advantage of the SbSeI nanosensor is very short measurement time (less than 1 ms) what allows to accomplish the real time inspection of the ultrasonic reactor operation. It is important benefit comparing to the pyroelectric sensors which the response times are of the order of a few seconds [38].

4.2 Triboelectric Nanogenerators

A novel hybrid piezo/triboelectric SbSeI nanogenerator was presented in Ref. [39]. The nanowires of SbSeI were synthesized under ultrasonic irradiation and compressed under high pressure (120 MPa) into the bulk pellet. These technologies were described extensively in the Chaps. 2 (Sect. 2.5) and 3 (Sect. 3.5) of this book, respectively. The gold electrode was sputtered on the side of the SbSeI sample with the largest area. The sample was attached to the epoxy laminate. The gold layer covered with the Kapton film served as the second electrode. The average roughness of SbSeI sample surface was 1.78 (13) μm indicating its suitability for application in the high-performance triboelectric generator. The Kapton electrode was moved periodically over the surface of the SbSeI pellet. It was accomplished using a shaker which was driven by square or triangular signal. The frequency of the shaker oscillations was adjusted in the range from 0.5 Hz to 200 Hz, whereas the amplitude of the vibrations remained constant (5 mm). The air gap between the Kapton film and SbSeI sample was changed from 0 mm up to 10 mm.

The voltage output of the SbSeI/Kapton nanogenerator originated from the both triboelectric and piezoelectric effects. The first mentioned phenomenon occurred when position of the Kapton electrode in respect to the SbSeI surface was changed in perpendicular direction (vertical contact/separation triboelectric mode) or in a parallel direction (lateral sliding triboelectric mode). The piezoelectric effect was observed when the SbSeI pellet was compressed or released due to the vertical movement of the Kapton electrode. When the SbSeI surface was in a contact with the Kapton film, the piezoelectric effect prevailed. In this case, the peak-to-peak voltage of 1.27 V was measured. After the air gap distance was increased to the 1.6 mm, the piezoelectric and triboelectric effects contributed to the nanogenerator response simultaneously. In result, peak-to-peak voltage attained the maximum value of 2.65 V. The output voltage drop was observed with the further increase of the air gap. It was attributed to the strong reduction of the piezoelectric effect and moderate decay of the friction induced triboelectricity. A huge decrease of the voltage signal occurred for air gap of 3.2 mm when the piezoelectric and friction effects were absent. When the air gap was larger than 5.5 mm, the peak-to-peak voltage was lower than 0.06 V due to limited electrostatic induction. An influence of the loaded mass on the nanogenerator response was examined. When the additional mass of 50 g was inserted onto the top electrode, the output voltage was enhanced approximately 3 times.

The piezo/triboelectric SbSeI/Kapton device generated the maximum output power of 3.35 mW per one cycle of its movement [39]. It corresponded to the surface power density of 9.57 mW/cm^2. The voltage generated per surface area and force unit (15.8 $V/(N \cdot cm^2)$) was found to be much higher than value of this parameter reported for other triboelectric nanogenerators based on the composites: PDMS/graphene oxide [40], PDMS/Ag-coated chinlon fabric [41], nylon/polytetrafluoroethylene (PTFE) [42], and cellulose/PVDF/$BaTiO_3$ [43]. The piezo/triboelectric SbSeI nanogenerator was used to power the LCD and LED connected with Graetz bridge. The aforementioned results demonstrated that SbSeI nanogenerator is suitable for its future application for mechanical energy harvesting.

Yu and coworkers developed a cutting-edge photo-enhanced triboelectric nanogenerator [44]. The bismuth oxyiodide (BiOI) and PDMS films were used as tribo-positive and tribo-negative materials, respectively. A layer of the BiOI nanoflakes was deposited on the fluorine doped indium tin oxide (FTO)/glass substrate via an acidic electrochemical method at room temperature. The sample was annealed at 623 K for 2 h in air to obtain the high crystallinity of the BiOI film. The mixture of PDMS prepolymer and crosslinker was degassed and spin-coated on the FTO/glass substrate. The thickness of the PDMS film was changed in the range from 35 μm to 110 μm by adjusting the spin-coating velocity. The PDMS layers were cured in at 378 K for 60 min. The electrical response of BiOI/PDMS nanogenerator was measured using a dynamic fatigue tester which allowed to control the contact force and frequency in the ranges of 1–50 N and 1–5 Hz, accordingly. An influence of light illumination on the nanogenerator response was examined. The BiOI/PDMS device generated peak-to-peak voltages of 59 V and 73 V under dark condition and under light illumination (AM 1.5 G), respectively. Similarly, the short-circuit current density of the BiOI/PDMS nanogenerator registered under light illumination (3.3 mA/m^2) was higher than this measured under dark condition (2.25 mA/m^2). A fluctuation of the electrical response of the BiOI/PDMS nanogenerator did not exceed 5% during the 24,000 contact cycles. It proved outstanding stability of the device. A strong enhancement of the surface potential was observed due to light illumination. It was attributed to a presence of internal electric field responsible for an effective separation of the electron–hole pairs. The photo-enhanced surface potential of the BiOI decayed after a long relaxation time of 3.5 h. It was suggested that this effect should not primarily result from the recombination of photogenerated carriers alone, but from uncompensated charges with significantly longer lifetimes or other physical properties, e.g. large dielectric constant of the material, slow charge exchange with the ambient, or an existence of deep surface states and deep traps [44]. The power surface density of the BiOI/PDMS nanogenerator attained the maximum value of 0.25 W/m^2 in the dark condition, whereas this parameter was enhanced up to 0.44 W/m^2 under light illumination. It was concluded that BiOI nanoflakes are promising for application in self-powered photodetectors and triboelectric energy harvesters with gained mechanical to electrical efficiency.

4.3 Pyroelectric Nanogenerators

Pyroelectric generators are devices that convert the thermal fluctuations into the electric energy [45]. They are promising for a recovery of a low-temperature waste heat which still suffers from low efficiency using conventional technologies [46]. Recently, a great interest in fabrication and investigation of the low-dimensional pyroelectric nanomaterials has been observed. Since the quantum confinement exists in the pyroelectric nanomaterials [47], the increased efficiency of thermal energy conversion to electrical power is expected. Until now, numerous different nanomaterials have been applied in pyroelectric nanogenerators, including non-ferroelectric nanostructures [48, 49], pristine ferroelectric polymers [50, 51], inorganic ferroelectric thin films [52], ferroelectric nanowires [53], and ferroelectric composites [54–56].

The first papers on the investigation of the pyroelectric effect in the chalcohalide materials were published in the 60 s of the twentieth century. Imai and coworkers measured pyroelectric current in the bulk single crystal of SbSI in the temperature range from 103 to 313 K [57]. The pyroelectric properties of the SbSeI crystals were studied in Ref. [58, 59]. The huge pyroelectric coefficient of 1.2 μC/(cm^2·K) was determined for SbSI single crystal at temperature close to the ferroelectric phase transition temperature [60]. The temperature dependences of the pyroelectric coefficient of the $SbSe_xS_{1-x}I$ mixed crystals were shown in Ref. [61, 62]. The pyroelectric coefficient of 180 nC/(cm^2·K) was reported for SbSI films in Ref. [63]. An existence of the giant electrocaloric effect at room temperature under a low electric field shift of 37 kV/m was revealed in Ref. [64].

The pyroelectric properties of the SbSeI nanowires were for the first time presented in Ref. [53]. Fabrication of the SbSeI pyroelectric nanogenerator involved several steps. At first, the SbSeI nanowires were grown sonochemically and dried to obtain the SbSeI xerogel. Then, the material was compressed into a bulk pellet under a pressure of 100 MPa at room temperature. The gold electrodes were sputtered on onto opposite sides of the sample. The SbSeI pellet was mounted on the plastic substrate and encapsulated in the silicone rubber to eliminate the influence of humidity changes on the electrical properties of the examined material. Finally, the sample was poled by applying the external electric field (12.8 kV/m) in order to align the electric dipoles in the SbSeI nanowires.

The SbSeI pyroelectric nanogenerator was subjected to the periodic heating–cooling cycles, as shown in Fig. 4.9a. The rate of the temperature change is depicted in Fig. 4.9b. It contained the sharp peaks corresponding to the fast rise or drop of temperature. The output voltage (Fig. 4.9c) and current (Fig. 4.9d) were measured as the responses of the device to the thermal input. The voltage and current waveforms were found to be highly correlated to the rate of the temperature changes, what confirmed the true pyroelectric effect in SbSeI nanowires. These experiments were repeated for temperature fluctuations in different ranges. The pyroelectric current attained the highest value of 11 nA upon exposure of the device to temperature variation from 324 to 334 K. The pyroelectric coefficient (p) was calculated using the well-known formula [65–67]

$$I = pA\frac{dT}{dt}, \tag{4.16}$$

where A means an area of the electrode, and dT/dt represents the rate of temperature change. The pyroelectric coefficient increased with the temperature rise reaching the large value of $p = 44(5)$ nC/(cm^2·K) at $T = 329$ K. The output power density of the SbSeI pyroelectric generator reached maximum value of 0.59(4) μW/m^2 [53].

An efficiency of the energy harvesting of the pyroelectric materials can be evaluated using the following figure of merit [68, 69]

$$F_E = \frac{p^2}{\varepsilon_r \cdot \varepsilon_0}, \tag{4.17}$$

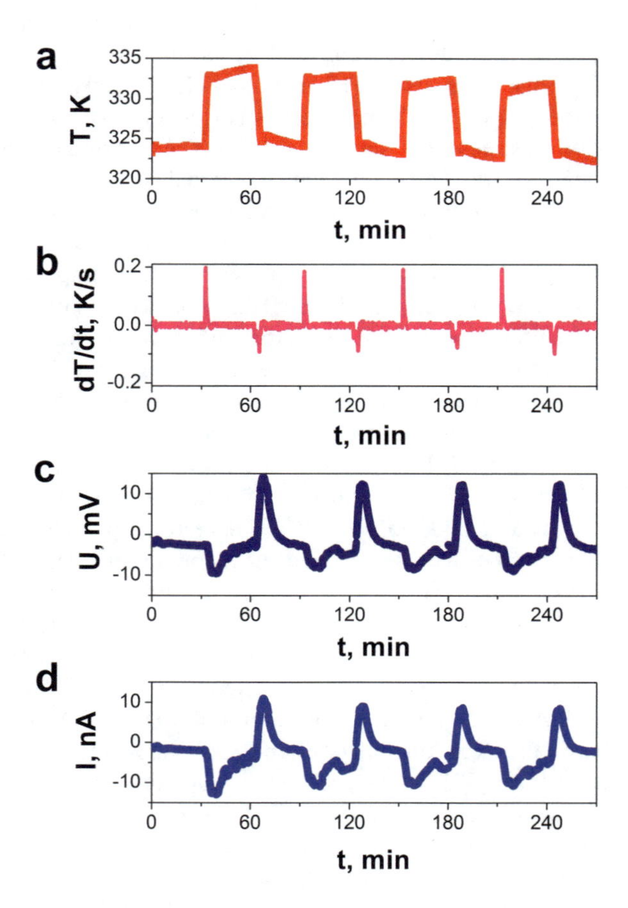

Fig. 4.9 **a** Typical periodic fluctuations of the temperature, **b** the corresponding rate of the temperature change, **c** the output voltage, and **d** current of the SbSeI pyroelectric nanogenerator

where p is pyroelectric coefficient, ε_r means a relative permittivity of the pyroelectric material, and ε_0 denotes vacuum permittivity. The energy-harvesting figure of merit $F_E = 24.30(6)$ J/(m^{3}K^2) was calculated for SbSeI pyroelectric nanogenerator at temperature of 297 K. It is higher than F_E values reported for other pyroelectric materials, e.g. Sn-doped $(Ba_{0.85}Ca_{0.15})(Zr_{0.1}Ti_{0.9})O_3$ ceramics [70], $BaSn_{0.05}Ti_{0.95}O_3$ [71], and $Ba_{0.85}Sr_{0.15}TiO_3$ [72]. A change of the energy stored in pyroelectric generator during one thermal cycle is described by the relation [69]

$$\Delta E = \frac{1}{2} \cdot \frac{p^2}{\varepsilon_r \cdot \varepsilon_0} \cdot A \cdot h \cdot (\Delta T)^2, \tag{4.18}$$

where h is thickness of the material, ΔT means temperature change. When the temperature variation was $\Delta T = 10$ K per one thermal cycle, the energy change in the SbSeI nanogenerator reached $\Delta E = 0.18(2)$ mJ.

The nanocomposite of the SbSI ferroelectric nanowires and titanium dioxide (TiO_2) non-ferroelectric nanoparticles was demonstrated for the first time in Ref. [56]. A facile drop-casting method was used to prepare a thin dense film of the SbSI-TiO_2 nanocomposite. TiO_2 nanoparticles played role the binders in the synthesized nanocomposite, which was crucial in reduction of voids between the nanowires in the porous structure of the SbSI xerogel. The film of the SbSI-TiO_2 nanocomposite was deposited on the polyethylene terephthalate (PET) substrate coated with indium tin oxide (ITO) layer. The Au electrode was sputtered on the top surface of the SbSI-TiO_2 film. The Curie temperature of the SbSI-TiO_2 was equal to 294(2) K. This value was in a good agreement with the phase transition temperatures published for ferroelectric SbSI [30, 60, 73]. The SbSI-TiO_2 device was poled to align the electric dipoles in the SbSI nanowires. Then, the device was examined as pyroelectric generator. It was inserted into the environmental test chamber. Temperature of the material was changed periodically (Fig. 4.10a, c), whereas the relative humidity was maintained at the constant level (50%). The rectangular waveform of the pyroelectric current was measured (Fig. 4.10b, d) as a response to the triangular transient characteristic of temperature. This effect was frequently observed in the case of other pyroelectric materials [49, 55, 74, 75]. The rectified pyroelectric current of the SbSI-TiO_2 nanocomposite was explained due to formation of the *pn* heterojunctions at the SbSI-ITO and SbSI-TiO_2 interfaces [56]. According to Ref. [76], a ferroelectric nanowire array can behave as the direct current generator due to possible rectification effect of the junction barrier. The rectified electric output was reported for numerous direct-current generators based on *pn* heterojunctions [77–79] and Schottky junctions [80, 81]. A potential barrier can occur at the SbSI-ITO and SbSI-TiO_2 heterojunctions, since the work function of p-type SbSI nanowires [82] is higher than these of n-type ITO layer [83] and n-type TiO_2 nanoparticles [84]. A positive pyroelectric potential was created when the SbSI-TiO_2 nanocomposite was heated ($dT/dt > 0$). The charge carriers flowed freely across the forward-biased ITO/SbSI and TiO_2/SbSI *pn* junctions (Fig. 4.10b, d). Almost zero pyroelectric current was measured when temperature was decreased ($dT/dt < 0$). It resulted from the potential barriers at the reversely biased ITO/SbSI and TiO_2/SbSI interfaces.

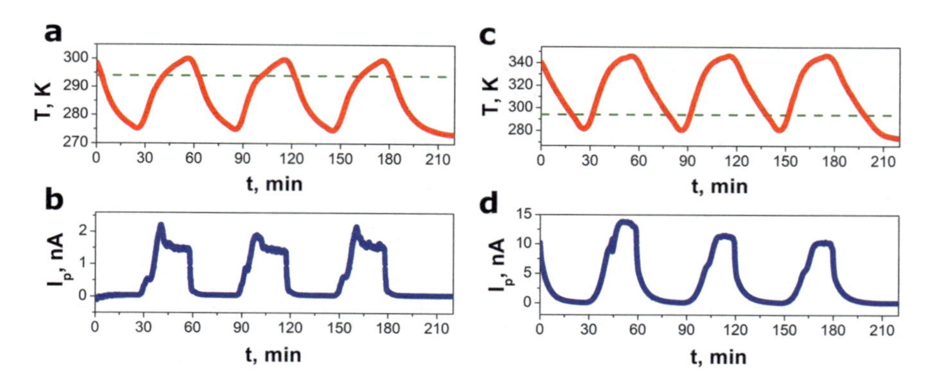

Fig. 4.10 The periodic temperature changes (**a**, **c**) and the corresponding output current (**b**, **d**) of the SbSI-TiO_2 pyroelectric nanogenerator. Dashed horizontal line shows the value of the Curie temperature (294(2) K). Reprinted from Mistewicz [56] under the terms of the Creative Commons Attribution 4.0 International License (CC BY 4.0). Copyright (2022) MDPI

According Eq. (4.16), the pyroelectric current density can be expressed as follows

$$J_p = p\frac{dT}{dt}. \tag{4.19}$$

Figure 4.11a presents the pyroelectric current of SbSI-TiO_2 nanocomposite as a linear function of the temperature change rate. Such behavior is typical for the pyroelectric effect. The theoretical formula (4.19) was best fitted to the experimental data in Fig. 4.11a to determine the pyroelectric coefficient ($p = 264(7)$ nC/(cm^2·K)) of the SbSI-TiO_2 nanocomposite. The peak values of the output voltage (U_{max}) and pyroelectric current ($I_{p\,max}$) were used to evaluate the maximum power density of the SbSI-TiO_2 nanogenerator

$$P_{S\,max} = \frac{U_{max} \cdot I_{p\,max}}{A}. \tag{4.20}$$

The average surface power density was calculated by integrating the product of transient pyroelectric voltage and current over the time.

$$P_{Savr} = \frac{1}{A \cdot \Delta t}\int_0^{\Delta t} U \cdot I_p dt, \tag{4.21}$$

where Δt is a time of the heating–cooling cycle. The values of P_{Smax} and P_{Savr} increased with a rise of temperature change rate (Fig. 4.11b). When the rate of the temperature change reached the highest value (62.5 mK/s), the maximum and average surface power densities were equal 8.39(2) μW/m^2 and 2.57(2) μW/m^2, respectively. The volume energy density of the SbSI-TiO_2 nanogenerator was determined using the following equation

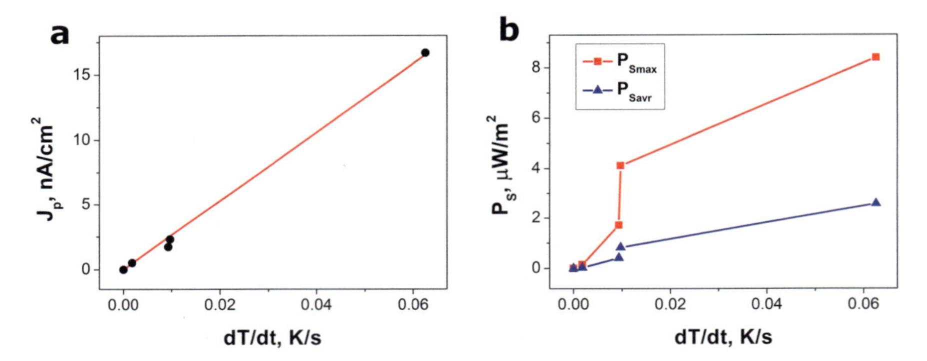

Fig. 4.11 The influence of the temperature change rate on the pyroelectric current density (**a**) and the surface power density (**b**) of the $SbSI\text{-}TiO_2$ pyroelectric nanogenerator. The solid line in (**a**) shows the theoretical formula (4.19) best fitted to the experimental data represented by the round points. The square and triangular points in (**b**) refer to the maximum and average surface power density, respectively. Reprinted from Mistewicz [56] under the terms of the Creative Commons Attribution 4.0 International License (CC BY 4.0). Copyright (2022) MDPI

$$\Delta E_V = \frac{1}{V}\int_0^t U \cdot I_p dt, \tag{4.22}$$

where V is volume of the $SbSI\text{-}TiO_2$ film. The $SbSI\text{-}TiO_2$ device generated volume energy density of 15.2 μJ/cm^3 during one heating–cooling cycle with temperature difference of 25 K.

The pyroelectric performance of the $SbSI\text{-}TiO_2$ nanogenerator was compared with values of the pyroelectric coefficients and output power densities measured for various pyroelectric materials (Table 4.5). The $SbSI\text{-}TiO_2$ nanocomposite possessed one of the highest pyroelectric coefficient among all compounds or composites presented in Table 4.5. It should be underlined that the high pyroelectric response of $SbSI\text{-}TiO_2$ nanocomposite originated from SbSI nanowires exclusively. The TiO_2 nanoparticles are expected to not contribute to the output voltage and current of the $SbSI\text{-}TiO_2$ nanogenerator since they do not exhibit pyroelectric properties. The chalcohalide nanowires have a substantial potential for application in pyroelectric detectors and devices for thermal energy harvesting. However, additional experiments are needed to increase the output power of the pyroelectric nanogenerators based on SbSI and SbSeI nanowires. The ternary pnictogen chalcohalides are highly anisotropic compounds which display the largest ferroelectric and pyroelectric properties along their lengths. Future alignment of the chalcohalide nanowires should result in an enhancement of their pyroelectric response. The pyroelectric coefficients and output power densities of the SbSI and SbSeI nanogenerators were averaged over randomly oriented nanowires (Table 4.5). These values do not represent the exact pyroelectric figures of merit of the individual nanowire. Furthermore, one can anticipate that the pyroelectric performance of the chalcohalide nanogenerator can be improved by a future selection of an appropriate polymer matrix and adjusting nanowires concentration in the nanocomposite.

4.4 Supercapacitors

Electrochemical energy conversion systems can be divided into the following groups: fuel cells, combustion engines, gas turbines, batteries, traditional capacitors, and supercapacitors [98]. The lithium-ion battery is one the most common used electrochemical energy storage technology. It is characterized by a high energy density and relatively long life cycle [99]. However, the applications of the rechargeable batteries are limited by their large internal resistances and low power densities. Supercapacitors can provide a few orders of magnitude higher power volume density in comparison to standard batteries [100]. Another advantages of the supercapacitors are very short charging time, high specific capacitance, long life cycle, and reduced memory effect [98]. The main drawback of the supercapacitors is their lower capability to store the electric charge comparing to the traditional batteries [100]. The supercapacitors are categorized into the two types: symmetrical and asymmetrical (hybrid) devices. The symmetrical supercapacitors consist of two electrodes prepared from the same material. The asymmetrical supercapacitors contain the electrodes that one of them exhibits electrochemical double-layer capacitance (EDLC) and the second electrode displays a pseudocapacitance [101]. Different materials have been used so far as the supercapacitor electrodes, including carbon materials [102–104], metal oxides [105, 106], conducting polymers [107–109], chalcogenides [110, 111], and chalcohalides [112–114]. Recent efforts toward improvement the electrochemical charge storage efficiency of the supercapacitors are focused on the optimization of electrical conductivity, structural flexibility, band gap and charge carrier mobility of the electrode material [115].

First report on application of bismuth chalcohalide, $Bi_{13}S_{18}I_2$, as electrode material in EDLC-type symmetric supercapacitor was provided in Ref. [112]. The device consisted of the $Bi_{13}S_{18}I_2$ electrode, carbon cloth, PTFE separator, and aqueous $NaClO_4$ electrolyte (Fig. 4.12a). The $Bi_{13}S_{18}I_2$ was prepared via thermal decomposition of a precursor solution containing bismuth xanthate ($Bi(xt)_3$) and BiI_3. A mixture of $Bi(xt)_3$, BiI_3, activated charcoal, and PTFE powder was dispersed in DMF. This heterogeneous solution was deposited on a carbon cloth using drop-casting method and heated at 423 K to obtain a high mass loading. A SEM analysis confirmed good coverage of the underlying carbon cloth with the rod-shaped $Bi_{13}S_{18}I_2$ (Fig. 4.12b). The electrochemical performance of the $Bi_{13}S_{18}I_2$ electrode was investigated using cyclic voltammetry (Fig. 4.12c), galvanostatic charge–discharge (Fig. 4.12d), and electrochemical impedance spectroscopy (Fig. 4.12f). The areal capacitance of $Bi_{13}S_{18}I_2$ EDLC-type symmetric supercapacitor was 210.68 mF/cm^2 [112]. A strong capacitance retention of 99.7% after 5000 cycles revealed a superior cycle stability of the device (Fig. 4.12e).

Table 4.5 An overview of the pyroelectric coefficients and surface power densities determined for ternary chalcohalides and other pyroelectric materials. The used abbreviations are as follows: BCs—bulk ceramics; BNT—$Bi_{0.5}Na_{0.5}TiO_3$; NPs—nanoparticles; NRs—nanorods; NWs—nanowires; *p*—pyroelectric coefficient; P_S—surface power density; PVC—poly(vinyl chloride); PVDF—polyvinylidene difluoride; P(VDF–TrFE)—poly(vinylidenefluoride-co-trifluoroethylene); PZT—lead zirconate titanate; SC—single crystal; TFs—thin films. The values of the maximum and average surface power densities are denoted by "max" and "avr" superscripts, respectively. The data was taken from Mistewicz [56] under the terms of the Creative Commons Attribution 4.0 International License (CC BY 4.0). Copyright (2022) MDPI

Group of materials	Material	p, nC/(cm^2·K)	P_S, μW/m^2	References
Non-ferroelectric materials	ZnO NWs	1.5		[48]
	ZnO TFs	1.0–1.4		[85]
	CdS NRs	470		[49]
Inorganic ferroelectric bulk crystals or ceramics	PZT BCs	53.3	3700^{max}	[86]
	PZT BCs	20	13.6^{avr}	[87]
	$BaTiO_3$ BCs	10	2240^{max}	[88]
	$BaTiO_3$ BCs	16		[89]
	$LiNbO_3$ SC	5–8	219^{max}	[90]
	SbSI SC	1200		[60]
Pure ferroelectric polymers	PVDF	1.94		[74]
	PVDF	4	108^{max}	[50]
	PVDF		0.13^{max}	[91]
	P(VDF-TrFE)	2.4		[92]
	P(VDF-TrFE)	4.39	128^{max}	[51]
Ferroelectric thin films or nanomaterials	$Ba_{0.8}Sr_{0.2}TiO_3$ TFs	25		[52]
	$KNbO_3$ NWs	0.8		[93]
	SbSI TFs	0.008		[94]
	SbSI TFs	180		[63]
	SbSeI NWs	44(5)	$0.59(4)^{max}$	[53]
Ferroelectric composites	$BaTiO_3$-PVC	10.6		[95]
	PVDF-diamond NPs	8.7		[96]
	PVDF-TiO_2	2.45		[74]
	PVDF-ZnO NPs	~2.9		[54]
	PVDF-$CH_3NH_3PbI_3$	0.004	1.75^{max}	[97]
	P(VDF-TrFE)-BNT NPs	5		[75]
	P(VDF-TrFE)-$PbTiO_3$ NPs	4		[55]
	SbSI NWs-TiO_2 NPs	264(7)	$8.39(2)^{max}$ $2.57(2)^{avr}$	[56]

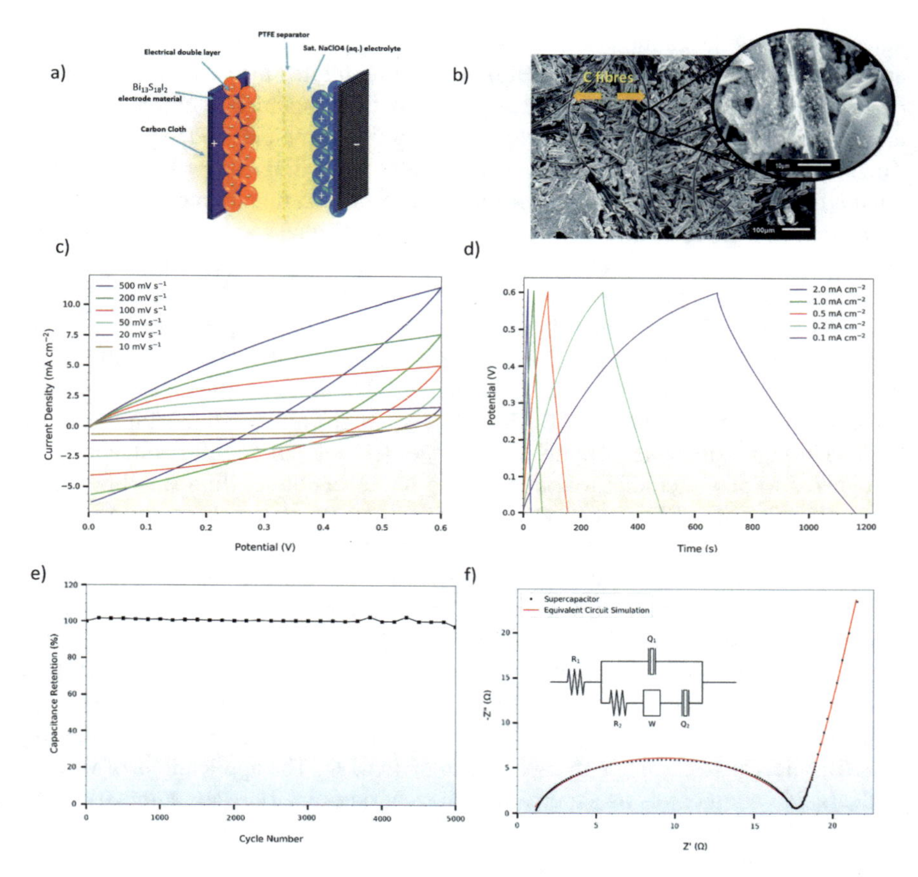

Fig. 4.12 **a** Scheme of the EDLC-type supercapacitor constructed from $Bi_{13}S_{18}I_2$ electrode material, carbon cloth, PTFE separator, and aqueous $NaClO_4$ electrolyte. **b** SEM micrograph of the $Bi_{13}S_{18}I_2$ electrode on carbon cloth. The cyclic voltammograms (**c**), galvanostatic charge–discharge curves (**d**), capacitance retention (**e**), and Nyquist plot (**f**) of the EDLC-type supercapacitor based on $Bi_{13}S_{18}I_2$. An inset in (**f**) shows the equivalent circuit of the supercapacitor. The black points and red solid curve represent experimental and simulated data, respectively. Reprinted from Adams et al. [112] under the terms of the Creative Commons Attribution 3.0 International License (CC BY 3.0). Copyright (2019) Royal Society of Chemistry.

Sun and coworkers presented a controlled and facile synthesis of two bismuth chalcohalides, i.e. BiSI and $Bi_{13}S_{18}I_2$ from solution in a single step [113]. These compounds were examined as electrode materials for supercapacitors. The BiSI and $Bi_{13}S_{18}I_2$ were mixed with activated charcoal powder and PTFE powder in ethanol to prepare working electrodes for both the three electrodes system and two-electrode EDLC system. The mixtures were homogenized under ultrasonic irradiation and drop-coated onto a conductive carbon paper. The electrodes, covered with BiSI and $Bi_{13}S_{18}I_2$, were separated by a thin polymer separator pre-soaked in a 3 M KOH aqueous electrolyte solution. The, the samples were inserted into the symmetrically-assembled capacitor test cells. The BiSI electrode exhibited the specific capacitance of 128 F/g at a current density of 2 A/g and good capacitance retention of 78.5% after 2000 cycles at a current density of 10 A/g in a three-electrode system [113]. The areal capacitance of the $Bi_{13}S_{18}I_2$ electrode of the EDLC-type supercapacitor attained 247 mF/cm^2 at a current density of 5 mA/cm^2. A retention of the $Bi_{13}S_{18}I_2$ capacitance achieved large value of 98.4% after 5000 cycles at a current density of 50 mA/cm^2. A preparation of the reduced graphene oxide (rGO) uniformly coated on BiSI composite and its use as a supercapacitor electrode was described in Ref. [116]. The BiSI-rGO electrode showed a maximum specific capacity of 234 C/g at the current density of 1 A/g and high capacity retention of 92.4% after 2000 cycles. It was found that the novel rGO coating route leads to improvement of the specific capacity and cycling stability of BiSI-rGO in comparison to pristine BiSI electrode [113]. The non-oxidizable bismuth covering the electrode surface after reduction was recognized to play a crucial role in affecting the reversible cycling stability [116]. An application of BiSI as electrode material in a photo-chargeable charge storage device was demonstrated in Ref. [117]. Zero-bias photocurrent of 0.1 μA/cm^2 was measured under 1 sun illumination. The specific capacitances of the BiSI supercapacitor were equal to 12 mF/g and 19 mF/g in dark condition and under light illumination, respectively [117]. Ultrasound irradiation mediated synthesis of SbSI nanorods was utilized to construct electrochemical energy storage devices [114]. Two different electrolytes were tested: 1 M NaOH and 1 M tetraethylammonium tetrafluoroborate ($TEABF_4$). The symmetric supercapacitor, based on the SbSI electrode and $TEABF_4$ electrolyte, had a high capacitance of 161 F/g. The applications of various chalcohalide compounds as supercapacitor electrode materials are summarized in Table 4.6. It should be concluded that the outstanding electrochemical performance, high stability, and facile synthesis of the chalcohalide materials prove their promising potential for supercapacitor applications.

Table 4.6 A comparison of the different supercapacitors based on the chalcohalide materials (used abbreviations: PVA—poly(vinyl alcohol); rGO – reduced graphene oxide; $TEABF_4$—tetraethylammonium tetrafluoroborate)

Electrode material	Electrolyte	Potential window	Specific capacity or capacitance	Current density	Capacitance retention	References
BiSI	3 M KOH	From –0.4 V to 0.4 V	128 F/g	2 A/g	78.5% after 2000 cycles	[113]
	1 M PVA-KOH		12 mF/g	0.15 μA/cm^2 at 0.4 V		[117]
BiSI-rGO	3 M KOH	From – 1.2 V to 0 V versus Ag/AgCl	234 C/g	1 A/g	92.4% after 2000 cycles	[116]
$Bi_{13}S_{18}I_2$	$NaClO_4$	From 0 V to 0.6 V	6.58 F/g	0.1 mA/cm^2	99.7% after 5000 cycles	[112]
	3 M KOH	From – 0.4 V to 0.4 V	86 F/g	2 A/g	98.4% after 5000 cycles	[113]
SbSI	1 M $TEABF_4$	From – 3 V to 3 V	161.16 F/g		91.4% after 3000 cycles	[114]

References

1. B. Yaghootkar, S. Azimi, B. Bahreyni, A high-performance piezoelectric vibration sensor. IEEE Sens. J. **17**, 4005 (2017)
2. Z. Wang, X. Pan, Y. He, Y. Hu, H. Gu, Y. Wang, Piezoelectric nanowires in energy harvesting applications. Adv. Mater. Sci. Eng. **2015**, 165631 (2015)
3. A. Erturk, D.J. Inman, *Appendix A: Piezoelectric Constitutive Equations*, in *Piezoelectric Energy Harvesting* (2011)
4. L.B. Kong, H. Huang, S. Li, *Fundamentals of Ferroelectric Materials*, in *Ferroelectric Materials for Energy Applications* (2018)
5. Z.L. Wang, On Maxwell's displacement current for energy and sensors: the origin of nanogenerators. Mater. Today **20**, 74 (2017)
6. D. Berlincourt, H. Jaffe, W.J. Merz, R. Nitsche, Piezoelectric effect in the ferroelectric range in SbSI. Appl. Phys. Lett. **4**, 61 (1964)
7. K. Hamano, T. Nakamura, Y. Ishibashi, T. Ooyane, Piezoelectric property of SbSI single crystal. J. Phys. Soc. Japan **20**, 1886 (1965)
8. A.S. Bhalla, R.E. Newnham, T.R. Shrout, L.E. Cross, Piezoelectric Sbsi: polymer composites. Ferroelectrics **41**, 207 (1982)
9. Y. Ren, M. Wu, J.M. Liu, Ultra-high piezoelectric coefficients and strain-sensitive curie temperature in hydrogen-bonded systems. Natl. Sci. Rev. **8**, nwaa203 (2021)

10. B. Peng, K. Xu, H. Zhang, Z. Ning, H. Shao, G. Ni, J. Li, Y. Zhu, H. Zhu, C.M. Soukoulis, 1D SbSeI, SbSI, and SbSBr with high stability and novel properties for microelectronic, optoelectronic, and thermoelectric applications. Adv. Theory Simulations **1**, 1700005 (2018)
11. S. Bestley Joe, S. Maflin Shaby, Performance analysis on the electrical behaviour of zinc oxide based nanowire for energy harvesting applications, in *Proceedings of the 5th International Conference on Electronics, Communication and Aerospace Technology, ICECA 2021* (2021), pp. 230–236
12. K. Mistewicz, M. Jesionek, H.J. Kim, S. Hajra, M. Kozioł, Ł Chrobok, X. Wang, Nanogenerator for determination of acoustic power in ultrasonic reactors. Ultrason. Sonochem. **78**, 105718 (2021)
13. B. Toroń, K. Mistewicz, M. Jesionek, M. Kozioł, D. Stróż, M. Zubko, Nanogenerator for dynamic stimuli detection and mechanical energy harvesting based on compressed SbSeI nanowires. Energy **212**, 118717 (2020)
14. W.Z. Xiao, L. Xu, G. Xiao, L.L. Wang, X.Y. Dai, Two-dimensional hexagonal chromium chalco-halides with large vertical piezoelectricity, high-temperature ferromagnetism, and high magnetic anisotropy. Phys. Chem. Chem. Phys. **22**, 14503 (2020)
15. K. Mistewicz, M. Nowak, D. Stróz, R. Paszkiewicz, SbSI nanowires for ferroelectric generators operating under shock pressure. Mater. Lett. **180**, 15 (2016)
16. V. Agrawal, K. Bhattacharya, Impact induced depolarization of ferroelectric materials. J. Mech. Phys. Solids **115**, 142 (2018)
17. Z. Liu et al., Lead-Free (Ag,K)NbO_3 materials for high-performance explosive energy conversion. Sci. Adv. **6**, eaba0367 (2020)
18. Z. Liu, W. Ren, H. Nie, P. Peng, Y. Liu, X. Dong, F. Cao, G. Wang, Pressure driven depolarization behavior of $Bi_{0.5}Na_{0.5}TiO_3$ based lead-free ceramics. Appl. Phys. Lett. **110**, 212901 (2017)
19. Z. Gao et al., Giant power output in lead-free ferroelectrics by shock-induced phase transition. Phys. Rev. Mater. **3**, 35401 (2019)
20. P.C. Lysne, C.M. Percival, Electric energy generation by shock compression of ferroelectric ceramics: normal-mode response of PZT 95/5. J. Appl. Phys. **46**, 1519 (1975)
21. S.I. Shkuratov, E.F. Talantsev, J. Baird, H. Temkin, L.L. Altgilbers, A.H. Stults, Longitudinal shock wave depolarization of $Pb(Zr_{52}Ti_{48})O_3$ polycrystalline ferroelectrics and their utilization in explosive pulsed power. AIP Conf. Proc. **845 II**, 1169 (2006)
22. S. Shkuratov, E. Talantsev, J. Bair, Application of piezoelectric materials in pulsed power technology and engineering, in *Piezoelectric Ceramics*, ed. by E. Talantsev (IntechOpen, Rijeka, 2010), p. Ch. 14
23. D.D. Jiang, J.M. Du, Y. Gu, Y.J. Feng, Shock wave compression of poled $Pb_{0.99}[(Zr_{0.90}Sn_{0.10})_{0.96}Ti_{0.04}]_{0.98}Nb_{0.02}O_3$ ceramics: depoling currents in axial and normal modes. Chin. Sci. Bull. **57**, 2554 (2012)
24. S.T. Montgomery, Simulation of the effects of shock stress and electrical field strength on shock-induced depoling of normally poled PZT 95/5. AIP Conf. Proc. **620**, 201 (2003)
25. X. Wang, Piezoelectric nanogenerators-harvesting ambient mechanical energy at the nanometer scale. Nano Energy **1**, 13 (2012)
26. K. Mistewicz, Recent advances in ferroelectric nanosensors: toward sensitive detection of gas, mechanothermal signals, and radiation. J. Nanomater. **2018**, 2651056 (2018)
27. Y. Purusothaman, N.R. Alluri, A. Chandrasekhar, S.J. Kim, Photoactive piezoelectric energy harvester driven by antimony Sulfoiodide (SbSI): A $A_VB_{VI}C_{VII}$ class ferroelectric-semiconductor compound. Nano Energy **50**, 256 (2018)
28. B. Toroń, P. Szperlich, M. Nowak, D. Stróż, T. Rzychoń, Novel Piezoelectric paper based on SbSI nanowires. Cellulose **25**, 7 (2018)
29. M. Jesionek, B. Toroń, P. Szperlich, W. Biniaś, D. Biniaś, S. Rabiej, A. Starczewska, M. Nowak, M. Kępińska, J. Dec, Fabrication of a new PVDF/SbSI nanowire composite for smart wearable textile. Polymer (Guildf). **180**, 121729 (2019)
30. S. Surthi, S. Kotru, R.K. Pandey, Preparation and electrical properties of ferroelectric SbSI films by pulsed laser deposition. J. Mater. Sci. Lett. **22**, 591 (2003)

31. M. Nowak, T. Tański, P. Szperlich, W. Matysiak, M. Kępińska, D. Stróż, Bober, B. Toroń, Using of sonochemically prepared SbSI for electrospun nanofibers. Ultrason. Sonochem. **38**, 544 (2017)
32. D.G. Eskin, K. Al-Helal, I. Tzanakis, Application of a plate sonotrode to ultrasonic degassing of aluminum melt: acoustic measurements and feasibility study. J. Mater. Process. Technol. **222**, 148 (2015)
33. G. Csány, M.D. Gray, M. Gyöngy, Estimation of acoustic power output from electrical impedance measurements. Acoustics
34. M.E. Frijlink, L. Løvstakken, H. Torp, Investigation of transmit and receive performance at the fundamental and third harmonic resonance frequency of a medical ultrasound transducer. Ultrasonics **49**, 601 (2009)
35. T.L. Szabo, Ultrasonic exposimetry and acoustic measurements, in *Diagnostic Ultrasound Imaging: Inside Out*, ed. by T.L.E. Szabo (Academic Press, Boston, 2014), pp. 565–604
36. K. Yasuda, T.T. Nguyen, Y. Asakura, Measurement of distribution of broadband noise and sound pressures in sonochemical reactor. Ultrason. Sonochem. **43**, 23 (2018)
37. M.A. Margulis, I.M. Margulis, Calorimetric method for measurement of acoustic power absorbed in a volume of a liquid. Ultrason. Sonochem. **10**, 343 (2003)
38. B. Zeqiri, P.N. Gélat, J. Barrie, C.J. Bickley, A novel pyroelectric method of determining ultrasonic transducer output power: device concept, modeling, and preliminary studies. IEEE Trans. Ultrason. Ferroelectr. Freq. Control **54**, 2318 (2007)
39. B. Toroń, K. Mistewicz, M. Jesionek, M. Kozioł, M. Zubko, D. Stróż, A New Hybrid Piezo/Triboelectric SbSeI Nanogenerator. Energy **238**, 122048 (2022)
40. B. Yang, W. Zeng, Z.H. Peng, S.R. Liu, K. Chen, X.M. Tao, A Fully Verified theoretical analysis of contact-mode triboelectric nanogenerators as a wearable power source. Adv. Energy Mater. **6**, 1600505 (2016)
41. J. Song, L. Gao, X. Tao, L. Li, Ultra-flexible and large-area textile-based triboelectric nanogenerators with a sandpaper-induced surface microstructure. Materials **11**, 2120 (2018)
42. Z. Zhang, J. Zhang, H. Zhang, H. Wang, Z. Hu, W. Xuan, S. Dong, J. Luo, A portable triboelectric nanogenerator for real-time respiration monitoring. Nanoscale Res. Lett. **14**, 354 (2019)
43. Z. Sun, L. Yang, S. Liu, J. Zhao, Z. Hu, W. Song, A green triboelectric nano-generator composite of degradable cellulose, piezoelectric polymers of PVDF/PA_6, and nanoparticles of $BaTiO_3$. Sensors **20**, 506 (2020)
44. Z. Yu, H. Yang, N. Soin, L. Chen, N. Black, K. Xu, P.K. Sharma, C. Tsonos, A. Kumar, J. Luo, Bismuth oxyhalide based photo-enhanced triboelectric nanogenerators. Nano Energy **89**, 106419 (2021)
45. S. Korkmaz, A. Kariper, Pyroelectric nanogenerators (PyNGs) in converting thermal energy into electrical energy: fundamentals and current status. Nano Energy **84**, 105888 (2021)
46. Z. Varga, B. Palotai, Comparison of low temperature waste heat recovery methods. Energy **137**, 1286 (2017)
47. A.N. Morozovska, E.A. Eliseev, M.D. Glinchuk, H.V. Shevliakova, G.S. Svechnikov, M.V. Silibin, A.V. Sysa, A.D. Yaremkevich, N.V. Morozovsky, V.V. Shvartsman, Analytical description of the size effect on pyroelectric and electrocaloric properties of ferroelectric nanoparticles. Phys. Rev. Mater. **3**, 104414 (2019)
48. Y. Yang, W. Guo, K.C. Pradel, G. Zhu, Y. Zhou, Y. Zhang, Y. Hu, L. Lin, Z.L. Wang, Pyroelectric nanogenerators for harvesting thermoelectric energy. Nano Lett. **12**, 2833 (2012)
49. M. Zhang, Q. Hu, K. Ma, Y. Ding, C. Li, Pyroelectric effect in CdS nanorods decorated with a molecular co-catalyst for hydrogen evolution. Nano Energy **73**, 104810 (2020)
50. T. Zhao et al., An infrared-driven flexible pyroelectric generator for non-contact energy harvester. Nanoscale **8**, 8111 (2016)
51. J. Kim, J.H. Lee, H. Ryu, J.H. Lee, U. Khan, H. Kim, S.S. Kwak, S.W. Kim, High-performance piezoelectric, pyroelectric, and triboelectric nanogenerators based on P(VDF-TrFE) with controlled crystallinity and dipole alignment. Adv. Funct. Mater. **27**, 1700702 (2017)

52. M. Mascot, D. Fasquelle, G. Velu, A. Ferri, R. Desfeux, L. Courcot, J.C. Carru, Pyro, ferro and dielectric properties of $Ba_{0.8}Sr_{0.2}TiO_3$ films deposited by Sol-Gel on platinized silicon substrates. Ferroelectrics **362**, 79 (2008)
53. K. Mistewicz, M. Jesionek, M. Nowak, M. Kozioł, SbSeI pyroelectric nanogenerator for a low temperature waste heat recovery. Nano Energy **64**, 103906 (2019)
54. K.S. Tan, W.C. Gan, T.S. Velayutham, W.H.A. Majid, Pyroelectricity enhancement of PVDF nanocomposite thin films doped with ZnO nanoparticles. Smart Mater. Struct. **23**, 125006 (2014)
55. M. Krause, I. Graz, S. Bauer-Gogonea, S. Bauer, B. Ploss, M. Zirkl, B. Stadlober, U. Helbig, $PbTiO_3$–P(VDF-TrFE)—nanocomposites for pressure and temperature sensitive skin. Ferroelectrics **419**, 23 (2011)
56. K. Mistewicz, Pyroelectric nanogenerator based on an SbSI-TiO_2 nanocomposite. Sensors **22**, 69 (2022)
57. K. Imai, S. Kawada, M. Ida, Anomalous pyroelectric properties of SbSI single crystals. J. Phys. Soc. Japan **21**, 1855 (1966)
58. B.P. Grigas, Pyroelectric effect in SbSeJ single crystals. Fiz. Tverd. Tela **9**, 2430 (1967)
59. B.P. Grigas, M. Mikalkevicius, Pyrocurrent in SbSI and SbSeI mono-crystals. Lith. J. Phys. **9**, 381 (1969)
60. W.A. Smith, J.P. Doughertyt, L.E. Cross, Pyroelectricity in SbSI. Ferroelectrics **33**, 3 (1981)
61. R. Chaves, H. Amaral, A. Levelut, S. Ziolkiewicz, M. Balkanski, M.K. Teng, J.F. Vittori, H. Stone, Tricritical point induced by atomic substitution in $SbSe_xS_{1-x}I$. Phys. Status Solidi **73**, 367 (1982)
62. A. Levelut, S. Ziolkienicz, Pyroelectric effect Of $SbSe_{0.50}S_{0.50}I$ in an applied electric field. Ferroelectrics **44**, 287 (1982)
63. S. Narayanan, R.K. Pandey, Physical vapor deposition of antimony Sulpho-Iodide (SbSI) thin films and their properties, in *IEEE International Symposium on Applications of Ferroelectrics* (1994), pp. 309–311
64. M.A. Hamad, Detecting giant electrocaloric properties of ferroelectric SbSI at room temperature. J. Adv. Dielectr. **03**, 1350008 (2013)
65. M. Sharma, R. Vaish, V.S. Chauhan, Development of figures of merit for pyroelectric energy-harvesting devices. Energy Technol. **4**, 843 (2016)
66. A. Thakre, A. Kumar, H.C. Song, D.Y. Jeong, J. Ryu, Pyroelectric energy conversion and its applications—flexible energy harvesters and sensors. Sensors **19**, 2170 (2019)
67. H. Ryu, S.W. Kim, Emerging pyroelectric nanogenerators to convert thermal energy into electrical energy. Small **17**, 1903469 (2021)
68. G. Sebald, L. Seveyrat, D. Guyomar, L. Lebrun, B. Guiffard, S. Pruvost, Electrocaloric and pyroelectric properties of $0.75Pb(Mg_{1/3}Nb_{2/3})O_3$–$0.25PbTiO_3$ single crystals. J. Appl. Phys. **100**, 124112 (2006)
69. C.R. Bowen, J. Taylor, E. Le Boulbar, D. Zabek, V.Y. Topolov, A Modified figure of merit for pyroelectric energy harvesting. Mater. Lett. **138**, 243 (2015)
70. S. Patel, Pyroelectric figures of merit and energy harvesting potential in ferroelectric cement composites. J. Mater. Sci. Mater. Electron. **31**, 16708 (2020)
71. K.S. Srikanth, V.P. Singh, R. Vaish, Enhanced pyroelectric figure of merits of porous $BaSn_{0.05}Ti_{0.95}O_3$ ceramics. J. Eur. Ceram. Soc. **37**, 3943 (2017)
72. K.S. Srikanth, V.P. Singh, R. Vaish, Pyroelectric performance of porous $Ba_{0.85}Sr_{0.15}TiO_3$ ceramics. Int. J. Appl. Ceram. Technol. **15**, 140 (2018)
73. M. Yoshida, K. Yamanaka, Y. Hamakawa, Semiconducting and dielectric properties of C-Axis oriented SbSI thin film. Jpn. J. Appl. Phys. **12**, 1699 (1973)
74. W.C. Gan, W.H.A. Abd Majid, Effect of TiO_2 on enhanced pyroelectric activity of PVDF composite. Smart Mater. Struct. **23**, 45026 (2014)
75. R.I. Mahdi, W.C. Gan, N.A. Halim, T.S. Velayutham, W.H.A. Majid, Ferroelectric and pyroelectric properties of novel lead-free polyvinylidenefluoride-trifluoroethylene-$Bi_{0.5}Na_{0.5}TiO_3$ nanocomposite thin films for sensing applications. Ceram. Int. **41**, 13836 (2015)

76. A.N. Morozovska, E.A. Eliseev, G.S. Svechnikov, S.V. Kalinin, Pyroelectric response of ferroelectric nanowires: size effect and electric energy harvesting. J. Appl. Phys. **108**, 42009 (2010)
77. L. Ren, A. Yu, W. Wang, D. Guo, M. Jia, P. Guo, Y. Zhang, Z.L. Wang, J. Zhai, P-n junction based direct-current triboelectric nanogenerator by conjunction of tribovoltaic effect and photovoltaic effect. Nano Lett. **21**, 10099 (2021)
78. Y. Lu, Z. Hao, S. Feng, R. Shen, Y. Yan, S. Lin, Direct-current generator based on dynamic PN junctions with the designed voltage output. IScience **22**, 58 (2019)
79. Y. Lu et al., Polarized water driven dynamic PN junction-based direct-current generator. Research **2021**, 1 (2021)
80. H. Zheng, R. Shen, H. Zhong, Y. Lu, X. Yu, S. Lin, Dynamic schottky diode direct-current generator under extremely low temperature. Adv. Funct. Mater. **31**, 2105325 (2021)
81. Y. Meng, L. Zhang, G. Xu, H. Wang, Direct-Current Generators Based on Conductive Polymers for Self-Powered Flexible Devices. Sci. Rep. **11**, 20258 (2021)
82. K. Mistewicz, M. Nowak, A. Starczewska, M. Jesionek, T. Rzychoń, R. Wrzalik, A. Guiseppi-Elie, Determination of electrical conductivity type of SbSI nanowires. Mater. Lett. **182**, 78 (2016)
83. M. Musztyfaga-Staszuk, Z. Starowicz, P. Panek, R. Socha, K. Gawlińska-Nęcek, The influence of material parameters on optical and electrical properties of Indium-Tin Oxide (ITO) layer. J. Phys. Conf. Ser. **1534**, 12001 (2020)
84. I. Iatsunskyi, M. Jancelewicz, G. Nowaczyk, M. Kempiński, B. Peplińska, M. Jarek, K. Załęski, S. Jurga, V. Smyntyna, Atomic layer deposition TiO_2 coated porous silicon surface: structural characterization and morphological features. Thin Solid Films **589**, 303 (2015)
85. T.J. Bukowski, K. McCarthy, F. McCarthy, G. Teowee, T.P. Alexander, D.R. Uhlmann, J.T. Dawley, B.J.J. Zelinski, Piezoelectric properties of Sol-Gel derived ZnO thin films. Integr. Ferroelectr. **17**, 339 (1997)
86. K. Zhang, Y. Wang, Z.L. Wang, Y. Yang, Standard and figure-of-merit for quantifying the performance of pyroelectric nanogenerators. Nano Energy **55**, 534 (2019)
87. T. Saito, Y. Kawai, T. Ono, Thermoelectric power generator based on a ferroelectric material combined with a bimetal thermostat, in *2013 Transducers and Eurosensors XXVII: The 17th International Conference on Solid-State Sensors, Actuators and Microsystems, TRANSDUCERS and EUROSENSORS 2013* (2013), pp. 2280–2283
88. F. Narita, M. Fox, K. Mori, H. Takeuchi, T. Kobayashi, K. Omote, Potential of energy harvesting in barium titanate based laminates from room temperature to cryogenic/high temperatures: measurements and linking phase field and finite element simulations. Smart Mater. Struct. **26**, 115027 (2017)
89. K. Zhao, B. Ouyang, Y. Yang, Enhancing photocurrent of radially polarized ferroelectric $BaTiO_3$ materials by Ferro-Pyro-phototronic effect. IScience **3**, 208 (2018)
90. H. Karim, M.R.H. Sarker, S. Shahriar, M.A.I. Shuvo, D. Delfin, D. Hodges, T.L. Tseng, D. Roberson, N. Love, Y. Lin, Feasibility study of thermal energy harvesting using lead free pyroelectrics. Smart Mater. Struct. **25**, 55022 (2016)
91. M.H. You, X.X. Wang, X. Yan, J. Zhang, W.Z. Song, M. Yu, Z.Y. Fan, S. Ramakrishna, Y.Z. Long, A self-powered flexible hybrid piezoelectric-pyroelectric nanogenerator based on non-woven nanofiber membranes. J. Mater. Chem. A **6**, 3500 (2018)
92. R.I. Mahdi, W.C. Gan, W.H. Abd Majid, Hot plate annealing at a low temperature of a thin ferroelectric P(VDF-TrFE) film with an improved crystalline structure for sensors and actuators. Sensors **14**, 19115 (2014)
93. Y. Yang, J.H. Jung, B.K. Yun, F. Zhang, K.C. Pradel, W. Guo, Z.L. Wang, Flexible pyroelectric nanogenerators using a composite structure of lead-free $KNbO_3$ nanowires. Adv. Mater. **24**, 5357 (2012)

94. T. Sudersena Rao, A. Mansinch, Electrical and optical properties of SbSI films. Jpn. J. Appl. Phys. **24**, 422 (1985)
95. M. Olszowy, C. Pawlaczyk, E. Markiewicz, J. Kułek, Dielectric and pyroelectric properties of $BaTiO_3$-PVC composites. Phys. Status Solidi Appl. Mater. Sci. **202**, 1848 (2005)
96. W. Li et al., High pyroelectric effect in Poly(Vinylidene Fluoride) composites cooperated with diamond nanoparticles. Mater. Lett. **267**, 127514 (2020)
97. A. Sultana et al., Methylammonium lead iodide incorporated Poly(Vinylidene Fluoride) nanofibers for flexible piezoelectric-pyroelectric nanogenerator. ACS Appl. Mater. Interfaces **11**, 27279 (2019)
98. Poonam, K. Sharma, A. Arora, S.K. Tripathi, Review of supercapacitors: materials and devices. J. Energy Storage **21**, 801 (2019)
99. L. Zhang, X. Hu, Z. Wang, F. Sun, D.G. Dorrell, A review of supercapacitor modeling, estimation, and applications: a control/management perspective. Renew. Sustain. Energy Rev. **81**, 1868 (2018)
100. A. González, E. Goikolea, J.A. Barrena, R. Mysyk, Review on supercapacitors: technologies and materials. Renew. Sustain. Energy Rev. **58**, 1189 (2016)
101. E.E. Miller, Y. Hua, F.H. Tezel, Materials for energy storage: review of electrode materials and methods of increasing capacitance for supercapacitors. J. Energy Storage **20**, 30 (2018)
102. Z. Li, K. Xu, Y. Pan, Recent Development of supercapacitor electrode based on carbon materials. Nanotechnol. Rev. **8**, 35 (2019)
103. S. Ghosh, R. Santhosh, S. Jeniffer, V. Raghavan, G. Jacob, K. Nanaji, P. Kollu, S.K. Jeong, A.N. Grace, Natural biomass derived hard carbon and activated carbons as electrochemical supercapacitor electrodes. Sci. Rep. **9**, 16315 (2019)
104. R. Vinodh, C.V.V.M. Gopi, V.G.R. Kummara, R. Atchudan, T. Ahamad, S. Sambasivam, M. Yi, I.M. Obaidat, H.J. Kim, A review on porous carbon electrode material derived from hypercross-linked polymers for supercapacitor applications. J. Energy Storage **32**, 101831 (2020)
105. C. An, Y. Zhang, H. Guo, Y. Wang, Metal oxide-based supercapacitors: progress and prospectives. Nanoscale Adv. **1**, 4644 (2019)
106. R. Liang, Y. Du, P. Xiao, J. Cheng, S. Yuan, Y. Chen, J. Yuan, J. Chen, Transition metal oxide electrode materials for supercapacitors: a review of recent developments, Nanomaterials **11**, 1248 (2021)
107. K.D. Fong, T. Wang, S.K. Smoukov, Multidimensional performance optimization of conducting polymer-based supercapacitor electrodes. Sustain. Energy Fuels **1**, 1857 (2017)
108. Q. Meng, K. Cai, Y. Chen, L. Chen, Research progress on conducting polymer based supercapacitor electrode materials. Nano Energy **36**, 268 (2017)
109. S. Banerjee, K.K. Kar, Conducting polymers as electrode materials for supercapacitors, in *Springer Series in Materials Science*, ed. by K.K. Kar, vol. 302 (Springer International Publishing, Cham, 2020), pp. 333–352
110. R.N.A. Raja Seman, M.A. Azam, Chalcogenide based 2D nanomaterials for supercapacitors, ed. by S. Thomas, A.B. Gueye, R.K. Gupta (Springer International Publishing, Cham, 2022), pp. 359–374
111. J. Theerthagiri, K. Karuppasamy, G. Durai, A. ul H.S. Rana, P. Arunachalam, K. Sangeetha, P. Kuppusami, H.S. Kim, Recent advances in metal chalcogenides (MX; X = S, Se) nanostructures for electrochemical supercapacitor applications: a brief review. Nanomaterials
112. K. Adams, A.F. González, J. Mallows, T. Li, J.H.J. Thijssen, N. Robertson, Facile synthesis and characterization of $Bi_{13}S_{18}I_2$ films as a stable supercapacitor electrode material. J. Mater. Chem. A **7**, 1638 (2019)
113. H. Sun, G. Yang, J. Chen, C. Kirk, N. Robertson, Facile Synthesis of BiSI and $Bi_{13}S_{18}I_2$ as stable electrode materials for supercapacitor applications. J. Mater. Chem. C **8**, 13253 (2020)
114. S. Manoharan, D. Kesavan, P. Pazhamalai, K. Krishnamoorthy, S.J. Kim, Ultrasound irradiation mediated preparation of antimony Sulfoiodide (SbSI) nanorods as a high-capacity electrode for electrochemical supercapacitors. Mater. Chem. Front. **5**, 2303 (2021)

115. S. Saha, P. Samanta, N.C. Murmu, T. Kuila, A Review on the heterostructure nanomaterials for supercapacitor application. J. Energy Storage **17**, 181 (2018)
116. H. Sun, X. Xiao, V. Celorrio, Z. Guo, Y. Hu, C. Kirk, N. Robertson, A novel method to synthesize BiSI uniformly coated with RGO by chemical bonding and its application as a supercapacitor electrode material. J. Mater. Chem. A **9**, 15452 (2021)
117. A.K. Pathak, A.C. Mohan, S.K. Batabyal, Bismuth Sulfoiodide (BiSI) for photo-chargeable charge storage device. Appl. Phys. A Mater. Sci. Process. **128**, 298 (2022)

Chapter 5
Photovoltaic Devices and Photodetectors

5.1 Ferroelectric-Photovoltaic Effect

The two fundamental processes are responsible for the photovoltaic effect. At first, the photons are absorbed by the semiconductor material leading to generation of the electrical charge carries, i.e. electrons and holes. Then, photogenerated electron–hole pairs are separated in built-in electric field arising from *pn* junction, Schottky junction or two electrodes with different work functions [1]. The energy band gap, the strength and space of built-in electric field of the semiconductor device are key factors affecting the photovoltaic performance which is usually evaluated by measuring the open circuit photovoltage (U_{OC}), short circuit photocurrent (I_{SC}) or short circuit photocurrent density (J_{SC}). The output photovoltage of the traditional photovoltaic device is limited by the bandgap of the semiconductor absorber. This problem can be overcome in the photovoltaic devices based on ferroelectric materials. It is possible to achieve a large above-bandgap photovoltage of the ferroelectric devices due to anomalous photovoltaic effect [2]. Thus, ferroelectric compounds have attracted in recent years considerable attention in the solar energy scavenging [3–6].

When a ferroelectric materials is illuminated using light with energy higher than its energy band gap, a steady photovoltage or photocurrent is generated along the polarization direction, as presented in Fig. 5.1a. The charge carriers are separated through the local electric fields originating from the electric polarization in the homogeneous bulk region of a ferroelectric with inherent non-centrosymmetry [7]. The charge carriers are moved toward device electrodes (Fig. 5.1b) and they are measured as the photovoltage or photocurrent. Phenomenon, described above, is called a ferroelectric-photovoltaic effect [8]. An orientation of the spontaneous polarization of the ferroelectric material determines a polarity of the photoresponse observed in the ferroelectric-photovoltaic effect. Therefore, the photovoltage or photocurrent can be switched from positive to negative (or in reversed direction) by applying an external electric field higher than coercive field of the ferroelectric material [9]. There have been investigated so far many different mechanisms responsible for ferroelectric-photovoltaic effect, including bulk photovoltaic effect [10–12], domain wall theory

K. Mistewicz, *Low-Dimensional Chalcohalide Nanomaterials*, NanoScience and Technology, https://doi.org/10.1007/978-3-031-25136-8_5

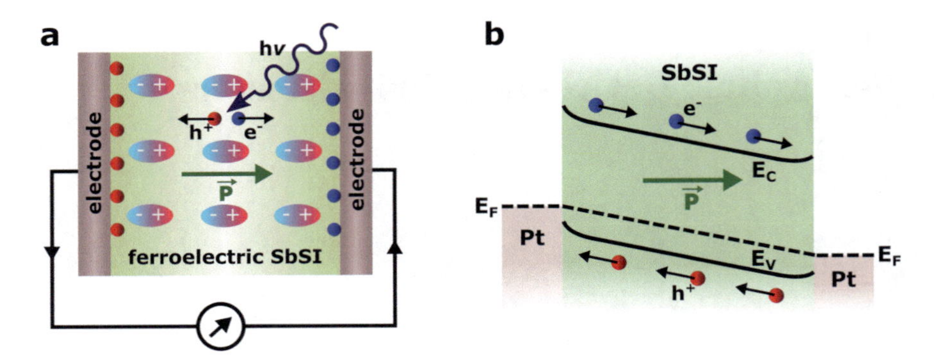

Fig. 5.1 A diagram presenting a photocurrent generation **a** and energy band diagram **b** of a poled Pt/SbSI/Pt ferroelectric-photovoltaic device. Used abbreviations and symbols: e^-—electron; h^+—hole; $h\nu$—photon; P—electric polarization; E_C—bottom of conduction energy band; E_V—top of valence energy band; E_F—Fermi level. Detailed description is given in the text. Reprinted from Mistewicz et al. [30] under the terms of the Creative Commons Attribution 4.0 International License (CC BY 4.0). Copyright (2019) MDPI

[13–15], Schottky junction theory [16, 17], and depolarization field effect [18, 19]. The current strategies of increasing photovoltaic performance of the ferroelectric based devices include flexo-photovoltaic effect application [20, 21], reducing the thickness of the ferroelectric film [16, 22], fabrication of ferroelectric-paraelectric superlattices [23], construction of layered ferroelectric/semiconductor heterostructure devices [18], using low or high work function metals as the electrodes in the case of n-type [1] or p-type [24] ferroeletrics, respectively. The ferroelectric-photovoltaic devices have a great potential in future application as solar cells [5, 25, 26], optically triggered memories [17, 27, 28], and optical transistors [29].

The photoconductivity of the ternary pnictogen chalcohalides was for the first time examined in Ref. [31]. The photovoltage and short circuit photocurrent of bulk single SbSI crystal were studied as a function of the light wavelength in Ref. [32]. Anomalous photovoltaic effect in bulk crystal of SbSI was described in Ref. [33, 34]. A linearly polarized light was used to illuminate the SbSI crystal. The photovoltaic current was measured as a function of the angle between the light polarization plane and the z-axis. The needle-shaped SbSI single crystals with four-terminal electrodes were used in Ref. [35] to investigate the local and non-local nature of bulk photovoltaic action under local photoexcitation. A shift current in SbSI single crystals was revealed. Shift current can be ascribed to the bulk photovoltaic phenomena. It comes from the geometric Berry phase of the constituting electron bands [36]. Nakamura and co-workers presented that the dissipation-less shift current dominated the current in the irradiated part of the sample, whereas the emerging internal electric field drove the spreading of current in the unirradiated part. A ultrafast evolution of the shift current in the SbSI single crystal was examined in Ref. [36] by detecting emitted terahertz electromagnetic waves. Since SbSI is usually regarded as a p-type semiconductor [37–39], the electrodes with large work functions are favorable to efficiently extract the shift current [24].

A ferroelectric-photovoltaic effect in the SbSI nanowires was for the first time revealed and studied comprehensively in Ref. [30]. The SbSI nanowires were fabricated from the elements in ethanol under ultrasonic irradiation. The details of this synthesis are described in the Chap. 2 of this book (Sect. 2.5). Afterwards, the SbSI nanowires were dispersed uniformly in toluene. Such suspension was dropped onto the alumina substrate equipped with platinum interdigitated electrodes separated by a gap of 250 μm. The SbSI nanowires were aligned perpendicularly to the electrodes by applying direct current (DC) electric field. Additional information on the electric field assisted alignment of nanowires is provided in the Chap. 3 in Sect. 3.7. Symmetric Pt electrodes were chosen to avoid an influence of different work functions and asymmetric Schottky contacts on the photovoltaic response. The photocurrent of the Pt/SbSI/Pt device was measured under vacuum in order to exclude the effect of humidity adsorption or photodesorption [40] on the electric properties of SbSI nanowires. The Pt/SbSI/Pt device was cooled down to temperature of 268 K which was much lower than Curie temperature of SbSI $T_C = 291(2)$ K [30]. The SbSI nanowires were poled by applying high external electric field of 10^6 V/m. The electric dipoles in ferroelectric SbSI nanowires were aligned in two different directions depending on whether positive (+*P*) or negative poling (–*P*) was used. The Pt/SbSI/Pt device was illuminated with monochromatic light ($\lambda = 488$ nm) emitted by argon laser. The optical power density (P_{opt}) was adjusted in the range from 0 to 127 mW/cm^2. Time dependences of short circuit photocurrent under zero bias voltage are shown in Fig. 5.2a, b. Positive and negative photocurrent responses were observed in the case of poling electric fields of -10^6 V/m and $+10^6$ V/m, respectively. It confirmed a true ferroelectric-photovoltaic effect for which photocurrent direction is opposite to the orientation of polarization vector [12]. When the Pt/SbSI/Pt device was illuminated, the photocurrent increased immediately, reached a maximum value, and decreased gradually to the steady level. The shape of the zero bias photocurrent relaxation was similar to those reported for other ferroelectric materials, e.g. lanthanum-modified lead zirconate titanate (PLZT) [1], $Pb(Zr_{20}Ti_{80})O_3$ [41], $BaTiO_3$ [42], and $BiFeO_3$ [43]. The transient characteristics of the short circuit photocurrent were least square fitted with a sum of the exponential functions [44]

$$I_{SC}(t) = I_S + I_1 \exp\left(-\frac{t - t_{on}}{\tau_1}\right) + I_2 \exp\left(-\frac{t - t_{on}}{\tau_2}\right), \tag{5.1}$$

where I_S means the constant value of short circuit photocurrent, I_1, I_2 are the pre-exponential factors, τ_1 and τ_2 are the time constants, t_{on} refers to time when the sample was illuminated. Two time constants τ_1, τ_2 resulted probably from surface recombination of photogenerated carriers and their capture in surface states. The values of the fitted parameters are listed in Table 5.1. The values of the open circuit photovoltage were calculated using following equation [10, 11, 23]

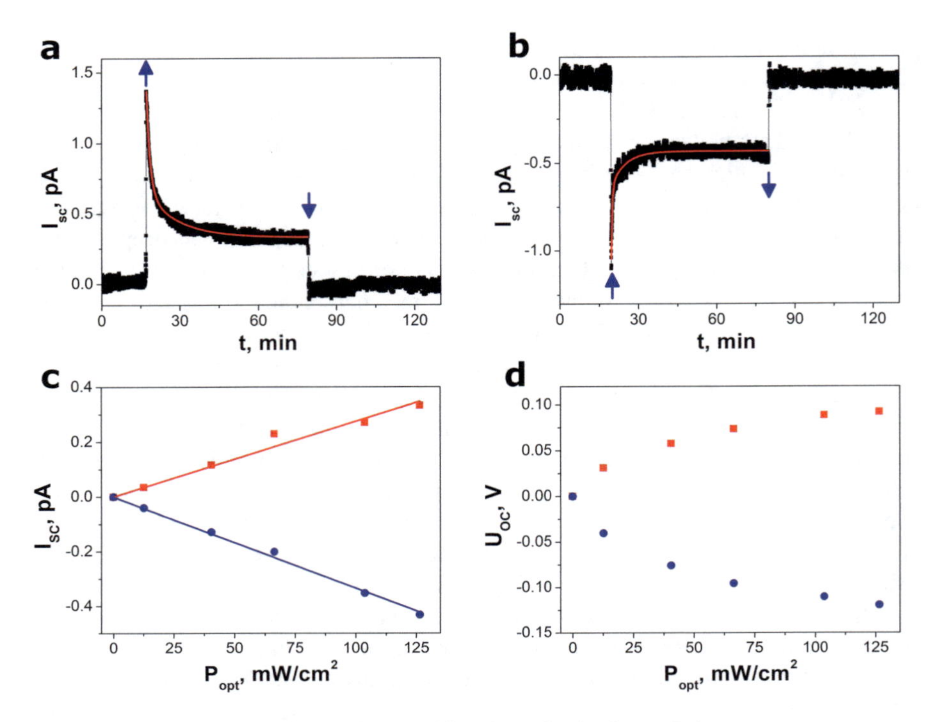

Fig. 5.2 Transient characteristics of a zero bias short circuit photovoltaic current responses on switching on (↑) and switching off (↓) illumination of the SbSI ferroelectric-photovoltaic device poled at two different electric fields: **a** $E = -10^6$ V/m and **b** $E = +10^6$ V/m. Influence of the optical power density on **c** zero bias short circuit photocurrent and **d** open circuit photovoltage of the SbSI ferroelectric-photovoltaic device poled at two different electric fields: $E = -10^6$ V/m (filled red square) and $E = + 10^6$ V/m (filled blue circle). Red solid curves in **a**, **b** show the best fitted dependences described by Eq. (5.1). The values of the fitted parameters of Eq. (5.1) are given in Table 5.1. The solid curves in **c** represent the best fitted dependences described by Eq. (5.3). The values of the fitted parameters are provided in the text. Adapted from Mistewicz et al. [30] under the terms of the Creative Commons Attribution 4.0 International License (CC BY 4.0). Copyright (2019) MDPI

$$U_{OC} = \frac{J_{SC}}{\sigma_d + \sigma_{ph}} d = R_{IL} \cdot I_{SC}, \tag{5.2}$$

where d is distance between electrodes, R_{IL} means electric resistance of SbSI nanowires under light illumination, σ_d and σ_{ph} mean dark conductivity and photoconductivity, respectively.

An increase of optical power density resulted in obvious enhancement of the photovoltaic response of the Pt/SbSI/Pt device (Fig. 5.2c, d). The short circuit density was found to be linear function of light intensity what was consistent with well-known Glass law [20, 45, 46]

$$I_{SC} = \kappa \alpha A P_{opt} = \gamma P_{opt}, \tag{5.3}$$

Table 5.1 Influence of poling electric field on parameters of dependence (5.1) best fitted to transient characteristic of the short circuit photocurrent of SbSI nanowires (Fig. 5.2a, b)

Parameter of Eq. (5.1)	Value of parameter	
	Device poled at $E = -10^6$ V/m	Device poled at $E = + 10^6$ V/m
I_S, pA	0.332(1)	– 0.432(1)
I_1, pA	0.716(7)	– 0.51(2)
I_2, pA	0.320(5)	– 0.200(5)
τ_1, s	85(2)	24(1)
τ_2, min	12.2(3)	5.5(2)

Data taken from Mistewicz et al. [30] under the terms of the

where α denotes absorption coefficient, A is the electrode area, κ is a Glass constant depending on the nature of the absorbing center and the wavelength [45]. The coefficient $\gamma = \kappa\alpha A$ was equal $-3.34(6) \cdot 10^{-16}$ m^2/V and $2.7(1) \cdot 10^{-16}$ m^2/V and in the case of positive and negative poling, respectively.

Usually, the photocurrent (or photovoltage) of the ferroelectric-photovoltaic cell consists of switchable component originating from a polarization-induced field and non-switchable component coming from a persistent built-in electric field [8].

$$I_{SC}^{p} = \frac{1}{2}\left|I_{SC}^{+} - I_{SC}^{-}\right|, \tag{5.4}$$

$$I_{SC}^{bi} = \frac{1}{2}\left|I_{SC}^{+} + I_{SC}^{-}\right|, \tag{5.5}$$

$$U_{OC}^{p} = \frac{1}{2}\left|U_{OC}^{+} - U_{OC}^{-}\right|, \tag{5.6}$$

$$U_{OC}^{bi} = \frac{1}{2}\left|U_{OC}^{+} + U_{OC}^{-}\right|, \tag{5.7}$$

where $I^p{}_{SC}$, $U^p{}_{OC}$ are short circuit photocurrent and open circuit photovoltage contributions related with the switchable ferroelectric polarization, $I^{bi}{}_{SC}$, $U^{bi}{}_{OC}$ are short circuit photocurrent and open circuit photovoltage components induced by the non-switchable internal bias field, the superscripts " + " and "–" denote the positive and negative poling, respectively. It was determined in Ref. [30] that switchable short circuit photocurrent and open circuit photovoltage of SbSI nanowires were respectively 8.5 and 7.7 times higher than those components originating from the non-switchable internal field. A weak built-in electric field in the Pt/SbSI/Pt device resulted probably from presence of iodine vacancies and heterogeneous distribution of SbSI nanowires. The photovoltaic performance of the Pt/SbSI/Pt device was compared with short circuit photocurrent and open circuit photovoltage generated by other ferroelectric-photovoltaic cells (Table 5.2). The output photocurrent and photovoltage of SbSI nanowires can be enhanced in the future by improvement of nanowires alignment and increase their coverage between electrodes.

Table 5.2 A summary of the ferroelectric-photovoltaic performance of different ferroelectric materials

Ferroelectric material	Morphology	Electrode materials	Ligh source	P_{opt}, mW/cm^2	U_{OC}, V	J_{SC}, nA/cm^2	References
BFO	Nanofibers	Au	AM 1.5G simulator	100	0.8		[47]
BFO doped with Pr	Nanotubes	Ag	AM 1.5G simulator	10	0.21		[48]
BIT	Thin film	Au and FTO	AM 1.5G simulator	100	0.02	180	[7]
BLFTO	Thin film	Al and Pt	solar simulator	100	0.2	$1.35 \cdot 10^6$	[49]
BTO doped with Ni and Nb	Bulk ceramic	Ag and ITO	AM 1.5G simulator	100		8	[50]
PIMN-PT	Bulk ceramic	Au and ITO	Laser, λ = 405 nm	120	23	6.16	[2]
PLZT	Thin film	LSMO and Nb:STO	UV, λ = 356 nm	0.86	0.71	2320	[22]
SbSI	Nanowires	Pt	Ar laser, λ = 488 nm	127	0.119(2)	> 7.4	[30]
SPS	Ceramic	Au	Xe lamp	2100		10	[51]
	Bulk crystal	Au	Xe lamp	2100		100	[51]
TiO_2-buffered PZT	Thick film	Pt/Ti and Au	UV, λ = 325 nm	24	1.7		[44]

Used abbreviations and symbols: BFO—$BiFeO_3$; BLFTO—$[Bi_{0.9}La_{0.1}][Fe_{0.97}Ta_{0.03}]O_3$; BTO—$BaTiO_3$; BIT—$Bi_4Ti_3O_{12}$; ITO—indium tin oxide; J_{SC}—short circuit photocurrent density; LSMO—$La_{0.7}Sr_{0.3}MnO_3$; Nb:STO—$SrTiO_3$ doped with Nb; PIMN-PT—$Pb(In_{1/2}Nb_{1/2})O_3$-$Pb(Mg_{1/3}Nb_{2/3})O_3$-$PbTiO_3$; PLZT—$(Pb_{0.97}La_{0.03})$ $(Zr_{0.52}Ti_{0.48})O_3$; PZT—lead zirconate titanate; SPS—$Sn_2P_2S_6$; P_{opt}—optical power density; U_{OC}—open circuit photovoltage

5.2 Solar Cells

5.2.1 *The Chalcohalide Compounds as Promising Photovoltaic Absorbers*

The metal chalcohalides have been recognized as so-far overlooked semiconductors attractive for future use in next generation photovoltaic devices due to their relatively low bandgap energies, high charge carrier mobilities, and moderate toxicity [52]. Furthermore, many metal chalcohalides possess high dielectric constants and tunable energy band gaps, which are beneficial for solar cells [53]. The low-dimensional chalcohalide nanomaterials have outstanding optoelectronic properties similar to their metal halide perovskites counterparts [54]. Therefore, antimony or bismuth chalcohalides have received a gained attention of the scientific community as new inorganic photovoltaic absorbers [55–59]. The chalcohalide glasses [60, 61], mixed-metal chalcohalides [62], lead chalcohalides [63], and perovskite based chalcohalides [54, 64] were examined as materials suitable for solar cells construction. Some chalcohalides, including SbSI, SbSeI, and BiOI, are supposed to be "defect-tolerant" photovoltaic materials [65]. The density functional theory and machine-learning approaches were used in Ref. [66] to confirm high spectroscopic limited maximum efficiencies, the suitable convex-hull stabilities and effective carrier masses of the selected chalcohalide materials. Davies and coworkers performed first-principles calculations on a large number of the chalcohalide compounds in order to find new photoactive semiconductors [67]. They identified two new chalcohalides, $Cd_5S_4Cl_2$ and Cd_4SF_6, as promising materials for use in a tandem solar cell.

The energy band gaps of the bismuth and antimony chalcohalides are in wide range 1–5 eV (Fig. 5.3a). The band structures of these compounds may be tuned through chemical substitution. One can remember that an appropriate selection of materials for electron transporting layer (ETL) and hole transporting layer (HTL) is crucial for achieving the high efficiency of the solar cell. Usually, the hole effective masses of the chalcohalides are higher than their electron effective masses (Fig. 5.3b). The electron or hole effective masses are lower than the rest mass of electron, what proves a possibility of good carrier mobility [68]. Certain ternary chalcohalides (e.g. SbSI and BiSI) can act as interfacial layers in solar cells, contributing to enhanced charge transfer [53]. The chalcohalide absorbers, utilized in photovoltaic devices, can be divided into three main groups: ternary compounds (SbSI, SbSeI, BiOI, BiSCl, BiSI, $Bi_{13}S_{18}Br_2$, $Bi_{13}S_{18}Cl_2$, $Bi_{13}S_{18}I_2$), quaternary compounds ($Sb_{0.67}Bi_{0.33}SI$, $Sn_2SbS_2I_3$, $Pb_2SbS_2I_3$, $Ag_3BiI_{5.92}S_{0.04}$), and composites (SbSI-polyacrylonitrile (PAN)). Furthermore, the antimony and bismuth chalcohalides can be distinguished among the photovoltaic absorbers. In the next two sections, the recent achievements in fabrication and characterization of the solar cells based on the antimony and bismuth chalcohalides will be presented.

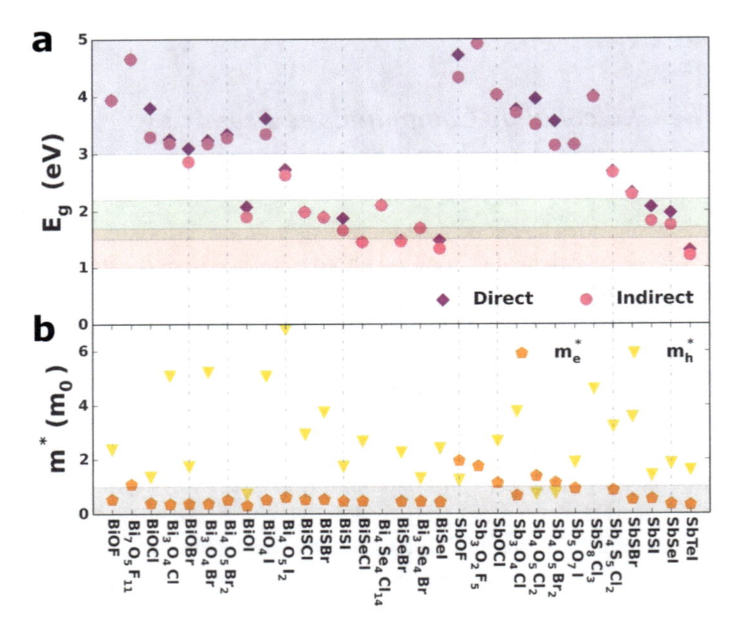

Fig. 5.3 The values of **a** energy band gaps and **b** effective masses of electron and hole calculated for various bismuth and antimony chalcohalides. The symbols $m_e{}^*$, $m_h{}^*$, and m_0 in **b** denote electron effective mass, hole effective mass, and the rest mass of electron, respectively. Adapted from Ran et al. [68] under the terms of the Creative Commons Attribution 4.0 International License (CC BY 4.0). Copyright (2018) Springer Nature

5.2.2 *Antimony Chalcohalide Based Solar Cells*

The thin films of antimony chalcohalides, used in solar cells, have been prepared till now via solution processing and spin-coating of polymer composites. These techniques were described detailly in Chap. 3 of this book. The solution processing of the chalcohalide films can be divided into two main groups: one-step [69] and two-step deposition [70]. Choi and coworkers presented for the first time fabrication of SbSI thin films in two stages [71]. At first, solution of $SbCl_3$ and thiourea was deposited on fluorine-doped tin oxide (FTO) glass substrate covered with TiO_2 blocking layer. In result, the amorphous Sb_2S_3 layer was formed. Then, SbI_3 solution was spin-coated on Sb_2S_3 layer and annealed at temperature of 473 K to obtain a dense SbSI film. A morphology of the SbSI was controlled by changing the input ratio of $SbCl_3$ to thiourea. The poly(3-hexylthiophene) (P3HT) and Au were chosen as hole transporting layer and top electrode, respectively. The energy level diagram of the FTO/TiO_2/SbSI/P3HT/Au device is depicted in Fig. 5.4a. The photovoltaic performance of the device was tested under standard light illumination with 100 mW/cm^2 optical power density (AM 1.5). A negligible hysteresis of the current density–voltage characteristic was observed (Fig. 5.4b). The open circuit voltage reached 0.548 V, whereas short circuit current density was equal to 5.45 mA/cm^2. A fill factor (*FF*)

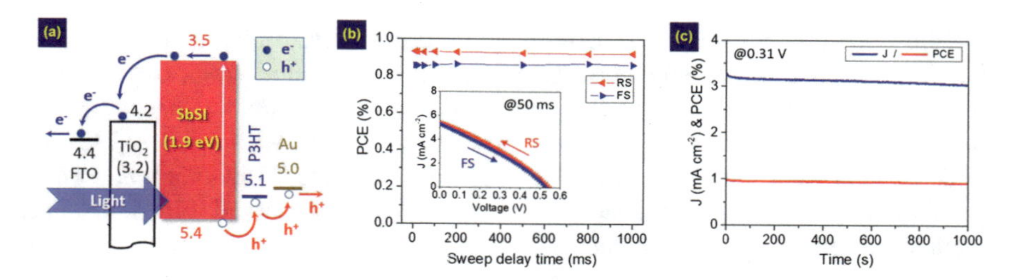

Fig. 5.4 The energy level diagram **a**, power conversion efficiency as a function of sweep delay time **b**, photocurrent density and power conversion efficiency stability **c** of FTO/TiO_2/SbSI/P3HT/Au solar cell. The inset in **b** presents current density–voltage characteristics of the device at a sweep delay time of 50 ms. The blue and red curves in **b** refer to forward scan (FS) and reverse scan (RS), respectively. Reprinted from Choi et al. [71] under the terms of the Creative Commons Attribution 4.0 International License (CC BY 4.0). Copyright (2018) American Institute of Physics (AIP)

of the solar cell is defined as a ratio of its maximum power (P_{max}) to a product of the open circuit voltage and short circuit current [72]

$$FF = \frac{P_{max}}{U_{OC} \cdot I_{SC}}. \tag{5.8}$$

The fill factor of the FTO/TiO_2/SbSI/P3HT/Au solar cell was 31% [71]. A power conversion efficiency (*PCE*) is described by the Scharber equation [73]

$$PCE = \frac{FF \cdot U_{OC} \cdot J_{SC}}{P_{in}}, \tag{5.9}$$

where P_{in} represents the input power density. The steady state photocurrent density of FTO/TiO_2/SbSI/P3HT/Au photovoltaic device was 3.12 mA/cm^2 at a voltage of 0.31 V, what corresponded to a stabilized *PCE* of 0.93% [71]. The short circuit current density and power conversion efficiency demonstrated good stability over time (Fig. 5.4c).

A first report on use of the SbSeI in solar cell was provided in Ref. [74]. The device was fabricated though several spin-coating cycles of SbI_3 solution on Sb_2Se_3 thin layer. This method was described comprehensively in Chap. 3 of this book (see Sect. 3.1). The mesoporous (mp)-TiO_2 film on FTO substrate was used as electron transporting layer. The poly(2,6-(4,4-bis-(2-ethylhexyl)-4H-cyclopenta[2,1-b;3,4-b′]dithiophene)-alt4,7(2,1,3-benzothiadiazole)) (PCPDTBT) and poly(3,4-ethylenedioxythiophene) doped with poly(4-styrenesulfonate) (PEDOT:PSS) were applied as the hole transporting material and hole transporting layer, accordingly. Figure 5.5a shows the current density–voltage curve of FTO/TiO_2/SbSeI/PCPDTBT/PEDOT:PSS/Au solar cell. The short circuit current density, open circuit voltage, fill factor, and power conversion efficiency were equal to 14.77 mA/cm^2, 0.473 V, 58.7%, and 4.1%, respectively [74]. The external quantum efficiency (EQE) spectrum is presented in Fig. 5.5b. The current density of 14.66 mA/cm^2 was calculated by integrating the overlap of the standard AM 1.5 G solar photon flux with the EQE spectrum. This value was close to short circuit current density of the SbSeI solar cell (J_{SC} = 14.77 mA/cm^2). When the bias voltage was maintained at the maximum power point (0.376 V), a stabilized JSC of 10.51 mA/cm^2 and a stabilized *PCE* of 3.95% were obtained (Fig. 5.5c). A normal distribution of the power conversion efficiency values was observed (Fig. 5.5d). The average power conversion efficiency reached approximately 2.5%, whereas solar cell with the highest photovoltaic performance exhibited the *PCE* of 4.1%.

An application of the chalcohalide composite in the photovoltaic device was presented for the first time in Ref. [75]. This device was fabricated in several steps. At first, the SbSI nanowires were grown under ultrasonic irradiation. A detailed description of the sonochemical synthesis of SbSI nanowires is given in Sect. 2.5 in Chap. 2 of this book. In the next step, a nanocomposite of the SbSI nanowires and PAN polymer was prepared and spin-coated on ITO substrate. This technology was described in Chap. 3 of this book (see Sect. 3.2). A need of high temperature treatment application was avoided. It was a significant advantage of the SbSI-PAN film fabrication in comparison to other known methods of SbSI photovoltaic structures

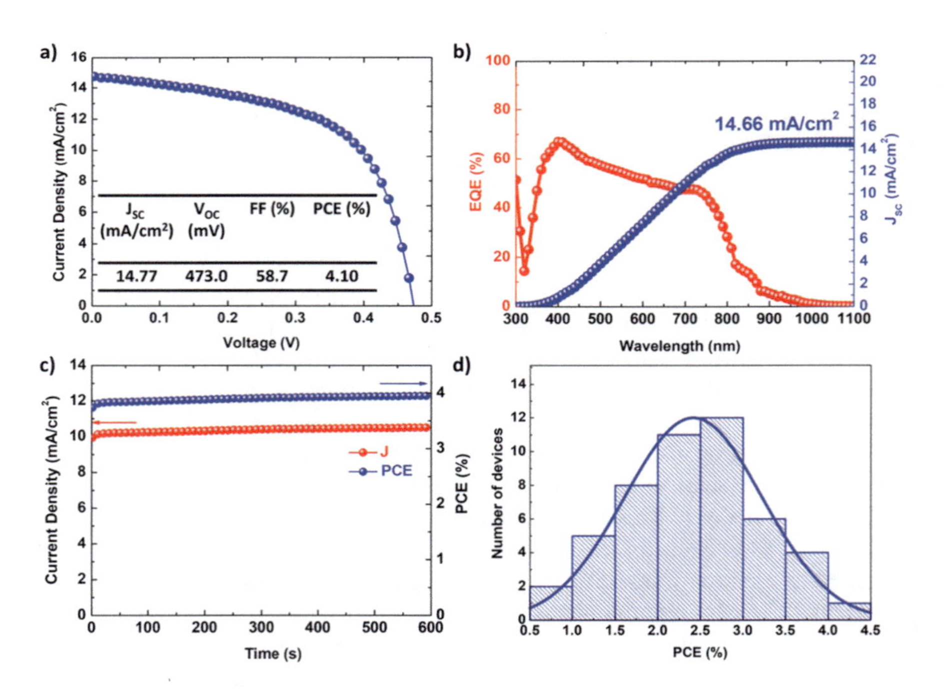

Fig. 5.5 The current density–voltage characteristics **a**, external quantum efficiency and short circuit current density spectra **b**, current density and power conversion efficiency stability **c** of the FTO/TiO_2/SbSeI/PCPDTBT/PEDOT:PSS/Au solar cell. Histogram **d** of power conversion efficiencies of from 49 individual SbSeI cells. Inset in **a** contains a table presenting values of device performance parameters. Reprinted from Nie et al. [74] under the terms of the Creative Commons Attribution 4.0 International License (CC BY 4.0). Copyright (2021) Wiley

preparation. The optical absorbance spectrum of SbSI-PAN film was measured to determine the indirect allowed energy band gap of 1.81 eV. The TiO_2 and P3HT were chosen as electron and hole transporting layers, respectively. In order to optimize the nanowires concentration in SbSI-PAN nanocomposite, over thirty ITO/TiO_2/SbSI-PAN/P3HT/Au devices containing various weight concentrations of SbSI nanowires (20, 40, and 60%) were tested under a standard illumination with power density of 100 mW/cm^2. An improvement of the photovoltaic performance was observed with increase of SbSI nanowires concentration in the nanocomposite (Fig. 5.6). When the weight concentration of SbSI nanowires was higher than 60%, the prepared SbSI-PAN films had poor quality. Thus, the best photovoltaic performance was achieved for devices based on nanocomposite containing 60% of SbSI nanowires. The average short circuit current density was equal to 1.84(20) μA/cm^2 and open circuit voltage reached 69(13) mV [75]. It should be underlined that the average short circuit current density of the ITO/TiO_2/SbSI-PAN/P3HT/Au devices was over two orders of magnitude higher than this parameter reported for Pt/SbSI/Pt ferroelectric-photovoltaic device [30]. However, the J_{SC} and U_{OC} of the photovoltaic device based on SbSI-PAN nanocomposite were much smaller than these for other SbSI solar cells (Table 5.3). It may result from the negative photoconductivity of SbSI nanowires [40] or weak charge transfer in the SbSI-PAN nanocomposite. Further studies are needed to eliminate aforementioned drawbacks of solar cells constructed from the composites of chalcohalide nanomaterials and polymers.

An interesting approach toward increasing charge transfer of the inorganic–organic hybrid solar cell was presented in Ref. [76]. In this paper, pure SbSI and SbSI-interlayered Sb_2S_3 solar cells were constructed by annealing an Sb_2S_3 film and SbI_3 powder in an inert gas atmosphere. The PCPDTBT was used as hole transporting material. An adjusting of SbSI morphology resulted in an intimate contact between SbSI and hole transporting layer. The power conversion efficiency of the FTO/TiO_2/SbSI/PCPDTBT/Au solar cell attained high value of 3.62% [76]. In the case of the FTO/TiO_2/Sb_2S_3/SbSI/PCPDTBT/Au solar cell the charge transfer was

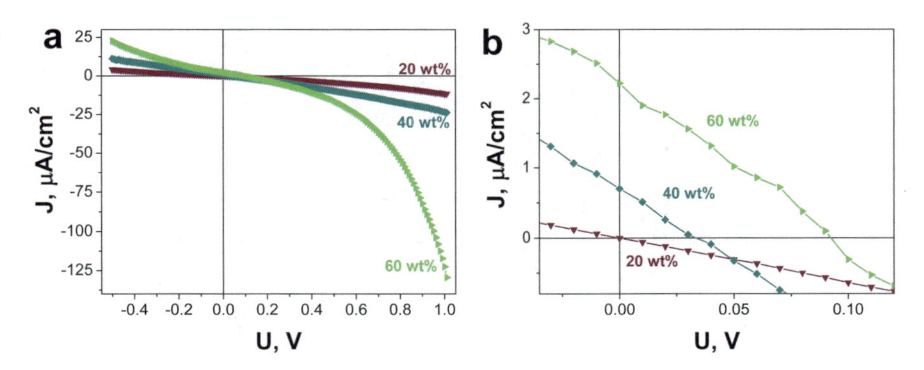

Fig. 5.6 The photovoltaic performance of ITO/TiO_2/SbSI-PAN/P3HT/Au devices for different concentrations of SbSI nanowires in SbSI-PAN composite. Reprinted from Mistewicz et al. [75] with permission from Elsevier. Copyright (2020) Elsevier

Table 5.3 A comparison of the photovoltaic performance of solar cells based on the antimony chalcohalides under a standard illumination with power density of 100 mW/cm^2

Material	Solar cell structure	U_{OC}, V	J_{SC}, mA/cm^2	*FF*, %	*PCE*, %	References
SbSI	FTO/TiO_2/SbSI/P3HT/Au	0.548	5.45	31	0.93	[71]
	FTO/TiO_2/SbSI/P3HT/Au	0.53	7.06	56.7	2.12	[39]
	FTO/TiO_2/SbSI/Spiro-MeOTAD/Au	0.55	6.42	56.6	2.00	[39]
	FTO/TiO_2/SbSI/PCPDTBT/Au	0.56	9.06	56.0	2.84	[39]
	FTO/TiO_2/SbSI/PCPDTBT/Au	0.60	9.26	65.2	3.62	[76]
	FTO/TiO_2/Sb_2S_3/SbSI/PCPDTBT/Au	0.62	14.92	66	6.08	[76]
SbSI-carbon	FTO/TiO_2/ZrO_2/SbSI-carbon	0.29	0.4		0.035	[77]
SbSI-PAN	ITO/TiO_2/SbSI-PAN/P3HT/Au	0.069	0.00184			[75]
SbSeI	FTO/CdS/SbSeI/Spiro-OMeTAD/Au	0.633	0.88	41.71	0.23	[69]
	FTO/TiO_2/SbSeI/PCPDTBT/PEDOT:PSS/Au	0.473	14.8	58.7	4.1	[74]
$Sb_{0.67}Bi_{0.33}SI$	FTO/TiO_2/Sb_2S_3/$Sb_{0.67}Bi_{0.33}SI$/PCPDTBT/PEDOT:PSS/Au	0.53	14.54	52.8	4.07	[78]
$Sn_2SbS_2I_3$	FTO/TiO_2/$Sn_2SbS_2I_3$/PCPDTBT/PEDOT:PSS/Au	0.440	16.1	57.0	4.04	[79]
$Pb_2SbS_2I_3$	FTO/TiO_2/$Pb_2SbS_2I_3$/PCPDTBT/Au	0.61	8.79	58.2	3.12	[80]
$(CH_3NH_3)SbSI_2$	FTO/TiO_2/$(CH_3NH_3)SbSI_2$/PCPDTBT/PEDOT:PSS/Au	0.65	8.12	58.5	3.08	[81]

Used abbreviations and symbols: *FF*—fill factor; FTO—fluorine-doped tin oxide; ITO—indium tin oxide; J_{SC}—short circuit photocurrent density; P3HT—poly(3-hexylthiophene); PAN—polyacrylonitrile; *PCE*—power conversion efficiency; PCPDTBT—poly(2,6-(4,4-bis-(2-ethylhexyl)-4H-cyclopenta[2,1-b;3,4-b′]dithiophene)-alt4,7(2,1,3-benzothiadiazole)); Spiro-MeOTAD—2,2′,7,7′-tetrakis(N,N-di-p-methoxyphenylamine)-9,9′-spirobifluorene; U_{OC}—open circuit photovoltage even more effective due to an existence of energetically favorable external driving force. Thus, the *PCE* of this solar cell reached 6.08%, which is the best result among the antimony chalcohalide photovoltaic devices developed so far (Table 5.3).

Different organic compounds have been employed so far as hole transporting materials in the antimony chalcohalide solar cells, such as poly(3-hexylthiophene) (P3HT) [71, 75], 2,2′,7,7′-tetrakis(N,N-di-p-methoxyphenylamine)-9,9′-spirobifluorene (Spiro-MeOTAD) [39, 69], and poly(2,6-(4,4-bis-(2-ethylhexyl)-4H-cyclopenta[2,1-b;3,4-b′]dithiophene)-alt4,7(2,1,3-benzothiadiazole)) (PCPDTBT) [76, 79, 81]. An effect of the hole transporting material on photovoltaic performance of the SbSI solar cell was studied in Ref. [39]. When the PCPDTBT was applied as hole transporting layer, the power conversion efficiency of the SbSI solar cell reached 2.84%. It was higher than the values of *PCE* of the SbSI solar cells containing P3HT and Spiro-MeOTAD as hole transporting materials. It was explained due to lower band gap of PCPDTBT in comparison to this for SbSI. The PCPDTBT absorbed light with long wavelength that could not be absorbed by the SbSI. Thus, the extra electric carriers were photogenerated contributing to enhanced short circuit photocurrent. This effect was recognized as a panchromatic photon extraction by hole transporting material [82]. It should be concluded that a selection of the hole transporting material is a crucial factor determining efficiency of the inorganic–organic hybrid solar cell.

The highest power conversion efficiencies of the antimony chalcohalide solar cells have been reported till now for photovoltaic structures based on SbSI-interlayered Sb_2S_3 [76] and ternary compounds, i.e. $Sb_{0.67}Bi_{0.33}SI$ [78] and $Sn_2SbS_2I_3$ [79]. However, the largest $PCE = 6.08\%$, achieved for the FTO/TiO_2/Sb_2S_3/SbSI/PCPDTBT/Au device, is still small in comparison to values of this parameter reported for perovskite solar cells [83–85]. The future efforts toward improvement of the photovoltaic properties of chalcohalide materials can be focused on bandgap engineering through transition-metal and chalcogenide doping [54]. Other promising strategy for increasing of *PCE* is dimensional reduction for tuning the electronic structure of the semiconductor material. The stability of the chalcohalide solar cells needs also future examination.

5.2.3 Bismuth Chalcohalide Based Solar Cells

Recently, bismuth-based materials have been revealed as emerging earth-abundant solar absorbers due to their low costs and non-toxicity [86–88]. According to [87], some bismuth chalcohalide materials (BiSI and BiSeI) have properties beneficial for solar energy applications, like low electron effective masses, suitable electronic structures, and a high level of defect tolerance. The bismuth chalcohalide photovoltaic absorbers can be divided into the two main groups: ternary compounds (BiOI, BiSCl, BiSI, $Bi_{13}S_{18}Br_2$, $Bi_{13}S_{18}Cl_2$, $Bi_{13}S_{18}I_2$) and quaternary compounds ($Sb_{0.67}Bi_{0.33}SI$, $Ag_3BiI_{5.92}S_{0.04}$). The thin films of the bismuth chalcohalides for photovoltaic applications can be fabricated using the solution processing method [89–91], chemical bath deposition [78, 92], oriented solvothermal growth [93, 94], and chemical vapor transport [95].

The bismuth oxyiodide (BiOI) thin films are grown by chemical vapor transport in a two-zone horizontal tube furnace and used in solar cell [95]. The ZnO and NiO_x were selected as electron and hole transporting materials, respectively. Only inorganic compounds were used to construct the solar cell. SEM micrograph of the ITO/NiO_x/BiOI/ZnO/Al device is presented in Fig. 5.7a. The BiOI had a textured surface morphology (Fig. 5.7b). The BiOI grains were grown along the (012) planes. The ZnO was conformal to BiOI (Fig. 5.7a–c). When thickness of the BiOI was below 600 nm, the large external quantum efficiency up to 80% was measured (Fig. 5.7d). The short circuit density was in the range from 0.4 mA/cm^2 to 6.3 mA/cm^2 depending on the BiOI layer thickness. The open circuit voltage was equal to 0.75 V. The power conversion efficiency of the BiOI solar cell reached 1.79%. A stability of the normalized efficiency of BiOI solar cell was demonstrated to be better than the $CH_3NH_3PbI_3$ photovoltaic device with similar structure (Fig. 5.7e). It was found in Ref. [95] that increase of the light intensity led to decrease of fill factor, shunt, and series resistances (Fig. 5.7f). This effect was explained due to rise of the BiOI film conductivity and enhancement of photogenerated carrier recombination when the light intensity was gained.

An application of the BiOI thin films in the solar cells was also demonstrated in Ref. [92]. The fabrication technology of the solar cell and its structure were similar to those presented in Ref. [95]. The effect of the BiOI films annealing on the photovoltaic performance of the ITO/NiO_x/BiOI/ZnO/Cr/Ag devices was examined. BiOI films were heated at a temperature of 373 K under vacuum. A relative reduction in the surface atomic fraction of iodine by over 40%, reduction in the surface bismuth fraction by over 5%, and an increase in the surface oxygen fraction by over 45% was calculated [92]. However, no significant influence of annealing time on power conversion efficiency (Fig. 5.8a), short circuit density (Fig. 5.8b), open circuit voltage (Fig. 5.8c), and fill factor (Fig. 5.8d) of the BiOI solar cell was observed. It was concluded in Ref. [92] that BiOI seems to be robust against processing conditions that lead to surface defects influenced by the concentration changes of the chemical elements in the BiOI film.

Among ternary bismuth chalcohalide compounds the BiSI is most frequently investigated as potential photovoltaic absorber [96]. The two steps solution processing technology was proposed in [89] to fabricate the BiSI films. In the first stage, the Bi_2S_3 was prepared using the Bi_2O_3-thiourea thiol-amine solution. The Bi_2S_3 and BiI_3 reacted leading to formation of BiSI in the second step. Bismuth to sulfur molar ratio of the Bi_2O_3-thiourea solution strongly affected the BiSI growth. Despite interesting technology of the material fabrication, the efficiency of the BiSI solar cell was very low. A single-precursor solution method of preparation of pure and stoichiometric BiSI thin films was shown in Ref. [91]. A single molecular precursor consisted of $Bi(NO_3)_3 \cdot 5H_2O$, thiourea and NH_4I in 2-methoxyethanol and acetylacetone. The BiSI thin film was sandwiched between layers of SnO_2 and poly(9,9-di-noctylfluorenyl-2,7-diyl) (F8), which served as electron and hole transporting materials, respectively. The BiSI solar cell was tested under AM 1.5 illumination

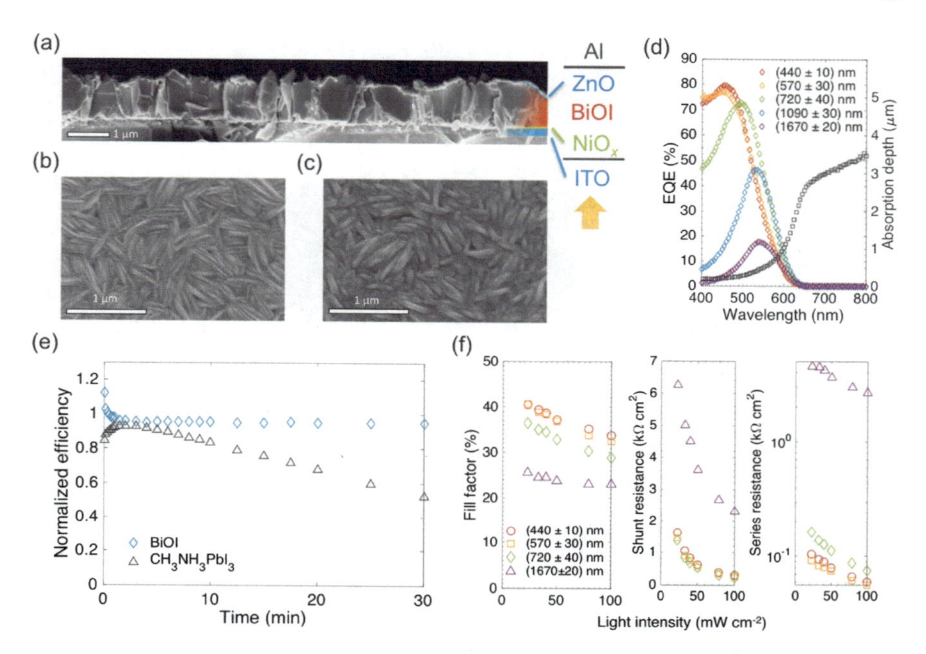

Fig. 5.7 The cross-sectional SEM image **a** of the ITO/NiO$_x$/BiOI/ZnO/Al solar cell. The SEM micrographs of **b** BiOI film and **c** ZnO layer deposited on BiOI film. An influence of BiOI thickness on external quantum efficiency absorption depth of BiOI **d**. The time dependence of normalized efficiencies of BiOI and $CH_3NH_3PbI_3$ solar cells **e**. An influence of light intensity on fill factor, shunt resistance, and series resistance of the solar cells based on BiOI films with different thickness **f**. Reprinted from Hoye et al. [95] under the terms of the Creative Commons Attribution 4.0 International License (CC BY 4.0). Copyright (2017) Wiley

(100 mW/cm^2) exhibiting open circuit voltage of 0.445 V, short circuit current density of 8.44 mA/cm^2, and power conversion efficiency of 1.32%. Yoo and coworkers presented that the application of the BiSI interlayer between the electron transport layer and the BiI_3 absorber supports photoinduced charge separation and electron extraction in the BiI_3 solar cells [90]. A solvothermal growth of BiSI nanorods along the [001] direction on a tungsten substrate was described in Ref. [94]. The copper thiocyanate (CuSCN) was chosen as the hole transporting material and deposited on BiSI film. The ITO layer was spattered on the top of the device to serve as the transparent electrode. The solar cell with structure of ITO/CuSCN/BiSI/W demonstrated a limited power conversion efficiency of 0.66% [94]. The photovoltaic performance of solar cells based on BiSI and other bismuth chalcohalides is summarized in Table 5.4.

A significant progress in fabrication and characterization of the bismuth chalcohalide solar cells has been observed in last years. However, the power conversion efficiencies of these devices (Table 5.4) are still relatively low in comparison to the solar cells based on the lead perovskites. The highest values of *PCE* have been obtained till now for quaternary compounds, i.e. $Sb_{0.67}Bi_{0.33}SI$ [78] and $Ag_3BiI_{5.92}S_{0.04}$ [99]. In order to improve photovoltaic performances of the bismuth chalcohalide solar cells,

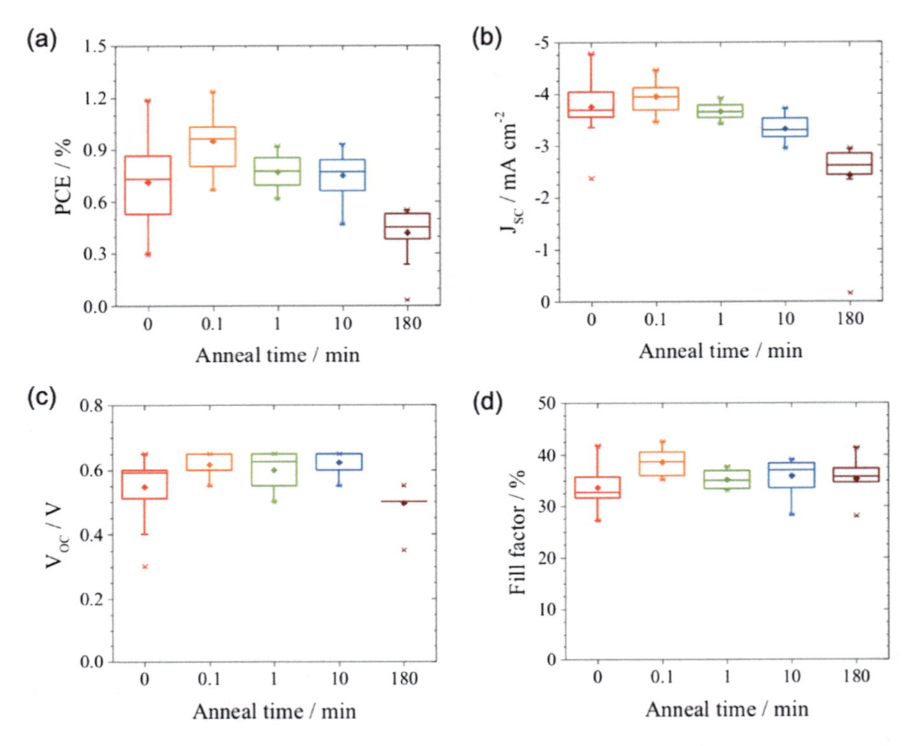

Fig. 5.8 An influence of annealing time on the power conversion efficiency **a**, short circuit current density **b**, open circuit voltage **c**, and fill factor **d** of the solar cells based on the BiOI thin films. Reprinted from Huq et al. [92] under the terms of the Creative Commons Attribution 4.0 International License (CC BY 4.0). Copyright (2020) Wiley

a special attention has to be paid to crystal orientation, material morphology, an existence of defects in the material, and design of the device architecture. The most promising strategies toward increasing the efficiencies of the bismuth chalcohalide solar cells are defect engineering [86], energy band gap tuning through adjusting of chemical composition [101], and band alignment against the properly chosen hole transporting and buffer layers [87]. A novel direction of efficient solar cells development is combining the wide bandgap chalcohalides with well-established silicon junctions in tandem photovoltaic devices [52]. It is expected that these solar cells will be constructed and investigated in a near future.

5.3 Photodetectors

A photodetector is an optoelectronic device that converts incoming electromagnetic radiation into the electrical signal. The photodetectors are commonly applied in telecommunication, biomedical imaging, motion detection, and environmental

Table 5.4 A comparison of the photovoltaic performance of solar cells based on the bismuth chalcohalides under a standard illumination with power density of 100 mW/cm^2

Material	Solar cell structure	U_{OC}, V	J_{SC}, mA/cm^2	*FF*, %	*PCE*, %	References
BiOI	ITO/NiO_x/BiOI/ZnO/Al	0.75	6.3	39	1.79	[95]
	ITO/NiO_x/BiOI/ZnO/Cr/Ag	0.62	4	39	0.95	[92]
BiSCl	FTO/TiO_2/BiSCl/(I_3^-/I^-)/Pt	0.54	9.87	25.5	1.36	[93]
BiSI	FTO/BiSI/CuSCN/Pt/FTO	0.38	2.5	28	0.25	[97]
	FTO/TiO_2/BiSI/P3HT/Au	0.4	0.1	11.1	$4.4 \cdot 10^{-6}$	[89]
	FTO/SnO_2/BiSI/F8/Au	0.445	8.44	35.14	1.32	[91]
	ITO/SnO_2/BiSI/BiI_3/Spiro-MeOTAD/Au	0.33	12.6	29	1.21	[90]
	ITO/CuSCN/BiSI/W	0.46	2.73	52.81	0.66	[94]
$Bi_{13}S_{18}Br_2$	FTO/TiO_2/$Bi_{13}S_{18}Br_2$/(I_3^-/I^-)/Pt	0.42	7.50	36	1.12	[98]
$Bi_{13}S_{18}Cl_2$	FTO/TiO_2/$Bi_{13}S_{18}Cl_2$/(I_3^-/I^-)/Pt	0.42	6.92	31	0.91	[98]
$Bi_{13}S_{18}I_2$	FTO/TiO_2/$Bi_{13}S_{18}I_2$/(I_3^-/I^-)/Pt	0.39	6.41	30	0.75	[98]
$Sb_{0.67}Bi_{0.33}SI$	FTO/TiO_2/Sb_2S_3/$Sb_{0.67}Bi_{0.33}SI$/PCPDTBT/PEDOT:PSS/Au	0.53	14.54	52.8	4.07	[78]
$Ag_3BiI_{5.92}S_{0.04}$	FTO/TiO_2/$Ag_3BiI_{5.92}S_{0.04}$/PTAA/Au	0.573	14.7	65.9	5.56	[99]
$MABiI_2S$	FTO/TiO_2/$MABiI_2S$/PCPDTBT/Au	0.22	1.96	30	0.13	[100]

Used abbreviations and symbols: F8—poly(9,9-di-n-octylfluorenyl-2,7-diyl); *FF*—fill factor; FTO—fluorine-doped tin oxide; ITO—indium tin oxide; J_{SC}—short circuit photocurrent density; *PCE*—power conversion efficiency; PTAA—poly[bis(4-phenyl)(2,4,6-trimethylphenyl)amine]; Spiro-MeOTAD—2,2′,7,7′-tetrakis(N,N-di-p-methoxyphenylamine)-9,9′-spirobifluorene; U_{OC}—open circuit photovoltage sensing. An incident light is absorbed in the photodetector and measured as electric current or voltage signal generated in the device. Generally, a sensing mechanism of a photodetector can be relied on various phenomena: photoconductive, photogating, and photovoltaic effects [102]. The low-dimensional semiconductors are attractive for use in photodetectors due to a wide spectral range of light absorption, hot-electron generation, a high surface-to-volume ratio and Debye length comparable to their sizes [103].

Several figures of merit have been proposed in order to characterize the sensing performance of the photodetectors. The responsivity (R_λ) is one of the key parameters which describes the ratio between photocurrent generation and incident light energy [102, 104, 105]

$$R_\lambda = \frac{I_{PC}}{P_{in} \cdot A}, \tag{5.10}$$

where I_{PC} means a photocurrent signal, P_{in} is a power density of incident light, A denotes the area of the sensing material. Another important factor is noise equivalent power (*NEP*)

$$NEP = \frac{I_n}{R_\lambda}, \tag{5.11}$$

where I_n means noise current, which is related with the dark current. A detectivity (D^*) of a photodetector is defined as [102, 106, 107]

$$D^* = R_\lambda \sqrt{\frac{A}{2eI_D}}, \tag{5.12}$$

where e denotes the elementary electronic charge, I_D is the dark current of the device. The external quantum efficiency (*EQE*) is a next significant parameter used to evaluate the photodetection performance of the device. It is defined as the ratio of number of the electron–hole pairs, that contribute to the photocurrent, to the total number of photons impinging on the device [108]. The external quantum efficiency can be calculated by applying following formula

$$EQE = \frac{R_\lambda hc}{e\lambda} \cdot 100\%, \tag{5.13}$$

where h is the Planck's constant, c is light velocity, and λ means wavelength of the incident light.

Antimony sulfoiodide is the most frequently used in the photodetectors among the all chalcohalide semiconductors (Table 5.5). A hydrothermal method of the SbSI microrods fabrication was presented in Ref. [106]. The material was prepared from $SbCl_3$, thiourea, and NH_4I in HCl aqueous solution at 433 K for 4 h. An individual SbSI microrod was placed on the SiO_2/Si substrate and connected to the external measurement circuit using symmetric ITO electrodes. The current response of the photodetector was tested under monochromatic light illumination with different wavelengths: 420, 475, 520, 550, and 650 nm. The ITO/SbSI/ITO photodetector displayed a detectivity of $2.3 \cdot 10^8$ Jones and short response/recovery times below 0.3 s [106]. A fabrication of the SbSI micro-crystals film was proposed in Ref. [109]. An evaporation target was prepared by drop-casting of SbI_3 on a microscope slide. Then, a FTO substrate was covered with the amorphous Sb_2S_3 and inserted in front of SbI_3 target in the chamber with nitrogen atmosphere. After the evaporation target was heated up to 523 K, the SbI_3 sublimed and reacted with the Sb_2S_3 what led to formation of crystalline SbSI micro-needles film. In order to avoid the shorting between the top and the bottom electrodes an insulating buffer layer was deposited on SbSI film by spin-coating of poly(methyl methacrylate) (PMMA). The chlorobenzene was spin-coated to partly etch the PMMA layer and reveal the tips of the SbSI micro-needles. At the end, a gold metal electrode was evaporated on the top of the sample. The photocurrent of 40 nA was registered under zero bias voltage, what was attributed to a built-in potential arising from asymmetric electrodes of the device. The responsivity of FTO/SbSI/Au photodetector was equal to 10^{-5} A W^{-1} [109]. Its detectivity attained 10^9 Jones. The response and recovery times were lower than 8 ms and 34 ms, respectively [109]. In 2019, a mixed-dimensional heterostructure of an individual PbI_2 flake and an individual single crystal SbSI whisker was constructed and used as a photodetector [107]. This device had a higher larger light-harvesting cross section than a photodetector based on an SbSI whisker. The two-dimensional PbI_2 flake and one-dimensional SbSI whisker played roles of a current collector and a current channel, respectively. The dark current of the SbSI/PbI_2 photodetector was lower in comparison to this for the individual SbSI device. The responsivity and detectivity of the SbSI/PbI_2 photodetector attained the high values of $26.3 \cdot 10^{-3}$ A W^{-1} and $4.37 \cdot 10^{13}$ Jones, respectively, under illumination of monochromatic light with a wavelength of 650 nm. Moreover, the SbSI/PbI_2 photodetector exhibited very fast response with rise time of 12 ms [107]. A flexible photodetector based on SbSI microrod was described in Ref. [110]. The SbSI microrods were synthesized hydrothermally from sulfur, iodine, and $SbCl_3$ at 453 K for 20 h. The photodetector was constructed by transferring a single SbSI microrod on the flexible polyimide (PI) substrate. The symmetric electrical contacts were fabricated using a sliver paste. The device showed responsivity of 13.74 mA W^{-1} and detectivity of $5.43 \cdot 10^{10}$ Jones under illumination with monochromatic light (635 nm). The response of the flexible Ag/SbSI/Ag photodetector remained almost unchanged after more than 2000 bending cycles.

An application of a few SbSI nanowires array as a photodetector was presented in Ref. [116]. The SbSI nanowires were synthesized under ultrasonic treatment and aligned in electric field, what was described in Chap. 3 of this book (Sect. 3.7). The electrode distance (1 μm) was comparable to the lengths of SbSI nanowires. A photocurrent response of a few SbSI nanowires array to argon laser illumination (488 nm) is shown in Fig. 5.9a. The dependence of illumination intensity on photocurrent (Fig. 5.9b) was fitted with semiempirical power formula [117]

Table 5.5 A summary of the chalcohalide photodetectors

Light absorbing material	Material morphology	λ, nm	R_λ, A W^{-1}	D^*, Jones	t_r, s	t_d, s	EQE, %	References
BiOBr	Nanosheets	365	12.4	$1.6 \cdot 10^{13}$	0.551	2.416	$4.22 \cdot 10^3$	[111]
BiTeCl	Nanoplates	532						[112]
BiSeI	Single crystals	635	3.2	$7 \cdot 10^{10}$	0.145	0.098	$6.22 \cdot 10^2$	[113]
	Nano/micro-wires	515	$7 \cdot 10^4$	$2.5 \cdot 10^{14}$	0.6	1	$1.8 \cdot 10^7$	[104]
BiSI	Thin films	625	62.1	$2 \cdot 10^{13}$	0.571	0.112	$9.28 \cdot 10^3$	[105]
SbSI	Microcrystal	650	10^{-5}	10^9	0.008	0.034		[109]
	Microrod	650		$2.3 \cdot 10^8$	< 0.3	< 0.3		[106]
	Microrod	630	$4.5 \cdot 10^{-6}$		0.1	0.01		[114]
	Microrod	630			0.3	0.05		[115]
	Microrod	635	0.0185	$7.3 \cdot 10^{10}$	0.063	0.126		[110]
	Nanowires	488			< 4.5	< 4.5		[116]
	Nanowires	488	$2.1 \cdot 10^{-8}$	$3.9 \cdot 10^6$	0.60	1.22	$5.3 \cdot 10^{-8}$	this work
$SbSI/PbI_2$	SbSI whisker and PbI_2 flake heterostructure	650	0.0263	$4.4 \cdot 10^{13}$	0.012	0.008		[107]

Used abbreviations and symbols: D^*—detectivity; EQE—external quantum efficiency; R_λ—responsivity; t_d—decay (recovery) time; t_r—rise (response) time

$$I_{PC} = I_{PC0} \cdot I_L^{\gamma}, \tag{5.14}$$

where I_{PC0} is a constant, I_L means light intensity, and $\gamma = 0.766(61)$ is a determined power coefficient. The response and recovery times of the SbSI photodetector was lower than 4.5 s.

The SbSI photodetector was fabricated via deposition of the SbSI nanowires on the Al_2O_3 substrate and their alignment between Pt interdigitated electrodes. More information on this technology can be found in Sect. 3.7 in Chap. 3. The electrode distance (250 μm) was much higher than the average length of SbSI nanowires. The current–voltage characteristics of this photodetector in dark condition and under monochromatic light illumination ($\lambda = 488$ nm, $P_{opt} = 127$ mW/cm^2) are presented in Fig. 5.10a. The electric current registered under light illumination was over one order of magnitude larger than this measured in dark condition. The photocurrent response exhibited short rise time ($t_r = 0.60$ s) and decay ($t_d = 1.22$ s) times (Fig. 5.10b). The responsivity of $2.06(3) \cdot 10^{-8}$ A W^{-1}, noise equivalent power of 1.87(2) mW, detectivity of $3.93(5) \cdot 10^6$ Jones, and external quantum efficiency of $5.25(7) \cdot 10^{-8}$% were calculated for SbSI photodetector using Eqs. (5.10), (5.11), (5.12), and (5.13), respectively. The values of R_λ, D^*, and EQE, determined for SbSI nanowires, are low in comparison for figures of merit reported for photodetectors based on other chalcohalide nanomaterials (Table 5.5).

Liu and coworkers demonstrated a way to change a photocurrent of SbSI single crystal by applying a pressure up to 23 GPa [118]. The chemical vapor deposition (CVD) method was used to grow the SbSI crystal from Sb_2S_3 and SbI_3 powders as solid precursors. The transient characteristics of photocurrent were registered at room temperature (300 K) under different pressures at zero bias voltage (Fig. 5.11a) and bias voltage of 5 V (Fig. 5.11b). The pressure-induced photocurrent was observed under the pressures of 6.7 and 14 GPa at zero bias voltage (Fig. 5.11a). It suggested

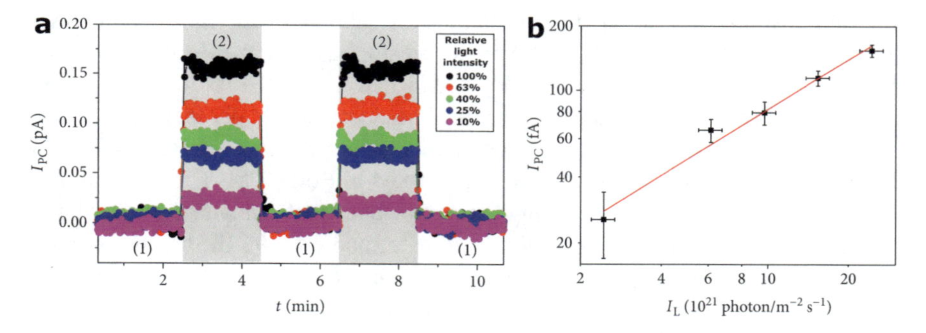

Fig. 5.9 **a** The time dependence of photocurrent of a few SbSI nanowires array measured in dark condition (1) and under light illumination (2) with different relative intensities (T = 298 K, p = 10^{-2} Pa). **b** The influence of illumination intensity on photocurrent of a few SbSI nanowires array. The red solid line in **b** represents the best fitted theoretical dependence (5.14). Adapted from Mistewicz [116] under the terms of the Creative Commons Attribution 4.0 International License (CC BY 4.0). Copyright (2018) Hindawi

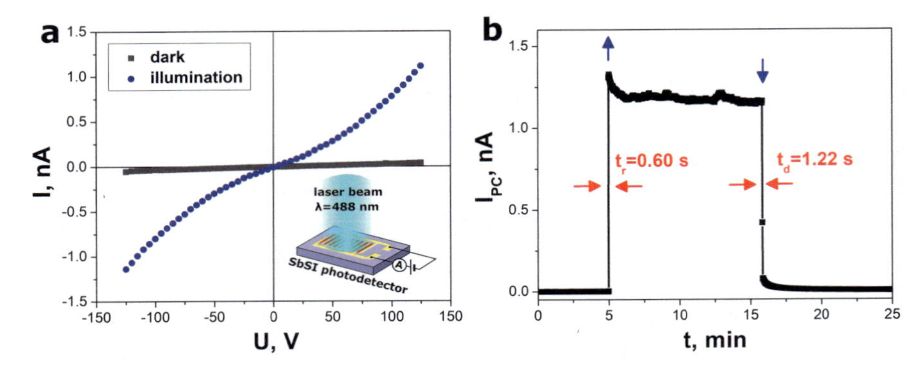

Fig. 5.10 Current–voltage characteristics **a** and time dependence of photocurrent **b** of the SbSI nanowires aligned between Pt electrodes on Al_2O_3 substrate ($T = 261$ K, $p = 10^{-3}$ Pa). Inset in **a** presents scheme of the SbSI photodetector. The blue vertical arrows in **b** indicate switching on (↑) and switching off (↓) illumination. The red horizontal arrows show rise time (t_r) and decay time (t_d) of the photocurrent

an existence of ferroelectric properties of SbSI under a high pressure. When non-zero bias voltage was applied to the SbSI crystal, the photocurrent attained the highest value under a pressure of 14 GPa (Fig. 5.11b, c). Similarly, the electrical resistivity of the SbSI crystal reached the maximum value under pressure of 12 GPa (Fig. 5.11d). These effects were explained due to an enhanced polarization originating from the stereochemical expression of lone-pair electrons [118]. Increased polarization led to improved efficiency of charge separation and rise of photocurrent. These results provided a new approach in understanding of what role may play the lone-pair electrons in a ferroelectric-photovoltaic effect.

Several report were published on using the bismuth chalcohalide materials in photodetectors (Table 5.5). The space-confined chemical vapor deposition (SCCVD) was demonstrated in Ref. [111] as method of fabrication of high-quality two-dimensional BiOBr nanocrystals. The photodetector based on the BiOBr nanosheets showed a high responsivity of 12.4 A W^{-1} and a detectivity of $1.6 \cdot 10^{13}$ Jones under monochromatic light illumination (365 nm). The electrical and photoelectrical properties of two-dimensional BiTeCl nanostructures were examined in Ref. [112]. The BiSI photodetector was for the first time presented in Ref. [105]. The BiI_3 thin film was converted into BiOI layer using a reactive bath that contained a solution of methanol and water. Then, the BiOI was transformed into the BiSI film through a sulfurization process in the H_2S gas. The BiSI film was transferred onto the Si/SiO_2 substrate equipped with lithographically patterned Au electrodes. Such prepared BiSI photodetector exhibited a good stability, high responsivity ($R_\lambda = 62.1$ A W^{-1}), and large detectivity ($D^* = 2 \cdot 10^{13}$ Jones). The best performance of the chalcohalide photodetectors has been achieved so far for device based on the BiSeI nano/micro-wires (Table 5.5). The one-dimensional BiSeI wires were exfoliated mechanically from two-dimensional material [104]. The needle-shaped crystals of BiSeI were peeled off with a tape and inserted in the SiO_2/Si substrate. A thicknesses of BiSeI

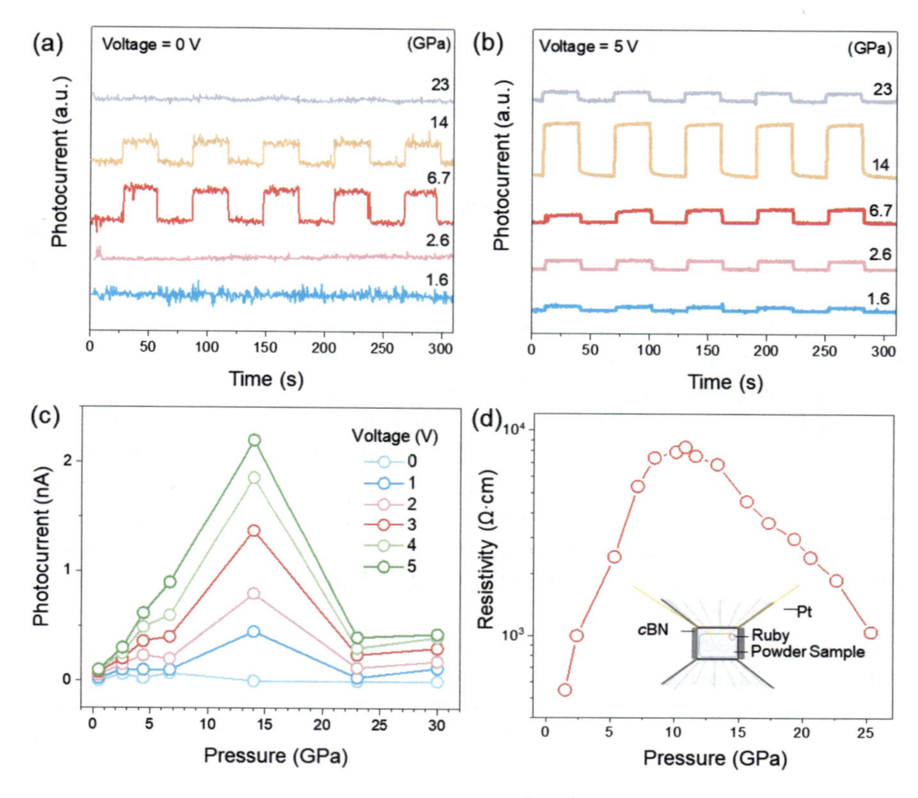

Fig. 5.11 Transient characteristics of photocurrent of SbSI single crystal at different pressures at bias voltages of **a** 0 V and **b** 5 V. The influence of pressure on photocurrent **c** and resistivity **d** of SbSI single crystal. The inset in **d** presents the schematic diagram for resistivity measurement. Reprinted from Liu et al. [118] under the terms of the Creative Commons Attribution 4.0 International License (CC BY 4.0). Copyright (2022) MDPI

wires were in the range from 70 nm to a few micrometers. The electrodes were fabricated using ultraviolet lithography and the thermal evaporation coating. The photocurrent of the 145 nm thick BiSeI wire was measured under monochromatic light illumination (515 nm). This device showed an excellent photodetection performance with enormous responsivity ($R_\lambda = 7 \cdot 10^4$ A W^{-1}), huge detectivity ($D^* = 2.5 \cdot 10^{14}$ Jones), and giant external quantum efficiency ($EQE = 1.8 \cdot 10^7\%$).

5.4 Detectors of Ionizing Radiation

The ionizing radiation detectors are commonly used for may purposes, including nuclear defense, medical imaging, and radiation monitoring in the nuclear reactors. In general, the radiation detectors can be divided into three main types: gas-filled, scintillation, and semiconductor devices [119]. Among semiconductor detectors there are devices, that can operate at room temperature [120]. It is a great advantage in comparison to high purity germanium (HPGe) detectors that have to be cooled to cryogenic temperatures. An ideal detector is made of a material that efficiently absorbs radiation and generates a spectral response that allows identification and characterization of radioactive sources [120]. The high average atomic number (Z) of the absorber and its high mass density is needed to successfully detect a high-energy radiation [121].

Usually, the bulk crystals of the chalcohalide compounds are examined as potential materials for use in radiation detectors (Table 5.6). Only BiSI based detectors [122, 123] were constructed from low-dimensional nanocrystals. The chalcohalides, listed in Table 5.6, consist of the anions coming from halogens with relatively low atomic numbers: chlorine ($Z = 17$), bromine ($Z = 35$), and iodine ($Z = 53$). However, they contain also the heavy-Z cations. Therefore, these compounds possess high electron high electron densities what is beneficial for efficient detection of the ionizing radiation. Furthermore, a sensing materials should have a substantial density (D), wide energy band gap, large electrical resistivity (ρ), and high mobility-lifetime products ($\mu\tau$) of the charge carriers.

The thallium compound, Tl_6SeI_4, was the first chalcohalide investigated as a semiconducting material for X- and γ-ray detection [128]. The bulk single crystal of Tl_6SeI_4 was obtained via a modified vertical Bridgman technique in a dual-zone furnace. The sample was cut perpendicularly to the [001] direction of the

Table 5.6 A comparison of the chalcohalide detectors of ionizing radiation

Material	Material morphology	D, g/cm^3	ρ, Ω cm	$\mu_e\tau_e$, cm^2/V	$\mu_h\tau_h$, cm^2/V	Detected radiation	S/N	References
BiSI	Nanorods and nanoparticles	6.41	$2 \cdot 10^{13}$			α particles, X-ray	1.4	[122]
BiSI/a-C	Nanocomposite	6.4	10^{11}			X-ray	7.44	[123]
$Hg_3S_2Cl_2$	Single crystal	6.80	10^{10}	$1.4 \cdot 10^{-4}$	$7.5 \cdot 10^{-5}$	X-ray, γ-ray		[124]
$Hg_3Se_2Br_2$	Single crystal	7.598	$5.2 \cdot 10^{11}$	$1.4 \cdot 10^{-4}$	$9.2 \cdot 10^{-5}$	X-ray		[125]
$Hg_3Se_2I_2$	Single crystal	7.38	10^{12}	$1.5 \cdot 10^{-5}$	10^{-6}	α particles, X-ray, γ-ray		[126]
SbSeI	Single crystal	5.80	10^8	$4.4 \cdot 10^{-4}$	$3.5 \cdot 10^{-4}$	X-ray, γ-ray		[127]
Tl_6SeI_4	Single crystal	7.38	$4 \cdot 10^{12}$	$7 \cdot 10^{-3}$	$6 \cdot 10^{-4}$	X-ray, γ-ray		[128]

Used abbreviations and symbols: a-C—morphous carbon; D—density; $\mu_e\tau_e$—mobility-lifetime product of electrons; S/N—signal-to noise ratio; $\mu_h\tau_h$—mobility-lifetime product of holes; ρ—electrical resistivity crystal, which was also growth direction. The high mobility-lifetime products of $7 \cdot 10^{-3}$ cm^2/V and $6 \cdot 10^{-4}$ cm^2/V were determined for electrons and holes, respectively. The large resistivity ($\rho = 4{\cdot}10^{12}$ Ω cm) of the Tl_6SeI_4 crystal was measured along the [001] direction [128]. The Co-57 source that emits radiation with characteristic energies of 14.4 keV, 122.1 keV, and 136.5 keV, was used to register a pulse height spectrum of the Tl_6SeI_4 detector. All expected major peaks were clearly resolved. Moreover, an energy resolutions of the pulse height spectra recorded by the Tl_6SeI_4 detector and the commercial detector based on the $Cd_{0.9}Zn_{0.1}Te$ (CZT) were close to each other. Antimony selenoiodide bulk single crystal was proposed in Ref. [127] as a suitable semiconductor for X-ray and γ-ray detection. The SbSeI crystal was grown using a vertical Bridgman method. Its electrical resistivity was approximately 10^8 Ω cm at room temperature. The mobility-lifetime products for electrons and holes were equal to $4.41{\cdot}10^{-4}$ cm^2/V and $3.52{\cdot}10^{-4}$ cm^2/V, accordingly. The SbSeI detector was irradiated with Ag Kα X-ray to measure photocurrent response. A long recovery time of the SbSeI detector was explained due to a slow thermal release of charge carriers from deep trap states [127]. The mercury chalcohalide compounds ($Hg_3Se_2Br_2$ [125], $Hg_3S_2Cl_2$ [124], and $Hg_3Se_2I_2$ [126]) were also examined as potential materials for room temperature X- and γ-ray detection. A solvothermal synthesis of BiSI nanorods and nanoparticles for application in ionizing radiation detectors was reported in Ref. [122]. The BiSI nanorods with average width of 150 nm were compacted into the pellet. Then, the electrodes were deposited on the top and bottom sides of the BiSI pellet. The electrical resistivity in the order of 10^{13} Ω cm was calculated [122]. The BiSI detector exhibited a significant response to X- and γ-ray radiation emitted by ^{241}Am source. Frutos and coworkers presented fabrication of the nanocomposite consisting of BiSI nanorods and amorphous carbon (aC) structures and its application for X and gamma ray detection [123]. The BiSI-aC nanocomposite was synthesized solvothermally using Bi_2S_3 and iodine in mono ethylene glycol at 453 K for 20 h. A pellet was prepared by compression of the BiSI-aC nanocomposite under uniaxial and isostatic press. Afterwards, the BiSI-aC pellet was annealed at temperature of 523 K for 3.5 h under Ar atmosphere. The gold electrodes were thermally evaporated on the both sides of the pellet. The high resistivity of $10^{11}{\cdot}\Omega$ cm was determined for the BiSI-aC nanocomposite [123]. The BiSI-aC detector was exposed to the radiation emitted by ^{241}Am source. A substantial influence of irradiation on the current–voltage characteristics of the BiSI-aC detector was observed. A high signal-to noise ratio ($S/N = 7.44$) of the BiSI-aC detector response was achieved. It should be concluded that investigations of chalcohalide nanomaterials for their use in the ionizing radiation detectors are still at early stage. Thus, these studies should be continued in the future to improve devices performances. A special attention should be paid to the response kinetics, stability, and sensitivity of the detectors based on low-dimensional chalcohalide nanostructures.

References

1. J. Zhang, X. Su, M. Shen, Z. Dai, L. Zhang, X. He, W. Cheng, M. Cao, G. Zou, Enlarging photovoltaic effect: combination of classic photoelectric and ferroelectric photovoltaic effects. Sci. Rep. **3**, 2109 (2013)
2. X. Qi, K. Li, E. Sun, B. Song, D. Huo, J. Li, X. Wang, R. Zhang, B. Yang, W. Cao, Large photovoltaic effect with ultrahigh open-circuit voltage in relaxor-based ferroelectric $Pb(In_{1/2}Nb_{1/2})O_3$-$Pb(Mg_{1/3}Nb_{2/3})O_3$-$PbTiO_3$ ceramics. J. Mater. Sci. Technol. **104**, 119 (2022)
3. X. Zhao, K. Song, H. Huang, W. Han, Y. Yang, Ferroelectric materials for solar energy scavenging and photodetectors. Adv. Opt. Mater. **10**, 2101741 (2022)
4. X. Han, Y. Ji, Y. Yang, Ferroelectric photovoltaic materials and devices. Adv. Funct. Mater. **32**, 2109625 (2022)
5. H. Li, F. Li, Z. Shen, S.T. Han, J. Chen, C. Dong, C. Chen, Y. Zhou, M. Wang, Photoferroelectric perovskite solar cells: principles advances and insights. Nano Today **37**, 101062 (2021)
6. I.E. Castelli, T. Olsen, Y. Chen, Towards photoferroic materials by design: recent progress and perspectives. JPhys Energy **2**, 11001 (2020)
7. H. He, Z. He, Z. Jiang, J. Wang, T. Liu, N. Wang, A controllable photoresponse and photovoltaic performance in $Bi_4Ti_3O_{12}$ ferroelectric thin films. J. Alloys Compd. **694**, 998 (2017)
8. S. Pal, N.V. Sarath, K.S. Priya, P. Murugavel, A review on ferroelectric systems for next generation photovoltaic applications. J. Phys. D. Appl. Phys. **55**, 283001 (2022)
9. L. Huang, M. Wei, C. Gui, L. Jia, Ferroelectric photovoltaic effect and resistive switching behavior modulated by ferroelectric/electrode interface coupling. J. Mater. Sci. Mater. Electron. **31**, 20667 (2020)
10. Y. Li, J. Fu, X. Mao, C. Chen, H. Liu, M. Gong, H. Zeng, Enhanced bulk photovoltaic effect in two-dimensional ferroelectric $CuInP_2S_6$. Nat. Commun. **12**, 5896 (2021)
11. A. Zenkevich, Y. Matveyev, K. Maksimova, R. Gaynutdinov, A. Tolstikhina, V. Fridkin, Giant Bulk Photovoltaic Effect in Thin Ferroelectric $BaTiO_3$ Films. Phys. Rev. B—Condens. Matter Mater. Phys. **90**, 161409 (2014)
12. W. Ji, K. Yao, Y.C. Liang, Bulk photovoltaic effect at visible wavelength in epitaxial ferroelectric $BiFeO_3$ thin films. Adv. Mater. **22**, 1763 (2010)
13. R. Inoue, S. Ishikawa, R. Imura, Y. Kitanaka, T. Oguchi, Y. Noguchi, M. Miyayama, Giant photovoltaic effect of ferroelectric domain walls in perovskite single crystals. Sci. Rep. **5**, 14741 (2015)
14. C.X. Qian, H.J. Feng, Q. Zhang, J. He, Z.X. Chen, M.Z. Wang, X.C. Zeng, Domain wall conduction in calcium-modified lead titanate for polarization tunable photovoltaic devices. Cell Reports Phys. Sci. **1**, 100043 (2020)
15. J. Seidel, D. Fu, S.Y. Yang, E. Alarcón-Lladó, J. Wu, R. Ramesh, J.W. Ager, Efficient photovoltaic current generation at ferroelectric domain walls. Phys. Rev. Lett. **107**, 126805 (2011)
16. Z. Tan et al., Thinning ferroelectric films for high-efficiency photovoltaics based on the Schottky barrier effect. NPG Asia Mater. **11**, 20 (2019)
17. A. Rivera-Calzada, F. Gallego, Y. Kalcheim, P. Salev, J. del Valle, I. Tenreiro, C. León, J. Santamaría, I.K. Schuller, Switchable optically active Schottky barrier in $La_{0.7}Sr_{0.3}MnO_3$/$BaTiO_3$/ITO ferroelectric tunnel junction. Adv. Electron. Mater. **7**, 2100069 (2021)
18. J.M. Yan, K. Wang, Z.X. Xu, J.S. Ying, T.W. Chen, G.L. Yuan, T. Zhang, H.W. Zheng, Y. Chai, R.K. Zheng, Large ferroelectric-polarization-modulated photovoltaic effects in bismuth layered multiferroic/semiconductor heterostructure devices. J. Mater. Chem. C **9**, 3287 (2021)
19. R. Gao, W. Cai, G. Chen, X. Deng, X. Cao, C. Fu, Enhanced ferroelectric photovoltaic effect based on converging depolarization field. Mater. Res. Bull. **84**, 93 (2016)

20. Y. Gong, C. Chen, F. Zhang, X. He, H. Zeng, Q. Yang, Y. Li, Z. Yi, Ferroelectric photovoltaic and flexo-photovoltaic effects in $(1 - x)(Bi_{0.5}Na_{0.5})TiO_3$-$xBiFeO_3$ systems under visible light. J. Am. Ceram. Soc. **103**, 4363 (2020)
21. J. Jiang, Z. Chen, Y. Hu, Y. Xiang, L. Zhang, Y. Wang, G.C. Wang, J. Shi, Flexo-photovoltaic effect in MoS_2. Nat. Nanotechnol. **16**, 894 (2021)
22. M. Qin, K. Yao, Y.C. Liang, High efficient photovoltaics in nanoscaled ferroelectric thin films. Appl. Phys. Lett. **93**, 122904 (2008)
23. Y. Yun, L. Mühlenbein, D.S. Knoche, A. Lotnyk, A. Bhatnagar, Strongly enhanced and tunable photovoltaic effect in ferroelectric-paraelectric superlattices. Sci. Adv. **7**, eabe4206 (2021)
24. M. Nakamura, H. Hatada, Y. Kaneko, N. Ogawa, Y. Tokura, M. Kawasaki, Impact of electrodes on the extraction of shift current from a ferroelectric semiconductor SbSI. Appl. Phys. Lett. **113**, 232901 (2018)
25. B. Chen, J. Shi, X. Zheng, Y. Zhou, K. Zhu, S. Priya, Ferroelectric solar cells based on inorganic-organic hybrid perovskites. J. Mater. Chem. A **3**, 7699 (2015)
26. K.T. Butler, J.M. Frost, A. Walsh, Ferroelectric materials for solar energy conversion: photoferroics revisited. Energy Environ. Sci. **8**, 838 (2015)
27. R. Guo, L. You, Y. Zhou, Z.S. Lim, X. Zou, L. Chen, R. Ramesh, J. Wang, Non-volatile memory based on the ferroelectric. Nat. Commun. **4**, 1990 (2013)
28. D. Li, D. Zheng, C. Jin, W. Zheng, H. Bai, High-performance photovoltaic readable ferroelectric nonvolatile memory based on La-doped $BiFeO_3$ films. ACS Appl. Mater. Interfaces **10**, 19836 (2018)
29. A. Makhort, R. Gumeniuk, J.F. Dayen, P. Dunne, U. Burkhardt, M. Viret, B. Doudin, B. Kundys, Photovoltaic-ferroelectric materials for the realization of all-optical devices. Adv. Opt. Mater. **10**, 2102353 (2022)
30. K. Mistewicz, M. Nowak, D. Stróż, A ferroelectric-photovoltaic effect in SbSI nanowires. Nanomaterials **9**, 580 (2019)
31. R. Nitsche, W.J. Merz, Photoconduction in ternary V-VI-VII compounds. J. Phys. Chem. Solids **13**, 154 (1960)
32. K. Irie, Photovoltaic effect in ferroelectric and photoconductive SbSI. J. Phys. Soc. Japan **30**, 1506 (1971)
33. V.M. Fridkin, A.I. Rodin, Anomalous photovoltaic effect in ferroelectric SbSI and cubic piezoelectric ZnS. Phys. Status Solidi **61**, 123 (1980)
34. D.R. Akopov, A.I. Rodin, A.A. Grekov, Anomalous photovoltaic effect in $A^{V}B^{VI}C^{VII}$ ferroelectrics. Ferroelectrics **26**, 855 (1980)
35. M. Nakamura, H. Hatada, Y. Kaneko, N. Ogawa, M. Sotome, Y. Tokura, M. Kawasaki, Non-local photocurrent in a ferroelectric semiconductor SbSI under local photoexcitation. Appl. Phys. Lett. **116**, 122902 (2020)
36. M. Sotome et al., Spectral dynamics of shift current in ferroelectric semiconductor SbSI. Proc. Natl. Acad. Sci. U. S. A. **116**, 1929 (2019)
37. K. Mistewicz, M. Nowak, A. Starczewska, M. Jesionek, T. Rzychoń, R. Wrzalik, A. Guiseppi-Elie, Determination of electrical conductivity type of SbSI nanowires. Mater. Lett. **182**, 78 (2016)
38. T. Sudersena Rao, A. Mansinch, Electrical and Optical Properties of Sbsi Films. Jpn. J. Appl. Phys. **24**, 422 (1985)
39. R. Nie, H.S. Yun, M.J. Paik, A. Mehta, B.W. Park, Y.C. Choi, S. Il Seok, Efficient solar cells based on light-harvesting antimony sulfoiodide. Adv. Energy Mater. **8**, 1701901 (2018)
40. M. Nowak, K. Mistewicz, A. Nowrot, P. Szperlich, M. Jesionek, A. Starczewska, Transient characteristics and negative photoconductivity of SbSI humidity sensor. Sensors Actuators A Phys. **210**, 32 (2014)
41. C. Wang, D. Cao, F. Zheng, W. Dong, L. Fang, X. Su, M. Shen, Photocathodic behavior of ferroelectric $Pb(Zr, Ti)O_3$ films decorated with silver nanoparticles. Chem. Commun. **49**, 3769 (2013)
42. J.E. Spanier et al., Power conversion efficiency exceeding the Shockley-Queisser limit in a ferroelectric insulator. Nat. Photonics **10**, 611 (2016)

43. S. Gupta, M. Tomar, V. Gupta, Ferroelectric photovoltaic properties of Ce and Mn codoped $BiFeO_3$ thin film. J. Appl. Phys. **115**, 14102 (2014)
44. R. Gupta, M. Tomar, R.P. Tandon, V. Gupta, Impact of TiO_2 buffer layer on the ferroelectric photovoltaic response of CSD grown PZT thick films. Appl. Phys. A Mater. Sci. Process. **127**, 427 (2021)
45. A.M. Glass, D. Von Der Linde, T.J. Negran, High-voltage bulk photovoltaic effect and the photorefractive process in $LiNbO_3$. Appl. Phys. Lett. **25**, 233 (1974)
46. M. Qin, K. Yao, Y.C. Liang, S. Shannigrahi, Thickness effects on photoinduced current in ferroelectric $(Pb_{0.97}La_{0.03})(Zr_{0.52}Ti_{0.48})O_3$ thin films. J. Appl. Phys. **101**, 14104 (2007)
47. L. Fei, Y. Hu, X. Li, R. Song, L. Sun, H. Huang, H. Gu, H.L.W. Chan, Y. Wang, Electrospun bismuth ferrite nanofibers for potential applications in ferroelectric photovoltaic devices. ACS Appl. Mater. Interfaces **7**, 3665 (2015)
48. G. Gopal Khan, R. Das, N. Mukherjee, K. Mandal, Effect of metal doping on highly efficient photovoltaics and switchable photovoltage in bismuth ferrite nanotubes. Phys. Status Solidi—Rapid Res. Lett. **6**, 312 (2012)
49. R.K. Katiyar, Y. Sharma, D. Barrionuevo, S. Kooriyattil, S.P. Pavunny, J.S. Young, G. Morell, B.R. Weiner, R.S. Katiyar, J.F. Scott, Ferroelectric photovoltaic properties in doubly substituted $(Bi_{0.9}La_{0.1})(Fe_{0.97}Ta_{0.03})O_3$ thin films. Appl. Phys. Lett. **106**, 82903 (2015)
50. L. Wu, A.R. Akbashev, A.A. Podpirka, J.E. Spanier, P.K. Davies, Infrared-to-ultraviolet light-absorbing $BaTiO_3$-based ferroelectric photovoltaic materials. J. Am. Ceram. Soc. **102**, 4188 (2019)
51. Y.W. Cho, S.K. Choi, Y.M. Vysochanskii, Photovoltaic effect of $Sn_2P_2S_6$ ferroelectric crystal and ceramics. J. Mater. Res. **16**, 3317 (2001)
52. F. Palazon, Metal chalcohalides: next generation photovoltaic materials? Sol. RRL **6**, 2100829 (2022)
53. Y.C. Choi, K.W. Jung, Recent progress in fabrication of antimony/bismuth chalcohalides for lead-free solar cell applications, Nanomaterials
54. S.J. Adjogri, E.L. Meyer, Chalcogenide perovskites and perovskite-based chalcohalide as photoabsorbers: a study of their properties, and potential photovoltaic applications. Materials
55. A. Zakutayev, J.D. Major, X. Hao, A. Walsh, J. Tang, T.K. Todorov, L.H. Wong, E. Saucedo, Emerging inorganic solar cell efficiency tables (version 2). J. Phys. Energy **3**, 32003 (2021)
56. C. Tablero, Optical properties of Sb(Se, Te)I and photovoltaic applications. J. Alloys Compd. **678**, 18 (2016)
57. R. Nishikubo, H. Kanda, I. García-Benito, A. Molina-Ontoria, G. Pozzi, A.M. Asiri, M.K. Nazeeruddin, A. Saeki, Optoelectronic and energy level exploration of bismuth and antimony-based materials for lead-free solar cells. Chem. Mater. **32**, 6416 (2020)
58. Y.T. Huang, S.R. Kavanagh, D.O. Scanlon, A. Walsh, R.L.Z. Hoye, Perovskite-inspired materials for photovoltaics and beyond-from design to devices. Nanotechnology **32**, 132004 (2021)
59. K.T. Butler, S. McKechnie, P. Azarhoosh, M. Van Schilfgaarde, D.O. Scanlon, A. Walsh, Quasi-particle electronic band structure and alignment of the V-VI-VII semiconductors SbSI, SbSBr, and SbSeI for solar cells. Appl. Phys. Lett. **108**, 112103 (2016)
60. I. Kebaili, I. Boukhris, Z.A. Alrowaili, M.M. Abutalib, M.S. Al-Buriahi, Characterization of physicochemical properties of As_2Se_3–GeTe–AgI chalcohalide glasses for solar cell and IR applications: influence of adding AgI. J. Mater. Sci. Mater. Electron. **33**, 800 (2022)
61. F. Huang, L. Chen, Y. Han, J. Tang, Q. Nie, P. Zhang, Y. Xu, Visible to near-infrared down-conversion in Tm3+/Yb3+ Co-doped chalcohalide glasses for solar spectra converter. Infrared Phys. Technol. **71**, 159 (2015)
62. S.R. Kavanagh, C.N. Savory, D.O. Scanlon, A. Walsh, Hidden spontaneous polarisation in the chalcohalide photovoltaic absorber $Sn_2SbS_2I_3$. Mater. Horizons **8**, 2709 (2021)
63. S. Toso et al., Nanocrystals of lead chalcohalides: a series of kinetically trapped metastable nanostructures. J. Am. Chem. Soc. **142**, 10198 (2020)
64. T. Li, X. Wang, Y. Yan, D.B. Mitzi, Phase stability and electronic structure of prospective sb-based mixed sulfide and iodide 3D perovskite $(CH_3NH_3)SbSI_2$. J. Phys. Chem. Lett. **9**, 3829 (2018)

65. R.E. Brandt et al., Searching for "defect-tolerant" photovoltaic materials: combined theoretical and experimental screening. Chem. Mater. **29**, 4667 (2017)
66. K. Choudhary, M. Bercx, J. Jiang, R. Pachter, D. Lamoen, F. Tavazza, Accelerated discovery of efficient solar cell materials using quantum and machine-learning methods. Chem. Mater. **31**, 5900 (2019)
67. D.W. Davies, K.T. Butler, J.M. Skelton, C. Xie, A.R. Oganov, A. Walsh, Computer-aided design of metal chalcohalide semiconductors: from chemical composition to crystal structure. Chem. Sci. **9**, 1022 (2018)
68. Z. Ran, X. Wang, Y. Li, D. Yang, X.G. Zhao, K. Biswas, D.J. Singh, L. Zhang, Bismuth and antimony-based oxyhalides and chalcohalides as potential optoelectronic materials. Npj Comput. Mater. **4**, 14 (2018)
69. Y.C. Choi, K.W. Jung, One-step solution deposition of antimony selenoiodide films via precursor engineering for lead-free solar cell applications. Nanomaterials **11**, 3206 (2021)
70. K.W. Jung, Y.C. Choi, Compositional engineering of antimony chalcoiodides via a two-step solution process for solar cell applications. ACS Appl. Energy Mater. **5**, 5348 (2022)
71. Y.C. Choi, E. Hwang, D.H. Kim, Controlled growth of SbSI thin films from amorphous Sb_2S_3 for low-temperature solution processed chalcohalide solar cells. APL Mater. **6**, 121108 (2018)
72. D. Rusirawan, I. Farkas, Identification of model parameters of the photovoltaic solar cells. Energy Procedia **57**, 39 (2014)
73. A. Eibeck, D. Nurkowski, A. Menon, J. Bai, J. Wu, L. Zhou, S. Mosbach, J. Akroyd, M. Kraft, Predicting power conversion efficiency of organic photovoltaics: models and data analysis. ACS Omega **6**, 23764 (2021)
74. R. Nie, M. Hu, A. M. Risqi, Z. Li, S. Il Seok, Efficient and stable antimony selenoiodide solar cells. Adv. Sci. **8**, 2003172 (2021)
75. K. Mistewicz et al., A simple route for manufacture of photovoltaic devices based on chalcohalide nanowires. Appl. Surf. Sci. **517**, 146138 (2020)
76. R. Nie, S. Il Seok, Efficient antimony-based solar cells by enhanced charge transfer. Small Methods **4**, 1900698 (2020)
77. A.K. Pathak, M.D. Prasad, S.K. Batabyal, One-dimensional SbSI crystals from Sb, S, and I mixtures in ethylene glycol for solar energy harvesting. Appl. Phys. A Mater. Sci. Process. **125**, 213 (2019)
78. R. Nie, J. Im, S. Il Seok, Efficient solar cells employing light-harvesting $Sb_{0.67}Bi_{0.33}SI$. Adv. Mater. **31**, 1808344 (2019)
79. R. Nie, K.S. Lee, M. Hu, M.J. Paik, S. Il Seok, Heteroleptic tin-antimony sulfoiodide for stable and lead-free solar cells. Matter **3**, 1701 (2020)
80. R. Nie, B. Kim, S.T. Hong, S. Il Seok, Nanostructured heterojunction solar cells based on $Pb_2SbS_2I_3$: linking lead halide perovskites and metal chalcogenides. ACS Energy Lett. **3**, 2376 (2018)
81. R. Nie, A. Mehta, B.W. Park, H.W. Kwon, J. Im, S. Il Seok, Mixed sulfur and iodide-based lead-free perovskite solar cells. J. Am. Chem. Soc. **140**, 872 (2018)
82. J.A. Chang, S.H. Im, Y.H. Lee, H.J. Kim, C.S. Lim, J.H. Heo, S. Il Seok, Panchromatic photon-harvesting by hole-conducting materials in inorganic-organic heterojunction sensitized-solar cell through the formation of nanostructured electron channels. Nano Lett. **12**, 1863 (2012)
83. J.Y. Kim, J.W. Lee, H.S. Jung, H. Shin, N.G. Park, High-efficiency perovskite solar cells. Chem. Rev. **120**, 7867 (2020)
84. D.I. Kim, J.W. Lee, R.H. Jeong, J.H. Boo, A high-efficiency and stable perovskite solar cell fabricated in ambient air using a polyaniline passivation layer. Sci. Rep. **12**, 697 (2022)
85. M. Jeong et al., Stable perovskite solar cells with efficiency exceeding 24.8% and 0.3-V voltage loss. Science **369**, 1615 (2020).
86. A.M. Ganose, S. Matsumoto, J. Buckeridge, D.O. Scanlon, Defect engineering of earth-abundant solar absorbers BiSI and BiSeI. Chem. Mater. **30**, 3827 (2018)
87. A.M. Ganose, K.T. Butler, A. Walsh, D.O. Scanlon, Relativistic electronic structure and band alignment of BiSI and BiSeI: candidate photovoltaic materials. J. Mater. Chem. A **4**, 2060 (2016)

88. L.C. Lee, T.N. Huq, J.L. Macmanus-Driscoll, R.L.Z. Hoye, Research update: bismuth-based perovskite-inspired photovoltaic materials. APL Mater. **6**, 84502 (2018)
89. Y.C. Choi, E. Hwang, Controlled growth of BiSi nanorod-based films through a two-step solution process for solar cell applications. Nanomaterials
90. B. Yoo, D. Ding, J.M. Marin-Beloqui, L. Lanzetta, X. Bu, T. Rath, S.A. Haque, Improved charge separation and photovoltaic performance of BiI_3 absorber layers by use of an in situ formed BiSI interlayer. ACS Appl. Energy Mater. **2**, 7056 (2019)
91. D. Tiwari, F. Cardoso-Delgado, D. Alibhai, M. Mombrú, D.J. Fermín, Photovoltaic performance of phase-pure orthorhombic BiSI thin-films. ACS Appl. Energy Mater. **2**, 3878 (2019)
92. T.N. Huq et al., Electronic structure and optoelectronic properties of bismuth oxyiodide robust against percent-level iodine-, oxygen-, and bismuth-related surface defects. Adv. Funct. Mater. **30**, 1909983 (2020)
93. S. Li, L. Xu, X. Kong, T. Kusunose, N. Tsurumach, Q. Feng, Enhanced photovoltaic performance of BiSCl solar cells through nanorod array. Chemsuschem **14**, 3351 (2021)
94. J. Xiong, Z. You, S. Lei, K. Zhao, Q. Bian, Y. Xiao, B. Cheng, Solution growth of BiSI nanorod arrays on a tungsten substrate for solar cell application. ACS Sustain. Chem. Eng. **8**, 13488 (2020)
95. R.L.Z. Hoye et al., Strongly enhanced photovoltaic performance and defect physics of air-stable bismuth oxyiodide (BiOI). Adv. Mater. **29**, 1702176 (2017)
96. N.T. Hahn, J.L. Self, C.B. Mullins, BiSI micro-rod thin films: efficient solar absorber electrodes? J. Phys. Chem. Lett. **3**, 1571 (2012)
97. N.T. Hahn, A.J.E. Rettie, S.K. Beal, R.R. Fullon, C.B. Mullins, N-BiSI thin films: selenium doping and solar cell behavior. J. Phys. Chem. C **116**, 24878 (2012)
98. S. Li, L. Xu, X. Kong, T. Kusunose, N. Tsurumachi, Q. Feng, $Bi_{13}S_{18}X_2$-based solar cells (X=Cl, Br, I): photoelectric behavior and photovoltaic performance. Phys. Rev. Appl. **15**, 34040 (2021)
99. N. Pai et al., Silver bismuth sulfoiodide solar cells: tuning optoelectronic properties by sulfide modification for enhanced photovoltaic performance. Adv. Energy Mater. **9**, 1803396 (2019)
100. C. Zhang, S. Teo, Z. Guo, L. Gao, Y. Kamata, Z. Xu, T. Ma, Development of a mixed halide-chalcogenide bismuth-based perovskite $MABiI_2S$ with small bandgap and wide absorption range. Chem. Lett. **48**, 249 (2019)
101. H. Kunioku, M. Higashi, R. Abe, Lowerature synthesis of bismuth chalcohalides: candidate photovoltaic materials with easily, continuously controllable band gap. Sci. Rep. **6**, 32664 (2016)
102. P.V.K. Yadav, B. Ajitha, Y.A. Kumar Reddy, A. Sreedhar, Recent advances in development of nanostructured photodetectors from ultraviolet to infrared region: a review. Chemosphere **279**, 130473 (2021)
103. Z. Li, K. Xu, F. Wei, Recent progress in photodetectors based on low-dimensional nanomaterials. Nanotechnol. Rev. **7**, 393 (2018)
104. H.J. Hu, W.L. Zhen, S.R. Weng, Y.D. Li, R. Niu, Z.L. Yue, F. Xu, L. Pi, C.J. Zhang, W.K. Zhu, Enhanced optoelectronic performance and photogating effect in quasi-one-dimensional BiSeI wires. Appl. Phys. Lett. **120**, 201101 (2022)
105. S. Farooq, T. Feeney, J.O. Mendes, V. Krishnamurthi, S. Walia, E. Della Gaspera, J. van Embden, High gain solution-processed carbon-free BiSI chalcohalide thin film photodetectors. Adv. Funct. Mater. **31**, 2104788 (2021)
106. G. Chen, W. Li, Y. Yu, Q. Yang, Fast and low-temperature synthesis of one-dimensional (1D) single-crystalline SbSI microrod for high performance photodetector. RSC Adv. **5**, 21859 (2015)
107. L. Sun, C. Wang, L. Xu, J. Wang, X. Liu, X. Chen, G.C. Yi, SbSI Whisker/PbI_2 flake mixed-dimensional van Der Waals heterostructure for photodetection. CrystEngComm **21**, 3779 (2019)
108. S. Ravishankar, C. Aranda, P.P. Boix, J.A. Anta, J. Bisquert, G. Garcia-Belmonte, Effects of frequency dependence of the external quantum efficiency of perovskite solar cells. J. Phys. Chem. Lett. **9**, 3099 (2018)

109. K.C. Gödel, U. Steiner, Thin film synthesis of SbSI micro-crystals for self-powered photodetectors with rapid time response. Nanoscale **8**, 15920 (2016)
110. J. Shen, X. Liu, C. Wang, J. Wang, B. Wu, X. Chen, X. Chen, G.C. Yi, Sbsi microrod based flexible photodetectors. J. Phys. D. Appl. Phys. **53**, 345106 (2020)
111. P. Liu, L. Yin, L. Feng, Y. Sun, H. Sun, W. Xiong, C. Xia, Z. Wang, Z. Liu, Controllable preparation of ultrathin 2D BiOBr crystals for high-performance ultraviolet photodetector. Sci. China Mater. **64**, 189 (2021)
112. Y. Liu, J. Yin, Z. Tan, M. Wang, J. Wu, Z. Liu, H. Peng, Electrical and photoresponse properties of inversion asymmetric topological insulator BiTeCl nanoplates. ChemNanoMat **3**, 406 (2017)
113. X. Yan, W.L. Zhen, H.J. Hu, L. Pi, C.J. Zhang, W.K. Zhu, High-performance visible light photodetector based on BiSeI single crystal. Chinese Phys. Lett. **38**, 68103 (2021)
114. Y. Purusothaman, N.R. Alluri, A. Chandrasekhar, S.J. Kim, Photoactive piezoelectric energy harvester driven by antimony sulfoiodide (SbSI): A $A_V B_{VI} C_{VII}$ class ferroelectric-semiconductor compound. Nano Energy **50**, 256 (2018)
115. Y. Purusothaman, N.R. Alluri, S.J. Kim, Piezo-phototronic dependent enhanced charge transportation in SbSI micro rod photodetector, in 23rd Opto-electronics and Communications Conference. OECC 2018 (2018), pp. 1–2
116. K. Mistewicz, Recent advances in ferroelectric nanosensors: toward sensitive detection of gas, mechanothermal signals, and radiation. J. Nanomater. **2018**, 2651056 (2018)
117. M. Nowak, Bober, B. Borkowski, M. Kępińska, P. Szperlich, D. Stróz, M. Sozańska, Quantum efficiency coefficient for photogeneration of carriers in SbSI nanowires. Opt. Mater. (Amst). **35**, 2208 (2013)
118. T. Liu et al., Pressure-enhanced photocurrent in one-dimensional SbSI via lone-pair electron reconfiguration. Materials **15**, 3845 (2022)
119. W.D. Kulp, Ionizing radiation ionizing radiation detectors radioactivity detectors, in *Encyclopedia of Sustainability Science and Technology*. ed. by R.A. Meyers (Springer, New York, 2012), pp.5560–5572
120. P.M. Johns, J.C. Nino, Room temperature semiconductor detectors for nuclear security. J. Appl. Phys. **126**, 40902 (2019)
121. T. Yang, F. Li, R. Zheng, Recent advances in radiation detection technologies enabled by metal-halide perovskites. Mater. Adv. **2**, 6744 (2021)
122. I. Aguiar, M. Mombrú, M.P. Barthaburu, H.B. Pereira, L. Fornaro, Influence of solvothermal synthesis conditions in BiSI nanostructures for application in ionizing radiation detectors. Mater. Res. Express **3**, 25012 (2016)
123. M.M. Frutos, M.E.P. Barthaburu, L. Fornaro, I. Aguiar, Bismuth chalcohalide-based nanocomposite for application in ionising radiation detectors. Nanotechnology **31**, 225710 (2020)
124. A.C. Wibowo, C.D. Malliakas, H. Li, C.C. Stoumpos, D.Y. Chung, B.W. Wessels, A.J. Freeman, M.G. Kanatzidis, An unusual crystal growth method of the chalcohalide semiconductor, β-$Hg_3S_2Cl_2$: a new candidate for hard radiation detection. Cryst. Growth Des. **16**, 2678 (2016)
125. H. Li, F. Meng, C.D. Malliakas, Z. Liu, D.Y. Chung, B. Wessels, M.G. Kanatzidis, Mercury chalcohalide semiconductor $Hg_3Se_2Br_2$ for hard radiation detection. Cryst. Growth Des. **16**, 6446 (2016)
126. Y. He et al., Controlling the vapor transport crystal growth of $Hg_3Se_2I_2$ hard radiation detector using organic polymer. Cryst. Growth Des. **19**, 2074 (2019)
127. A.C. Wibowo, C.D. Malliakas, Z. Liu, J.A. Peters, M. Sebastian, D.Y. Chung, B.W. Wessels, M.G. Kanatzidis, Photoconductivity in the chalcohalide semiconductor, SbSeI: a new candidate for hard radiation detection. Inorg. Chem. **52**, 7045 (2013)
128. S. Johnsen, Z. Liu, J.A. Peters, J.H. Song, S. Nguyen, C.D. Malliakas, H. Jin, A.J. Freeman, B.W. Wessels, M.G. Kanatzidis, Thallium Chalcohalides for X-Ray and γ-Ray detection. J. Am. Chem. Soc. **133**, 10030 (2011)

Chapter 6
Gas Nanosensors

6.1 Conductometric Sensors

Gas sensors play the significant roles in the environmental pollution monitoring, industrial safety, detection of fire, leaks, and toxic gasses [1]. The principle of operation of a conductometric gas sensor is based on the variation of its electrical conductivity or resistance, which is correlated to the analyte concentration in the surrounding atmosphere [2]. Different effects can take place on the surface of the sensing layer, e.g. physisorption, chemisorption, diffusion, and catalysis [3]. An interaction between the analyte and the sensor surface leads to charge injection/extraction processes [4]. Therefore, the majority carrier density of the sensing layer is changed resulting in modulation of the overall electrical conductivity of the sensor. Usually, the conductometric sensor consists of a sensitive semiconducting layer equipped with the contact electrodes. The direct current (DC) bias voltage is applied to the sensor and the current flowing through the electrodes is monitored as the response [3]. The conductometric gas sensors possess many advantages, like simplicity of their fabrication and operation, high robustness as well as low size or volume [4]. A response of the conductometric sensor is influenced by numerous factors, including operating temperature, illumination, analyte concentration, a presence of other interfering gases [5]. For example, the gas sensing response can be tailored by adjusting the operating temperature of the conductometric sensor [6, 7]. A significant improvement of the performance of a gas sensor can be also achieved by its exposure to UV light in order to shorten the recovery time [8]. Another strategy toward increasing the sensitivity of the gas sensor is a selection of an appropriate grain size [9] or thickness [10] of the sensing film. It should be underlined that the semiconductor nanomaterials [11], especially one-dimensional nanostructures [12, 13], are attractive for use in the conductometric gas sensors due to their high surface to volume ratios.

Devetak and coworkers presented an application of the chalcohalide $Mo_6S_3I_6$ nanowires network as a gas sensor for detection of various analyte vapors [14]. The nanowires of $Mo_6S_3I_6$ were dispersed in isopropanol (IPA) and deposited by

K. Mistewicz, *Low-Dimensional Chalcohalide Nanomaterials*, NanoScience and Technology, https://doi.org/10.1007/978-3-031-25136-8_6

simple drop-casting method on an oxidized silicon wafer equipped with Au/Ti interdigital electrodes. The resistance response of the $Mo_6S_3I_6$ sensor was measured upon exposing the device to different analytes (methanol, ethanol, acetone, water, ammonia, chloroform, and pyridine) diluted in a flowing inert gas. The operating temperature of the sensor was equal to 296 K. It was revealed that the sensitivity of the individual nanowires is much lower than the sensitivity of the network that contained the large number of nanowires. It was explained due to a condensation of analyte molecules in the network contacts with the metal [14]. Tunneling rate in the metal-nanowire contact junctions was responsible for a change of resistance at the bundle-metal contacts, which affected the main sensing properties of the $Mo_6S_3I_6$ nanowires network. Moreover, the response of the $Mo_6S_3I_6$ sensor was strongly dependent on the dipole moment of the analyte. The $Mo_6S_3I_6$ sensor showed relatively good selectivity and fast response/recovery. A future optimization of the intrinsic $Mo_6S_3I_6$ resistance, network electrodes, and nanowire density was proposed to improve the sensor response and reduce external noise.

A suitability of the antimony sulfoiodide (SbSI) nanowires for a humidity sensing was shown in Ref. [15, 16]. In order to establish humidity sensing mechanism of the SbSI nanowires, two types of nanosensors, made of xerogel as well as single nanowires, were fabricated, examined, and compared in Ref. [16]. The SbSI nanowires were synthesized under ultrasonic irradiation in water directly from elements: antimony, sulfur, and iodine. This method was described comprehensively in Chap. 2 of this book (Sect. 2.5). After synthesis was completed, the material was dried in order to evaporate the remaining water and obtain the SbSI xerogel. A bulk sample was cut from the SbSI xerogel. The opposite sides of the sample were covered with a silver paint to ensure the electrical contacts. The second type of the humidity sensor was constructed from an array of a few SbSI nanowires according to the method developed in Ref. [17]. A suspension of the SbSI nanowires in toluene was drop-casted on Si/SiO_2 substrate equipped with Au microelectrodes. The distance between electrodes of 1 μm was comparable to the average length of the nanowires. The DC electric field-assisted technique was applied to align the nanowires perpendicularly to the electrodes. More information on this method can be found in the Chap. 3 of this book (Sect. 3.7). The SbSI nanowires were bonded ultrasonically to the microelectrodes [17]. Finally, the Si/SiO_2 substrate with the array of a few SbSI nanowires was mounted into a standardized package. These two types of devices were tested as conductometric sensors of humidity at a low operating temperature of 280 K, which is lower than Curie temperature of SbSI nanowires (291 K [18]). The experiments were conducted in dark condition. The typical responses of the SbSI xerogel and an array of a few SbSI nanowires to the relative humidity (RH) change are compared in Fig. 6.1a. Transient characteristics of the dark current responses (I_D) were least square fitted with a following empirical dependence [16]

$$I_D(t) = I_S + I_1 \exp\left(-\frac{t - t_{on}}{\tau_1}\right) - I_2 \exp\left(-\frac{t - t_{on}}{\tau_2}\right), \tag{6.1}$$

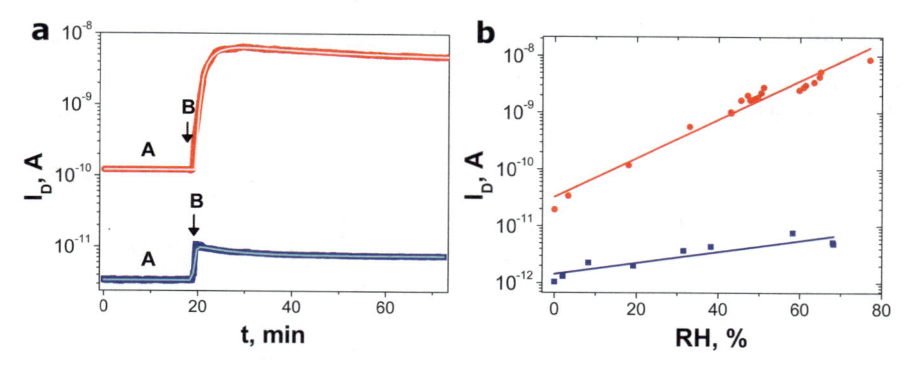

Fig. 6.1 The current responses **a** to a relative humidity change from (A) $RH = 0\%$ to (B) $RH = 59\%$ and influence of relative humidity on current responses **b** of two types of humidity sensors: constructed from SbSI xerogel (filled red circle) and array of a few SbSI nanowires (filled blue square). The solid lines in **a** and **b** represent the best fitted relations described by Eqs. (6.1) and (6.2), respectively. The values of the fitted parameters are given in Table 6.1. Reprinted from Mistewicz et al. [16] under the terms of the Creative Commons Attribution 4.0 International License (CC BY 4.0). Copyright (2017) Springer Nature

where I_S represents a stationary value of dark current, I_1 and I_2 denote the pre-exponential factors, τ_1 and τ_2 mean the time constants, t_{on} is the time stamp for which the relative humidity was changed. The values of fitted parameters are provided in Table 6.1. The response times of the array of a few SbSI nanowires to humidity change were lower than these parameters determined for SbSI xerogel. The current responses of the SbSI nanosensors were investigated as the functions of relative humidity (Fig. 6.1b). The experimental data was least square fitted with the formula given below

$$I_D(RH) = I_{D0} \exp(\alpha_H \cdot RH), \tag{6.2}$$

where I_{D0} is the pre-exponential factor which refers to the value of the dark current at $RH = 0\%$, α_H is a coefficient that describes the sensitivity of the humidity sensor. The values of fitted parameters of Eq. (6.2) are provided in Table 6.1. The α_H coefficient determined for SbSI xerogel was over three times higher than value of this parameter calculated for the array of a few SbSI nanowires. An exponential humidity dependence of electric conductance, similar to experimental data presented in Fig. 6.1b, was reported for graphite oxide [19], microcrystalline cellulose [20], Na_2O-MoO_3-P_2O_5 ceramic [21], sulfonated aromatic hydrocarbon polymers [22], and Nafion membranes [23].

The humidity sensing mechanism of the SbSI nanowires can be explained as follows. An adsorption of the water molecules on the ferroelectric SbSI nanowires was supported by van der Waals forces. Furthermore, the water as a polar molecule may interact with the electric polarization of the ferroelectric domains on the SbSI surface [24]. At a low values of *RH*, the H_2O molecules were adsorbed chemically on the SbSI surface by a dissociation mechanism [16]. The subsequent layers of the water

Table 6.1 The values of the parameters of Eqs. (6.1) and (6.2) fitted to the experimental data presented in Figs. 6.1a and b, respectively

Equation	Parameter	Value of the parameter determined for	
		Array of a few SbSI nanowires	SbSI xerogel
(6.1)	I_S, A	$7.532(4) \cdot 10^{-12}$	$4.51(2) \cdot 10^{-9}$
	I_1, A	$3.05(3) \cdot 10^{-12}$	$3.05(2) \cdot 10^{-9}$
	I_2, A	$10.4(1) \cdot 10^{-12}$	$7.4(1) \cdot 10^{-9}$
	τ_1, s	466(6)	1430(20)
	τ_2, s	25.9(4)	167(1)
(6.2)	I_{D0}, A	$1.43(23) \cdot 10^{-12}$	$32.5(54) \cdot 10^{-12}$
	α_H, $\%^{-1}$	$2.26(39) \cdot 10^{-2}$	$7.87(34) \cdot 10^{-2}$

Data taken from Mistewicz et al. [16] under the terms of the Creative Commons Attribution 4.0 International License (CC BY 4.0). Copyright (2017) Springer Nature

molecules were adsorbed physically on the SbSI nanowire via hydrogen bonding under medium or high humidity condition. An increase of *RH* led to formation of the water clusters on nanowires surfaces and at the nanowire boundaries. The high electrostatic fields coming from the chemisorbed layer were responsible for dissociation of the physisorbed water [21]

$$2H_2O \leftrightarrow H_3O^+ + OH^-, \tag{6.3}$$

where the H_3O^+ and OH^- are hydronium and hydroxyl ions, respectively. A proton transfer process was the main conduction mechanism of the SbSI nanosensors of humidity. The protons tunneled from one water molecule to the next via hydrogen bonding according to the Grotthuss mechanism [25–27]

$$H_2O + H_3O^+ \rightarrow H_3O^+ + H_2O. \tag{6.4}$$

The array of a few SbSI nanowires was much simpler system in respect to the SbSI xerogel which humidity sensing properties were affected by random distribution of contacts between separate nanowires and contacts between nanowires and electrodes. The connections between separate SbSI nanowires were eliminated in the sensor based on the array of a few SbSI nanowires. Taking into consideration a significantly lower sensitivity of the array of a few SbSI nanowires to humidity in comparison to SbSI xerogel (Fig. 6.1b), it should be concluded that the electrical conductance of assemblies of SbSI nanowires in moist environment was mainly caused by water clusters agglomerated on the nanowire boundaries and near the contacts between nanowires [16, 28].

An electrical conductivity type of the SbSI nanowires was determined in Ref. [29]. It was done by examining the influence of hydrogen and oxygen adsorption on electric response of an array of a few SbSI nanowires. This method seems be much simpler than Hall or thermoelectric measurements which require a special instrumentation

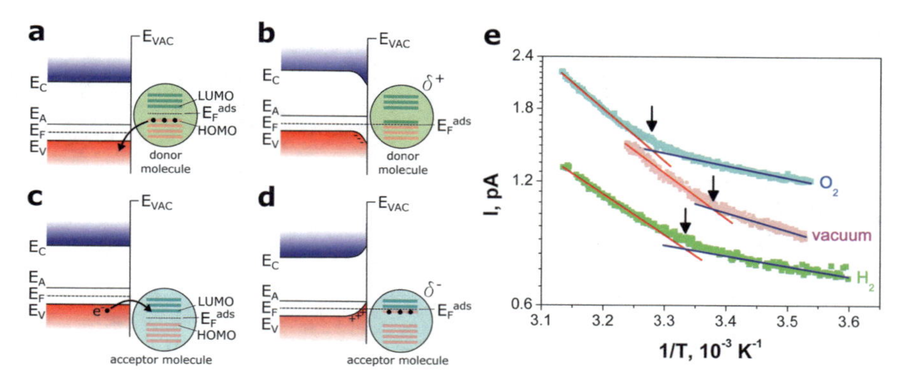

Fig. 6.2 Energy band diagram of p-type SbSI before **a**, **c** and after adsorption of donor **b** and acceptor **d** gas molecules. The Arrhenius plot of electric current **e** flowing through an array of a few SbSI nanowires in different ambient gas atmospheres: hydrogen, oxygen, and vacuum. The blue and red lines in **e** represent the best fitted relation (6.5) in ferroelectric and paraelectric phases, respectively. The arrows indicate the Curie temperatures. Used abbreviations and symbols: E_{VAC}—energy of electron in vacuum, E_C—bottom of conduction energy band, E_A—acceptor level, E_F—Fermi level of SbSI semiconductor, E_{Fads}—Fermi level of adsorbate, E_V—top of valence energy band, LUMO, HOMO—lowest unoccupied and highest occupied molecular orbitals, respectively. Reprinted from Mistewicz et al. [29] with permission from Elsevier. Copyright (2016) Elsevier

in the case of low-dimensional nanomaterials. The hydrogen molecule is known as an electron donor [30, 31]. Its adsorption on the semiconductor surface results in electron donation (Fig. 6.2a, b). The oxygen molecule is an electron acceptor [32, 33]. After its adsorption the electron is withdrawn from semiconductor surface (Fig. 6.2c, d). The electric conductance of the SbSI nanowires decreased and increased due to hydrogen and oxygen gases, respectively (Fig. 6.2e). Such behavior proved the p-type electrical conductivity of SbSI nanowires. It should be mentioned that p-type electrical conductivity was also determined for bulk crystals [34], films [35–37], and nanowires [15, 38] of SbSI.

The Arrhenius plot of an electric conductance of a ferroelectric material consists of two linear curves with different slopes that correspond to the paraelectric and ferroelectric phases [39–41]. This fact was used to determine the Curie temperatures of SbSI nanowires in different ambient gas atmospheres (Fig. 6.2e). Temperature dependences of electric current were least squares fitted using following formula

$$I(T) = I_0 \cdot \exp\left(-\frac{E_A}{k_B T}\right), \tag{6.5}$$

where I_0 is the pre-exponential coefficient of the electric current, E_{A} means the activation energy, and k_{B} denotes the Boltzmann constant. The Curie temperature values of 295.9(2) K, 300.0(2) K, and 305.0(2) K were determined for SbSI nanowires in the vacuum, hydrogen, and oxygen atmospheres, respectively. This effect can be ascribed to an influence of gas adsorption [42] or pressure [43, 44] on phase transition of ferroelectric material.

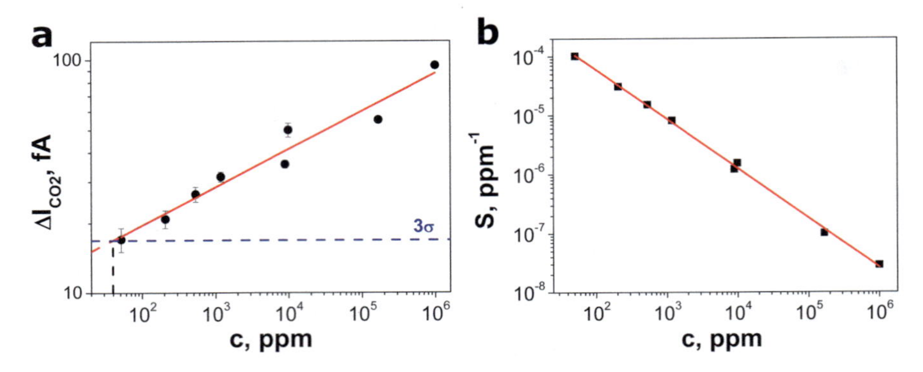

Fig. 6.3 An influence of CO_2 gas concentration on the current response **a** and sensitivity **b** of the SbSI nanosensor. The blue dashed line in **a** represents three times the standard deviation of the nanosensor noise. The red solid lines in **a** and **b** correspond to the best fitted dependences (6.6) and (6.8), respectively. The values of the fitted parameters are given in the text. Reprinted from Mistewicz et al. [16] under the terms of the Creative Commons Attribution 4.0 International License (CC BY 4.0). Copyright (2017) Springer Nature

The array of a few SbSI nanowires was tested as carbon dioxide gas sensor at the operating temperature of 304 K [16]. The CO_2 adsorption on SbSI nanowires surfaces resulted in an increase of electric current flowing through the sensor at a constant bias voltage (Fig. 6.3a). The current response (ΔI_{CO_2}) of a few SbSI nanowire array was least squares with the empirical relation

$$\Delta I_{CO_2} = \Delta I_0 \cdot c^{\varphi}, \tag{6.6}$$

where c is a gas concentration, $\Delta I_0 = 9.3(12)$ fA is the constant, φ is the power coefficient. The value of 0.162(15) was determined for φ coefficient in the case of CO_2 concentration expressed in parts per million (ppm) units. According to [45, 46], the limit of sensor detection (c_{min}) is the value of gas concentration that corresponds to a sensor sensitivity three times greater than the standard deviation of the noise signal. The CO_2 detection limit of 40(31) ppm was calculated for sensor based on array of a few SbSI nanowires [16]. The sensitivity of the current response (Fig. 6.3b) was determined using following equation [47]

$$S = \frac{1}{c} \cdot \frac{\Delta I_{CO_2}}{I_0}, \tag{6.7}$$

where I_0 is the electric current when gas concentration is equal to zero. The Eqs. (6.6) and (6.7) can be combined into one relation

$$S = S_0 \cdot c^{\varphi - 1}, \tag{6.8}$$

where $S_0 = \Delta I_{CO2}/I_0$ is a constant. The CO_2 molecule acts as an electron acceptor [48, 49]. Thus, its adsorption on the SbSI surface led to electron transfer from SbSI into the CO_2 molecule and formation of the partially charged species $CO_2^{\delta-}$ [50]. An increase of the electric current flowing through SbSI nanosensor due to the carbon dioxide adsorption (Fig. 6.3a) confirmed their p-type electric conductivity of the SbSI nanowires.

An ammonia gas detection with the ferroelectric SbSI nanosensor was for the first time demonstrated in Ref. [51]. A few SbSI nanowires were deposited on glass chip and aligned between Au microelectrodes in the external electric field. The responses of the SbSI nanosensor to different NH_3 concentrations were measured at two operating temperatures, i.e. below (280 K) and above (304 K) ferroelectric-paraelectric phase transition of SbSI (291 K [18]). The response was calculated using following equation

$$R = \frac{\Delta I}{I_0} \cdot 100\%, \tag{6.9}$$

where ΔI is the change of electric current due to ammonia adsorption. The increase of the ammonia concentration resulted in exponential rise of the sensor response (Fig. 6.4a). It followed the power law known for the conductometric gas sensors [52, 53]

$$R = R_0 \cdot c^{\varphi}. \tag{6.10}$$

Transient characteristic of the SbSI nanosensor response demonstrated good stability and reversibility. The comparable values of the ammonia detection limits of 6.0(24) ppm and 6.3(39) ppm were determined at the operating temperatures of 280 K and 304 K, respectively [51]. The SbSI nanosensor exhibited a meaningful selectivity to NH_3 against other possible interfering gases: H_2O, CO_2, O_2, H_2, N_2O, CO, and N_2 (Fig. 6.4b). The sensitivity of the device was calculated using Eq. (6.8). The sensor response and sensitivity at the operating temperature of 280 K were

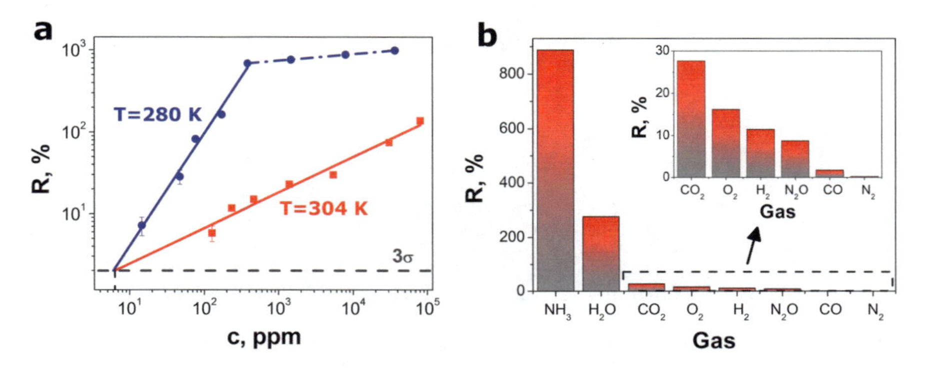

Fig. 6.4 Influence of ammonia concentration on response of the SbSI nanosensor **a** and its selectivity toward different interfering gases at operating temperature of 280 K **b**. Blue and red lines in **a** represent the best fitted dependence described by Eq. (6.10). The horizontal dashed line in **a** shows three times the standard deviation of the noise signal. Adapted from Mistewicz et al. [51] with permission from Elsevier. Copyright (2018) Elsevier

approximately one order of magnitude higher than those achieved at the operating temperature of 304 K. It proved that ferroelectric properties of the SbSI nanosensor are crucial for its large sensitivity.

A following ammonia sensing mechanism was proposed in Ref. [51]. The NH_3 molecule possesses the relatively high permanent electric dipole moment of 1.4 Debye. Therefore, it interacts strongly with the electric polarization of the ferroelectric domains (Fig. 6.5a). It favors a gas molecular adsorption the ferroelectric surface [42, 54, 55]. Since the SbSI surface is partially covered with H_2O molecules [56], NH_3 gas can be adsorbed (Fig. 6.5b) by a dissociation mechanism [57]

$$NH_3 + H_2O \leftrightarrow NH_4^+ + OH^-. \tag{6.11}$$

In the next step, a proton is transferred according to Grotthuss's mechanism [58, 59]. A vacant proton position is created when a proton tunnels (or hops) through

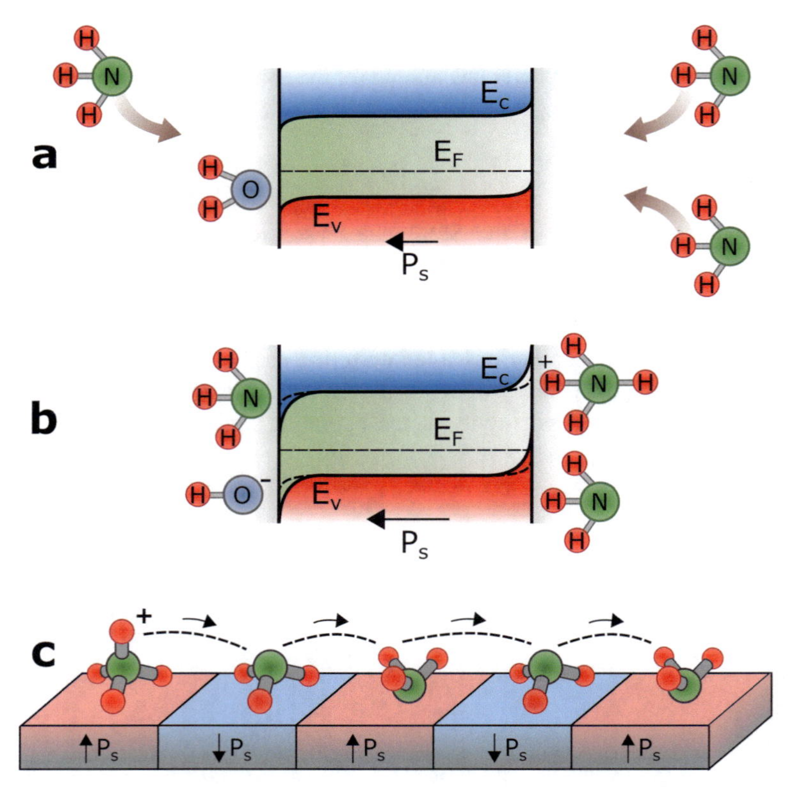

Fig. 6.5 Ammonia sensing mechanism of the SbSI nanosensor: **a** adsorption of NH_3 gas molecules on SbSI surface, **b** formation of ammonium ions, **c** the Grotthuss's chain reaction resulting in a proton transfer. Used abbreviations and symbols: E_C—bottom of conduction energy band, E_F—Fermi level of SbSI semiconductor, E_V—top of valence energy band, P_S—spontaneous polarization. Reprinted from Mistewicz et al. [51] with permission from Elsevier. Copyright (2018) Elsevier

the hydrogen bond from ammonium (NH_4^+) to neighboring ammonia molecule (Fig. 6.5c). Then, the neutral molecule NH_3 reorients itself and occupies the vacant proton position [51]. The ammonia clusters are assembled on the SbSI surface with increasing NH_3 concentration, what leads to formation of the ion-conductive layer and enhancement of the electric conductance of SbSI nanosensor. The Grotthuss's mechanism was also proposed to explain the ammonia detection using other sensing materials, including ferroelectric $Ba_{0.75}Sr_{0.25}TiO_3$ films [59], Ag-doped Fe_2O_3/SiO_2 composite films [57], and zeolites [60, 61].

The gas sensing performances of the conductometric detectors based on the low-dimensional chalcohalide nanomaterials are briefly summarized in Table 6.2. The majority of these sensors were constructed from SbSI nanowires. Further research is needed to reveal sensing properties of the other chalcohalide nanomaterials. Moreover, an investigation of the temperature dependences of the chalcohalide sensor responses should be studied in near future to provide a new insight into their sensing mechanisms. Especially the character of the gas adsorption (physi- or chemisorption) on the SbSI surface at relatively low temperatures is still controversial and requires further clarification.

Table 6.2 A summary on conductometric gas sensors based on chalcohalide nanomaterials

Sensing material	Device structure	Detected gas	T, K	Detection limit or lowest detected concentration	Reference
$Mo_6S_3I_6$	Network of nanowires	CH_3OH, C_2H_5OH, C_3H_6O, H_2O, NH_3, $CHCl_3$, C_5H_5N	296	$c_L = 0.5\%$	[14]
SbSI	Xerogel of randomly oriented nanowires	H_2O		$c_L \approx 3\%$	[15]
	Array of a few aligned nanowires	H_2, O_2	280		[29]
	Array of aligned nanowires	N_2O	298		[17]
	Array of a few aligned nanowires	H_2O	280	$c_L \approx 2\%$	[16]
	Array of a few aligned nanowires	CO_2	304	$c_{min} =$ 40(31) ppm	[16]
	Array of a few aligned nanowires	NH_3	280	$c_{min} =$ 6.0(24) ppm	[51]
			304	$c_{min} =$ 6.3(39) ppm	

Used symbols: c_L—lowest detected concentration of a gas; c_{min}—detection limit; T—operating temperature of a sensor

6.2 Photoconductive Sensors

An influence of humidity on the photoconductivity of the SbSI xerogel was shown in Ref. [62, 63]. The sample was illuminated using argon laser (488 nm). The positive photoconductivity occurred below some critical value of relative humidity (RH_C), whereas the negative photoconductivity was observed for $RH > RH_C$. The values of RH_C were equal to 39.8% [62] and 44% [63] at operating temperatures of 280 K and 304 K, accordingly. The negative photoconductivity was reported for many different low-dimensional materials, e.g. ZnO nanowires [64], InAs nanowires [65], Cd_3As_2 nanowires [66], Bi doped ZnSe nanowires [67], Si nanowire [68], WO_3 nanowire [69], MoS_2 monolayer [70, 71], and $PtSe_2$ film [72]. This phenomenon can originate from trap centers, surface gas molecules, and surface plasmons [73]. However, a photo-induced desorption of a gas is usually regarded as the main mechanism behind the negative photoconductivity [65, 67, 69, 71–73]. Similarly, it was suggested in Ref. [62], that photodesorption of water molecules from SbSI nanowires was responsible for observed negative photoconductivity of the SbSI xerogel in humid atmosphere. The same shape of electric response of the SbSI xerogel was achieved upon its heating, which resulted in thermal desorption of H_2O molecules. Thus, the photo-induced desorption of water from the SbSI surface was due to photonic as well as thermal action of light. These effects led to a changes of an electric polarization, surface band bending, and barriers at grain boundaries of ferroelectric SbSI [62].

A comparative study of the photoconductive sensors of humidity, constructed from SbSI xerogel and an array of a few SbSI nanowires, was presented in Ref. [16]. The nanosensors were illuminated using argon laser (488 nm). It emitted a light with the energy higher than energy band gap of the SbSI. The operating temperature of the sensor ($T = 280$ K) was lower than Curie temperature of the SbSI in order to ensure ferroelectric properties of the sensing material. Figure 6.6a depicts a difference between photocurrent (I_{PC}) responses of the SbSI nanosensors recorded at relative humidity of 53%. In both cases, the photocurrent increased fast after switching on illumination due to photogeneration of excess carriers in the semiconducting SbSI nanowires. This peak of the photocurrent can be also ascribed to the so-called hook anomaly effect which is commonly observed for infrared detectors [74]. The first pulse of the transient characteristic of the photocurrent was followed by its slow decrease to a steady value. The negative and positive photoconductivity was observed for of the SbSI xerogel and the array of a few SbSI nanowires, respectively. Figure 6.6b presents the influence of relative humidity on a steady value of photocurrent $\left(I_{PC_{const}}\right)$ response of the SbSI nanosensors under constant illumination. In the case of the array of a few SbSI nanowires, the values of $I_{PC_{const}}$ were positive in the entire range of relative humidity. The qualitatively different photocurrent behavior was demonstrated for the SbSI xerogel. The values of $I_{PC_{const}}$ were positive for RH lower than the critical value $RH_C = 39.8\%$. Above this threshold the steady photocurrent under constant illumination was negative (Fig. 6.6b).

A ratio of desorbed $\left(\Delta n_{H_2O}\right)$ to adsorbed water molecules $\left(n_{H_2O}\right)$ on surfaces of SbSI nanowires was calculated in Ref. [16]. The ratio $\Delta n_{H_2O}/n_{H_2O}$ is shown as a function of RH in Fig. 6.7a. In the case of the array of a few SbSI nanowires, this parameter was almost independent on relative humidity. It was in contrast to the sensor based

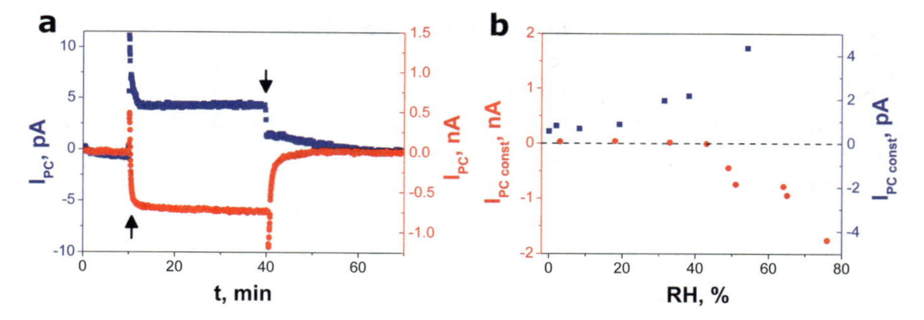

Fig. 6.6 Transient characteristic of photocurrent responses **a** to switching on (arrow up) and switching off (arrow down) illumination (488 nm) at $RH = 53\%$. The influence of relative humidity on a steady photocurrent **b** under constant illumination of SbSI xerogel (filled red circle) and array of a few SbSI nanowires (filled blue square). Reprinted from Mistewicz et al. [16] under the terms of the Creative Commons Attribution 4.0 International License (CC BY 4.0). Copyright (2017) Springer Nature

on the SbSI xerogel. When the *RH* was low, the H_2O molecules were strongly bound in the SbSI xerogel probably due to water chemisorption near electrode-nanowire and nanowire-nanowire contacts. It resulted in a slight $\Delta n_{H_2O}/n_{H_2O}$ ratio. When the relative humidity was high, the H_2O molecules were weakly physisorbed in the SbSI xerogel. Therefore, majority of them were easily desorbed under light illumination leading to a large the $\Delta n_{H_2O}/n_{H_2O}$ ratio. The increase of relative light illumination $\left(I_L/I_{L_{max}}\right)$ led to obvious rise of the number of water molecules photodesorbed from the SbSI xerogel (Fig. 6.7b).

The research results, described above, were measured at only one light wavelength (488 nm) and two values of the sensor operating temperature (280 and 304 K). In order

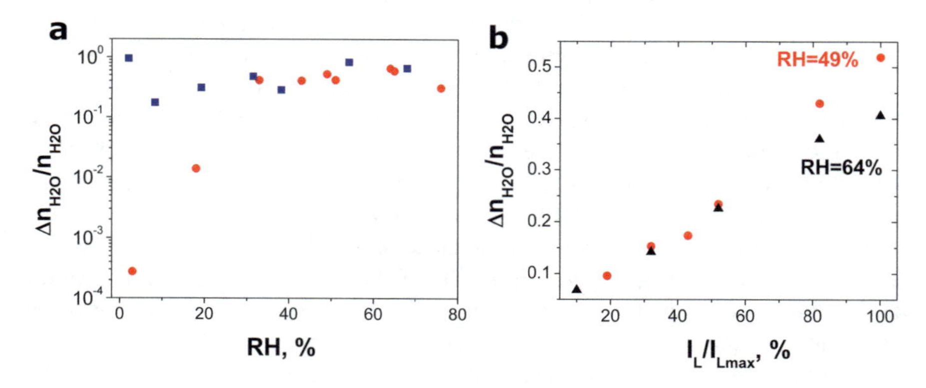

Fig. 6.7 The influence of relative humidity **a** and light intensity **b** on ratio of desorbed to adsorbed water molecules on the surfaces of SbSI xerogel (filled red circle, filled black triangle) and array of a few SbSI nanowires (filled blue square). Reprinted from Mistewicz et al. [16] under the terms of the Creative Commons Attribution 4.0 International License (CC BY 4.0). Copyright (2017) Springer Nature

to explore the humidity sensing mechanism of the SbSI nanowires in more detail, an influence of light wavelength and temperature on the response of photoconductive SbSI nanosensors has to be examined. Moreover, a detection of the other gases than water vapor using the photoconductive SbSI nanosensors can be demonstrated in the future. Since the other chalcohalide compounds (BiSI, SbSeI, BiSBr, BiOI, BiSCl) possess the relatively low energy bang gaps, they should be considered as sensing materials for a construction of photoconductive gas sensors.

6.3 Impedance Sensors

A working principle of the impedance sensor relies on the electric impedance change in a repeatable fashion that can be correlated to gas concentration [75]. The impedance spectroscopy (IS) can be used as an ultra-sensitive technique to identify and measure different toxic gases [76]. In this approach, the alternating current (AC) response of a sensor is measured over a range of frequencies, typically from subhertz to megahertz [77]. A sensitivity, selectivity, and response time of a sensor can be determined using DC measurements, whereas AC investigation provides the same information, but also enables the different contributions from the bulk, surfaces and interfaces, grain boundaries, electrode contacts, and even substrate to be quantified [78]. Furthermore, the impedance spectroscopy is a useful method that can be applied to study the nature of conduction processes, the mechanism of interactions between gas and sensing material, as well as processes with different time constants [79]. The electric equivalent circuit can be determined using IS technique, what ensures a better understanding of sensing performance in computational scheme [80].

The effect of water adsorption on the electrical impedance of the SbSI xerogel was explored in Ref. [81]. The experiments were performed for different temperatures in the range from 279 to 320 K. The increase of the relative humidity from 10 to 85% caused three orders of magnitude decrease of the total complex impedance. The Nyquist characteristics were least square fitted taking into consideration many possible theoretical models, like Voigt or Maxwell circuits, and their combinations. Two various equivalent circuits of the SbSI xerogel were determined at low (20%) and high (80%) relative humidity. Influence of temperature and humidity on relaxation time was presented. The polarization of water molecules strongly contributed to the temperature dependence of electric capacitance of the SbSI xerogel [81]. Thus, the ferroelectric-paraelectric phase transition of the SbSI xerogel could not be observed at low humidity. A dissociation of water molecules at low *RH* generated mobile protons responsible for the observed impedance patterns. When the relative humidity was high, the proton transfer according to Grotthuss chain reaction was recognized as the main sensing mechanism.

A first report on investigation of antimony selenoiodide (SbSeI) nanowires using the impedance spectroscopy and application of this material in a high performance humidity sensor was provided in Ref. [82]. The SbSeI nanowires were fabricated under ultrasonic irradiation. The sonochemical method of the chalcohalide nanomaterials preparation was elaborated in Chap. 2 of this book (Sect. 2.5). The SbSeI nanowires were compacted into the bulk pellet by applying a pressure of 100 MPa at room temperature. An application of this method enabled to increase an area of the electrodes and to decrease a sample thickness. It resulted in a reduction of an electric resistance of a sample, what is a great advantage in the case of the highly resistive one-dimensional chalcohalide nanomaterials. Moreover, a porosity of the SbSeI xerogel was maintained after compression at high pressure. It was calculated in Ref. [82], that SbSeI nanowires constituted only 44% of the total volume of the SbSeI pellet. The porous structure of the sample containing many voids favored water vapor adsorption and was crucial for a sensitive detection of humidity. More information on the high pressure compression of the chalcohalide nanowires can be found in Chap. 3 of this book (Sect. 3.5). The Au electrodes were sputtered on the opposite sides of the SbSeI sample. Impedance spectroscopy measurements were performed at different temperatures from 293 to 345 K and values of relative humidity from 30 to 80%. The impedance modulus response of the SbSeI nanosensor (Fig. 6.8a) was registered as a function of relative humidity from low to high *RH* and in the opposite direction, what corresponded to the adsorption and desorption processes, respectively. The low hysteresis of 3.7% *RH* was determined for humidity change rate of 0.083% min^{-1}. A sensitivity of the SbSeI nanosensor (Fig. 6.8b) was calculated as the absolute vale of impedance derivative with respect to relative humidity [83]

$$S_Z = \left|\frac{\mathrm{d}Z}{\mathrm{d}(RH)}\right|. \quad (6.12)$$

The sensitivity attained the maximum values of 33 kΩ/%*RH* at $RH = 41\%$ and 34 kΩ/%*RH* at $RH = 38\%$ in the case of adsorption and desorption, respectively [82]. These values are higher than impedance sensitivity reported for humidity sensors

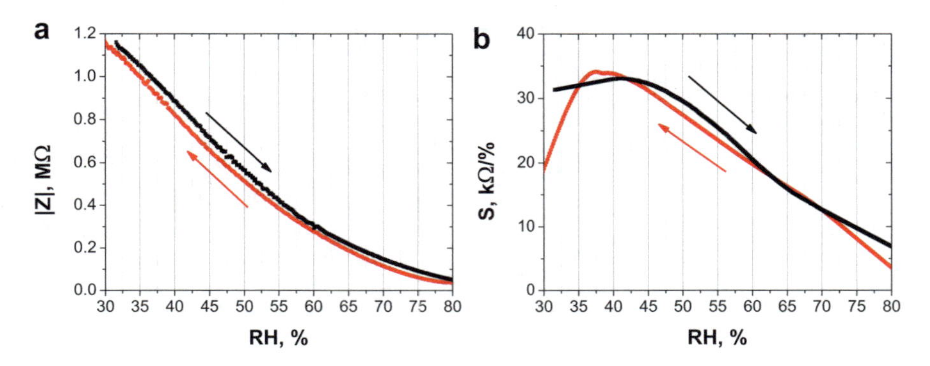

Fig. 6.8 The impedance response **a** and sensitivity **b** of the SbSeI nanosensor as functions of increasing (filled black square) and decreasing (filled red circle) relative humidity measured at operating temperature of 293 K. Reprinted from Mistewicz et al. [82] with permission from Elsevier. Copyright (2020) Elsevier

based on CuO doped ZnO nanoparticles [84], carbon nanotubes [85], and graphene–carbon nanotubes–silicone adhesive nanocomposite [86].

The SbSeI nanosensor was subjected to periodic *RH* fluctuations from 30 to 80% at a constant temperature of 293 K (Fig. 6.9a). The impedance modulus response of the SbSeI nanosensor was highly correlated to humidity input changes (Fig. 6.9b). It showed an excellent repeatability and stability. An existence of relaxation processes with a non-Debye type nature in SbSeI xerogel were confirmed from analysis of the Nyquist plots [82]. The equivalent electric model of SbSeI xerogel was established by least-square fitting of Nyquist characteristics. It consisted of three main branches corresponding to SbSeI nanowires/electrode interface, nanowire/nanowire interface, and nanowires with adsorbed water molecules. A humidity sensing mechanism of the SbSeI xerogel was explained in two regimes of relative humidity. When *RH* was low, the H_2O molecules were adsorbed chemically on sensor surface via dissociation mechanism. The H^+ ions were formed and contributed to electrical conductivity. They were jumping through interstitials sites and channels of the nanocrystalline structure and hopping between adjacent hydroxyl ions [87]. In the case of high relative humidity, the water molecules were adsorbed physically and dissociated into hydronium and hydroxyl ions according to Eq. (6.3). The excess protons were transferred along the hydrogen bonds via electrically neutral H_2O molecules in agreement with the Grotthuss mechanism (Eq. 6.4). Thus, the electrical impedance of the SbSeI xerogel decreased strongly with increasing relative humidity. The experimental data, presented in Ref. [82], proved a meaningful potential of SbSeI nanowires for a reliable humidity sensing.

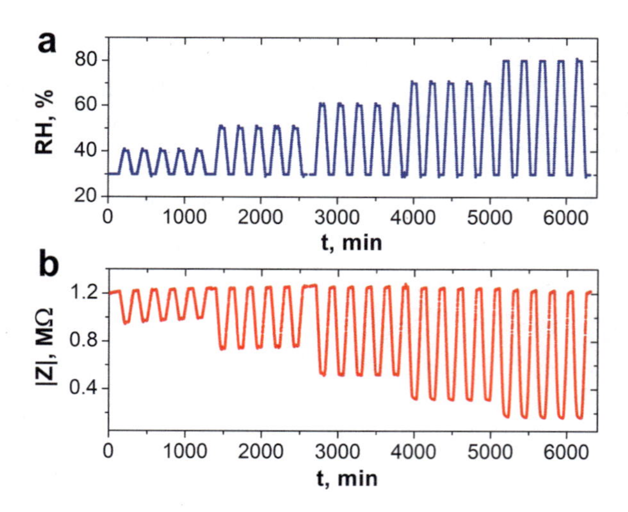

Fig. 6.9 The periodic changes of relative humidity **a** and impedance response of SbSeI nanosensor **b** measured at operating temperature of 293 K. Reprinted from Mistewicz et al. [82] with permission from Elsevier. Copyright (2020) Elsevier

6.4 Quartz Crystal Microbalance Sensors

In general, the quartz crystal microbalance (QCM) sensor consists of a thin piezoelectric crystal plate or cantilever with metal electrodes deposited on each side of the plate [88]. A sensing layer is deposited on the surface of the QCM sensor electrode [89, 90]. A working principle of the quartz crystal microbalance sensor of a gas can be briefly described as follows. The piezoelectric crystal plate of the QCM sensor is excited mechanically or electrically. Thickness shear acoustic waves undergo constructive interference such that resonances are observed at particular frequencies [91]. When the QCM sensor is exposed to an analyte, a molecular absorption of the gas on the sensing layer leads to a variation of the effective mass of the vibrating structure. In result, the resonant frequency of the QCM sensor is changed [88, 92, 93]. The shift of the fundamental frequency (Δf) can be calculated using Sauerbrey equation [90, 94]

$$\Delta f = -\frac{2 f_0^2}{A\sqrt{D_q \mu_q}} \Delta m, \tag{6.13}$$

where f_0 means a resonant frequency of the QCM, Δm is an additional mass attached on the surface of the QCM, A is the electrode surface area, D_q is the quartz density, and μ_q refers to the shear modulus of quartz. It should be noted that this device is a non-specific sensor and only detects the deposited mass changes [95]. Thus, the sorption properties of the sensing layer are crucial for an effective gas detection. The QCM sensors possess many advantages, including high stability, low cost, and sub-nanogram measurement resolution [94].

A first attempt to use the chalcohalide material in cantilever-based gas sensors was demonstrated in Ref. [93]. The reagents (Ge, S, and AgI) were heated in the quartz ampoule at a temperature of 1273 K for 48 h under vacuum (10^{-3} Pa). The bulk glasses of the ternary chalcohalide $(GeS_x)_{100-y}(AgI)_y$ were obtained, where $y = 5, 10, 15, 20$, and $x = 1.2, 1.5, 2, 3$. A thermal evaporation method was applied to deposit the Ge–S–AgI films on Si/SiO_2 cantilevers. The sorption properties of the Ge–S–AgI layer were studied by exposing the sensor to various gases, such as water vapor, ethanol, acetone, and ammonia. A dry nitrogen was used as a carrier gas in these experiments. The highest change of the resonance frequency of the cantilever sensor occurred after acetone adsorption. It was explained due to the highest molecular weight of this analyte [93]. The sensor exhibited a good reversibility of the response to C_3H_6O, H_2O, and C_2H_5OH, what proved that these analytes were adsorbed physically on the Ge–S–AgI surface. A qualitatively different response was observed when the sensor was exposed to ammonia gas, which was chemisorbed on the surface of the sensitive layer. Thus, the sensor baseline was shifted down as a result of an additional mass of NH_3 on the cantilever. Moreover, ammonia adsorption led to a modification of the Ge–S–AgI surface and an increase of the sensitivity of water vapor detection [93].

A synthesis of the chalcohalide compounds from the $GeSe_2$–Ag_2Se–AgI system was presented in Ref. [96]. The basic physicochemical properties of these materials were examined (glass-transition temperatures, density, and microhardness) in order to verify their suitability for future use in gas sensors.

A preparation and characterization of a novel QCM humidity sensor based on bismuth oxychloride (BiOCl) was shown in Ref. [97]. The BiOCl powder was fabricated via a hydrolysis method from bismuth nitrate ($Bi(NO_3)_3 \cdot 5H_2O$) and hydrochloric acid (HCl). A solution of the BiOCl in pure water was deposited to the electrode of QCM using a drop-casting method. The humidity sensing properties of BiOCl based QCM sensor were examined at room temperature (298 K). The resonant frequency decreased significantly with the *RH* increase from 11.3 to 97.3%. A low hysteresis of 2% was determined by measuring the sensor response from low *RH* to high *RH* and in the reverse direction. This slight hysteresis of the response was attributed to trapping of H_2O molecules in the deep layer of the BiOCl film what prevented their complete desorption. The short response and recovery times of 5.2 s and 4.5 s were achieved, respectively [97]. The shift of resonant frequency measured after 3 months was close to its initial value recorded the first day of the experiment. It confirmed an excellent stability of the sensor response. The BiOCl sensor exhibited good selectivity to water vapor at high RH against other possible interfering gases: acetone, methane, ethanol, ammonia, and nitrogen dioxide. The BiOCl sensor was used to detect the human nose breaths, since exhaled and inhaled gases have different levels of humidity during breathing. The humidity sensing mechanism was proposed. At first, the H_2O molecules were chemisorbed on the BiOCl surface. Then, they were adsorbed physically onto the hydroxyl layer by attaching with two neighboring hydroxyl groups through hydrogen double bonds. After water adsorption or desorption the mass of the QCM surface was changed leading to the shift of a resonant frequency.

Only a few chalcohalide compounds have been applied until now as sensing materials in QCM gas sensors. The sorption properties of many members of the chalcohalide family of materials remain unknown. Strategies for incorporation of chalcohalide nanomaterials into the functional devices, presented in the Chap. 3 of this book, seem to be promising for future preparation of novel QCM gas sensors. In order to obtain a high performance sensing layers, the recently developed methods of the chalcohalide films preparation should be considered, including solution processing, spin-coating deposition, drop-casting, and pressure assisted sintering.

References

1. R. Jana, S. Hajra, P.M. Rajaitha, K. Mistewicz, H.J. Kim, Recent advances in multifunctional materials for gas sensing applications. J. Environ. Chem. Eng. **10**, 1 (2022)
2. G. Di Francia, B. Alfano, E. Massera, M.L. Miglietta, T. Polichetti, Chemical sensors: conductometric gas sensors, in *Encyclopedia of Sensors and Biosensors* (Elsevier, 2023), pp. 189–208
3. G. Korotcenkov, V. Brinzari, B.K. Cho, Conductometric gas sensors based on metal oxides modified with gold nanoparticles: a review. Microchim. Acta **183**, 1033 (2016)

4. A. Vázquez-López, J. Bartolomé, A. Cremades, D. Maestre, High-Performance Room-Temperature Conductometric Gas Sensors: Materials and Strategies, Chemosensors **10**, 227 (2022)
5. H. Hashtroudi, I.D.R. MacKinnon, M. Shafiei, Emerging 2D hybrid nanomaterials: towards enhanced sensitive and selective conductometric gas sensors at room temperature. J. Mater. Chem. C **8**, 13108 (2020)
6. J.K. Radhakrishnan, M. Kumara, Geetika, Effect of temperature modulation, on the gas sensing characteristics of ZnO nanostructures, for gases O_2, CO and CO_2. Sensors Int. **2**, 100059 (2021)
7. F. Xu, C. Zhou, H.P. Ho, A rule for operation temperature selection of a conductometric VOC gas sensor based on ZnO nanotetrapods. J. Alloys Compd. **858**, 158294 (2021)
8. R. Wagner, D. Schönauer-Kamin, R. Moos, Novel Operation Strategy to Obtain a Fast Gas Sensor for Continuous Ppb-Level NO_2 Detection at Room Temperature Using ZnO—A Concept Study with Experimental Proof, Sensors **19**, 4104 (2019)
9. G. Korotcenkov, S.D. Han, B.K. Cho, V. Brinzari, Grain size effects in sensor response of nanostructured SnO_2- and In_2O_3-based conductometric thin film gas sensor. Crit. Rev. Solid State Mater. Sci. **34**, 1 (2009)
10. J.F. Chang, H.H. Kuo, I.C. Leu, M.H. Hon, The effects of thickness and operation temperature on ZnO: Al thin film CO gas sensor. Sensors Actuators, B Chem. **84**, 258 (2002)
11. X. Kang, S.P. Yip, Y. Meng, W. Wang, D. Li, C. Liu, J.C. Ho, High-performance electrically transduced hazardous gas sensors based on low-dimensional nanomaterials. Nanoscale Adv. **3**, 6254 (2021)
12. Z. Wang, L. Zhu, S. Sun, J. Wang, W. Yan, One-Dimensional Nanomaterials in Resistive Gas Sensor: From Material Design to Application, Chemosensors **9**, 198 (2021)
13. N. Kaur, M. Singh, E. Comini, One-dimensional nanostructured oxide chemoresistive sensors. Langmuir **36**, 6326 (2020)
14. M. Devetak, B. Berčič, M. Uplaznik, A. Mrzel, D. Mihailovic, $Mo_6S_3I_6$ nanowire network vapor pressure chemisensors. Chem. Mater. **20**, 1773 (2008)
15. M. Nowak et al., Fabrication and characterization of SbSI gel for humidity sensors. Sensors Actuators, A Phys. **210**, 119 (2014)
16. K. Mistewicz, M. Nowak, R. Paszkiewicz, A. Guiseppi-Elie, SbSI nanosensors: from gel to single nanowire devices. Nanoscale Res. Lett. **12**, 97 (2017)
17. K. Mistewicz, M. Nowak, R. Wrzalik, J. Śleziona, J. Wieczorek, A. Guiseppi-Elie, Ultrasonic processing of SbSI nanowires for their application to gas sensors. Ultrasonics **69**, 67 (2016)
18. K. Mistewicz, M. Nowak, D. Stróż, A ferroelectric-photovoltaic effect in SbSI nanowires, Nanomaterials **9**, 580 (2019)
19. V.A. Smirnov, N.N. Denisov, A.E. Ukshe, Y.M. Shul'Ga, Effect of humidity on the conductivity of graphite oxide during its photoreduction. High Energy Chem. **47**, 242 (2013)
20. M. Nilsson, Conductance phenomena in microcrystalline cellulose. Phys. Status Solidi C Conf. **3**, 251 (2006)
21. M. Foucaud, S. Renka, T. Klaser, J. Popović, Ž. Skoko, P. Mošner, L. Koudelka, A. Šantić, Sodium-ion conductivity and humidity-sensing properties of Na_2O-MoO_3-P_2O_5 glass-ceramics. Nanomaterials **12**, 240 (2022)
22. T. Higashihara, K. Matsumoto, M. Ueda, Sulfonated aromatic hydrocarbon polymers as proton exchange membranes for fuel cells. Polymer (Guildf). **50**, 5341 (2009)
23. S. Ma, Z. Siroma, H. Tanaka, Anisotropic conductivity over in-plane and thickness directions in nafion-117. J. Electrochem. Soc. **153**, A2274 (2006)
24. K. Mistewicz, Recent advances in ferroelectric nanosensors: toward sensitive detection of gas, mechanothermal signals, and radiation. J. Nanomater. **2018**, 2651056 (2018)
25. N. Agmon, The Grotthuss mechanism. Chem. Phys. Lett. **244**, 456 (1995)
26. Z. Li, M. Zhang, L. Yang, R. Wu, Z. Wu, Y. Jiang, L. Zhou, Y. Liu, The effect of surface hydroxyls on the humidity-sensitive properties of LiCl-doped $ZnSn(OH)_6$ sphere-based sensors. Nanomaterials **12**, 467 (2022)
27. Y. Lu, G. Yang, Y. Shen, H. Yang, K. Xu, Multifunctional flexible humidity sensor systems towards noncontact wearable electronics. Nano-Micro Lett. **14**, 150 (2022)

28. K. Mistewicz, M. Nowak, Prevention of food spoilage using nanoscale sensors, in *Nanobiosensors*, ed. by A. M. B. T.-N. Grumezescu (Academic Press, 2017), pp. 245–288
29. K. Mistewicz, M. Nowak, A. Starczewska, M. Jesionek, T. Rzychoń, R. Wrzalik, A. Guiseppi-Elie, Determination of electrical conductivity type of SbSI nanowires. Mater. Lett. **182**, 78 (2016)
30. L.Y. Guo, S. Liang, Z. Yang, L. Jin, Y. Tan, Z. Huang, Gas-sensing properties of dissolved gases in insulating material adsorbed on SnO_2–GeSe monolayer. Chemosensors **10**, 212 (2022)
31. K.I.M. Rojas, A.R.C. Villagracia, S.C. Narido, J.L.V. Moreno, N.B. Arboleda, First principles study of H_2 adsorption on Ni-decorated silicene. Mater. Res. Express **6**, 55509 (2019)
32. Z. Zhang, J.T. Yates, Effect of adsorbed donor and acceptor molecules on electron stimulated desorption: $O_2/TiO_2(110)$. J. Phys. Chem. Lett. **1**, 2185 (2010)
33. M. V. Manasa, G. Sarala Devi, P.S. Prasada Reddy, B. Sreedhar, High performance CO_2 gas sensor based on noble metal functionalized semiconductor nanomaterials for health and environmental safety. Mater. Res. Express **6**, 125041 (2019)
34. I. Ikemoto, X-ray photoelectron spectroscopic studies of SbSI. Bull. Chem. Soc. Jpn. **54**, 2519 (1981)
35. R. Nie, H.S. Yun, M.J. Paik, A. Mehta, B.W. Park, Y.C. Choi, S. Il Seok, Efficient solar cells based on light-harvesting antimony sulfoiodide. Adv. Energy Mater. **8**, 1701901 (2018)
36. A. Mansingh, T.S. Rao, I-V and C-V characteristics of ferroelectric SbSI(Film)-Si-metal. Ferroelectrics **50**, 263 (1983)
37. T. Sudersena Rao, A. Mansinch, Electrical and optical properties of SbSI films. Jpn. J. Appl. Phys. **24**, 422 (1985)
38. M. Nowak, E. Talik, P. Szperlich, D. Stróz, XPS analysis of sonochemically prepared SbSI ethanogel. Appl. Surf. Sci. **255**, 7689 (2009)
39. D. Bochenek, P. Niemiec, R. Skulski, M. Adamczyk, D. Brzezińska, Electrophysical properties of the multicomponent PBZT-type ceramics doped by Sn^{4+}. J. Electroceramics **42**, 17 (2019)
40. A. Mansingh, K.N. Srivastava, B. Singh, Effect of surface capacitance on the dielectric behavior of ferroelectric lead germanate. J. Appl. Phys. **50**, 4319 (1979)
41. K. Mistewicz, Pyroelectric nanogenerator based on an SbSI-TiO_2 nanocomposite. Sensors **22**, 69 (2022)
42. A.L. Cabrera, G. Tarrach, P. Lagos, G.B. Cabrera, Influence of crystallographic phase transitions in small ferroelectric particles on carbon dioxide adsorption. Ferroelectrics **281**, 53 (2002)
43. K. Gesi, Effect of hydrostatic pressure on ferroelectric and related phase transitions. Phase Transitions **40**, 187 (1992)
44. G.A. Samara, F. Bauer, The effects of pressure on the β molecular relaxation and phase transitions of the ferroelectric copolymer $P(VDF_{0.7}TrFe_{0.3})$. Ferroelectrics **135**, 385 (1992)
45. K.W. Kao, M.C. Hsu, Y.H. Chang, S. Gwo, J. Andrew Yeh, A sub-ppm acetone gas sensor for diabetes detection using 10 Nm thick ultrathin InN FETs. Sensors **12**, 7157 (2012)
46. H.P. Loock, P.D. Wentzell, Detection limits of chemical sensors: applications and misapplications. Sensors Actuators, B Chem. **173**, 157 (2012)
47. A. Qureshi, A. Altindal, A. Mergen, Electrical and gas sensing properties of Li and Ti Codoped NiO/PVDF thin film. Sensors Actuators, B Chem. **138**, 71 (2009)
48. A. Vimont, A. Travert, P. Bazin, J.C. Lavalley, M. Daturi, C. Serre, G. Férey, S. Bourrelly, P.L. Llewellyn, Evidence of CO_2 molecule acting as an electron acceptor on a nanoporous metal-organic-framework MIL-53 or $Cr^{3+}(OH)(O_2C\text{-}C_6H_4\text{-}CO_2)$. Chem. Commun. 3291 (2007)
49. H. Tang, L.N. Sacco, S. Vollebregt, H. Ye, X. Fan, G. Zhang, Recent advances in 2D/nanostructured metal sulfide-based gas sensors: mechanisms, applications, and perspectives. J. Mater. Chem. A **8**, 24943 (2020)
50. Y. Ye, H. Yang, J. Qian, H. Su, K.J. Lee, T. Cheng, H. Xiao, J. Yano, W.A. Goddard, E.J. Crumlin, Dramatic differences in carbon dioxide adsorption and initial steps of reduction between silver and copper. Nat. Commun. **10**, 1875 (2019)
51. K. Mistewicz, M. Nowak, D. Stróż, A. Guiseppi-Elie, Ferroelectric SbSI nanowires for ammonia detection at a low temperature. Talanta **189**, 225 (2018)

52. A. Tepore, A. Serra, D. Manno, L. Valli, G. Micocci, D.P. Arnold, Kinetic behavior analysis of porphyrin Langmuir-Blodgett films for conductive gas sensors. J. Appl. Phys. **84**, 1416 (1998)
53. T. Siciliano, M. Di Giulio, M. Tepore, E. Filippo, G. Micocci, A. Tepore, Ammonia sensitivity of Rf sputtered tellurium oxide thin films. Sensors Actuators, B Chem. **138**, 550 (2009)
54. L.C. TĂnase, N.G. Apostol, L.E. Abramiuc, C.A. Tache, L. Hrib, L. TrupinĂ, L. Pintilie, C.M. Teodorescu, Ferroelectric triggering of carbon monoxide adsorption on lead zirco-titanate (001) surfaces. Sci. Rep. **6**, 35301 (2016)
55. X. Tang, J. Shang, Y. Gu, A. Du, L. Kou, Reversible gas capture using a ferroelectric switch and 2D molecule multiferroics on the In_2Se_3 monolayer. J. Mater. Chem. A **8**, 7331 (2020)
56. A. Starczewska, R. Wrzalik, M. Nowak, P. Szperlich, Bober, J. Szala, D. Stróz, D. Czechowicz, Infrared spectroscopy of ferroelectric nanowires of antimony sulfoiodide. Infrared Phys. Technol. **51**, 307 (2008)
57. Y. Tang, Z. Li, X. Zu, J. Ma, L. Wang, J. Yang, B. Du, Q. Yu, Room-temperature NH_3 gas sensors based on Ag-doped γ-Fe_2O_3/SiO_2 composite films with sub-Ppm detection ability. J. Hazard. Mater. **298**, 154 (2015)
58. M. Meuwly, A. Bach, S. Leutwyler, Grotthus-type and diffusive proton transfer in 7-hydroxyquinoline·$(NH_3)_n$ clusters. J. Am. Chem. Soc. **123**, 11446 (2001)
59. C.E. Simion, A. Sackmann, V.S. Teodorescu, C.F. Ruşti, A. Stənoiu, Room temperature ammonia sensing with barium strontium titanate under humid air background. Sensors Actuators, B Chem. **220**, 1241 (2015)
60. M.E. Franke, U. Simon, R. Moos, A. Knezevic, R. Müller, C. Plog, Development and working principle of an ammonia gas sensor based on a refined model for solvate supported proton transport in zeolites. Phys. Chem. Chem. Phys. **5**, 5195 (2003)
61. M.E. Franke, U. Simon, Solvate-supported proton transport in zeolites. ChemPhysChem **5**, 465 (2004)
62. M. Nowak, K. Mistewicz, A. Nowrot, P. Szperlich, M. Jesionek, A. Starczewska, Transient characteristics and negative photoconductivity of SbSI humidity sensor. Sensors Actuators, A Phys. **210**, 32 (2014)
63. K. Mistewicz, M. Nowak, P. Szperlich, A. Nowrot, Humidity sensing using SbSI nanophotodetectors, in *Optics InfoBase Conference Papers* (Optica Publishing Group, Tucson, Arizona, 2014), p. JTu3A.31
64. Z. Fan, D. Dutta, C.J. Chien, H.Y. Chen, E.C. Brown, P.C. Chang, J.G. Lu, Electrical and photoconductive properties of vertical ZnO nanowires in high density arrays. Appl. Phys. Lett. **89**, 213110 (2006)
65. Y. Han, X. Zheng, M. Fu, D. Pan, X. Li, Y. Guo, J. Zhao, Q. Chen, Negative photoconductivity of InAs nanowires. Phys. Chem. Chem. Phys. **18**, 818 (2015)
66. K. Park, M. Jung, D. Kim, J.R. Bayogan, J.H. Lee, S.J. An, J. Seo, J. Seo, J.P. Ahn, J. Park, Phase controlled growth of Cd_3As_2 nanowires and their negative photoconductivity. Nano Lett. **20**, 4939 (2020)
67. X. Zhang, J. Jie, Z. Wang, C. Wu, L. Wang, Q. Peng, Y. Yu, P. Jiang, C. Xie, Surface induced negative photoconductivity in p-type ZnSe: Bi nanowires and their nano-optoelectronic applications. J. Mater. Chem. **21**, 6736 (2011)
68. E. Baek, T. Rim, J. Schütt, C.K. Baek, K. Kim, L. Baraban, G. Cuniberti, Negative photoconductance in heavily doped Si nanowire field-effect transistors. Nano Lett. **17**, 6727 (2017)
69. Y. Liu, P. Fu, Y. Yin, Y. Peng, W. Yang, G. Zhao, W. Wang, W. Zhou, D. Tang, Positive and negative photoconductivity conversion induced by H_2O molecule adsorption in WO_3 nanowire. Nanoscale Res. Lett. **14**, 144 (2019)
70. X. Xiao, J. Li, J. Wu, D. Lu, C. Tang, Negative photoconductivity observed in polycrystalline monolayer molybdenum disulfide prepared by chemical vapor deposition. Appl. Phys. A Mater. Sci. Process. **125**, 765 (2019)
71. J.K. Gustafson, D. Wines, E. Gulian, C. Ataca, L.M. Hayden, Positive and negative photoconductivity in monolayer MoS_2 as a function of physisorbed oxygen. J. Phys. Chem. C **125**, 8712 (2021)

72. A. Grillo, E. Faella, A. Pelella, F. Giubileo, L. Ansari, F. Gity, P.K. Hurley, N. McEvoy, A. DiBartolomeo, Coexistence of negative and positive photoconductivity in few-layer $PtSe_2$ field-effect transistors. Adv. Funct. Mater. **31**, 2105722 (2021)
73. B. Cui, Y. Xing, J. Han, W. Lv, W. Lv, T. Lei, Y. Zhang, H. Ma, Z. Zeng, B. Zhang, Negative photoconductivity in low-dimensional materials. Chinese Physics B **30**, 028507 (2021)
74. N. Sclar, Properties of doped silicon and germanium infrared detectors. Prog. Quantum Electron. **9**, 149 (1984)
75. J.M. Rheaume, A.P. Pisano, A review of recent progress in sensing of gas concentration by impedance change. Ionics (Kiel). **17**, 99 (2011)
76. V. Balasubramani, S. Chandraleka, T.S. Rao, R. Sasikumar, M.R. Kuppusamy, T.M. Sridhar, Review—recent advances in electrochemical impedance spectroscopy based toxic gas sensors using semiconducting metal oxides. J. Electrochem. Soc. **167**, 037572 (2020)
77. Y. Liu, Y. Lei, Pt-CeO_2 nanofibers based high-frequency impedancemetric gas sensor for selective CO and C_3H_8 detection in high-temperature harsh environment. Sensors Actuators, B Chem. **188**, 1141 (2013)
78. F. Schipani, D.R. Miller, M.A. Ponce, C.M. Aldao, S.A. Akbar, P.A. Morris, Electrical characterization of semiconductor oxide-based gas sensors using impedance spectroscopy: a review. Reviews in Advanced Sciences and Engineering **5**, 86 (2016)
79. A. Labidi, C. Jacolin, M. Bendahan, A. Abdelghani, J. Guérin, K. Aguir, M. Maaref, Impedance spectroscopy on WO_3 gas sensor. Sensors Actuators, B Chem. **106**, 713 (2005)
80. K. Dutta, Potential of impedance spectroscopy towards quantified analysis of gas sensors: a tutorial. IEEE Sens. J. **21**, 22220 (2021)
81. A. Starczewska, M. Nowak, P. Szperlich, B. Toroń, K. Mistewicz, D. Stróz, J. Szala, Influence of humidity on impedance of SbSI gel. Sensors Actuators, A Phys. **183**, 34 (2012)
82. K. Mistewicz, A. Starczewska, M. Jesionek, M. Nowak, M. Kozioł, D. Stróż, Humidity dependent impedance characteristics of SbSeI nanowires. Appl. Surf. Sci. **513**, 145859 (2020)
83. J. Wang, M.Y. Su, J.Q. Qi, L.Q. Chang, Sensitivity and complex impedance of nanometer zirconia thick film humidity sensors. Sensors Actuators, B Chem. **139**, 418 (2009)
84. T. Thiwawong, K. Onlaor, N. Chaithanatkun, B. Tunhoo, Preparation of copper doped zinc oxide nanoparticles by precipitation process for humidity sensing device. AIP Conf. Proc. **2010**, 20022 (2018)
85. H.S. Kim, J.H. Kang, J.Y. Hwang, U.S. Shin, Wearable CNTs-based humidity sensors with high sensitivity and flexibility for real-time multiple respiratory monitoring. Nano Converg. **9**, 35 (2022)
86. M.T.S. Chani, K.S. Karimov, A.M. Asiri, Impedimetric humidity and temperature sensing properties of the graphene-carbon nanotubes–silicone adhesive nanocomposite. J. Mater. Sci. Mater. Electron. **30**, 6419 (2019)
87. R. Álvarez, F. Guerrero, G. Garcia-Belmonte, J. Bisquert, Study of the humidity effect in the electrical response of the $KSbMoO_6$ ionic conductive ceramic at low temperature. Mater. Sci. Eng. B Solid-State Mater. Adv. Technol. **90**, 291 (2002)
88. M. Varga, A. Laposa, P. Kulha, J. Kroutil, M. Husak, A. Kromka, Quartz crystal microbalance gas sensor with nanocrystalline diamond sensitive layer. Phys. Status Solidi Basic Res. **252**, 2591 (2015)
89. E. Haghighi, S. Zeinali, Nanoporous MIL-101(Cr) as a sensing layer coated on a quartz crystal microbalance (QCM) nanosensor to detect volatile organic compounds (VOCs). RSC Adv. **9**, 24460 (2019)
90. F. Fauzi, A. Rianjanu, I. Santoso, K. Triyana, Gas and humidity sensing with quartz crystal microbalance (QCM) coated with graphene-based materials—a mini review. Sensors Actuators, A Phys. **330**, 112837 (2021)
91. K. Kanazawa, N.J. Cho, Quartz crystal microbalance as a sensor to characterize macromolecular assembly dynamics. J. Sensors **2009**, 824947 (2009)
92. H.M. Saraoglu, A.O. Selvi, M.A. Ebeoglu, C. Tasaltin, Electronic nose system based on quartz crystal microbalance sensor for blood glucose and Hba1c levels from exhaled breath odor. IEEE Sens. J. **13**, 4229 (2013)

93. B. Monchev, D. Filenko, N. Nikolov, C. Popov, T. Ivanov, P. Petkov, I.W. Rangelow, Investigation of the sorption properties of thin Ge-S-AgI films deposited on cantilever-based gas sensor. Appl. Phys. A Mater. Sci. Process. **87**, 31 (2007)
94. I.R. Jang, S.I. Jung, J. Park, C. Ryu, I. Park, S.B. Kim, H.J. Kim, Direct and controlled device integration of graphene oxide on quartz crystal microbalance via electrospray deposition for stable humidity sensing. Ceram. Int. **48**, 8004 (2022)
95. S.N. Songkhla, T. Nakamoto, Overview of quartz crystal microbalance behavior analysis and measurement. Chemosensors **9**, 350 (2021)
96. G. Vassilev, V. Vassilev, S. Boycheva, K. Petkov, New chalcohalide glasses from the $GeSe_2$-Ag_2Se-AgI system for nanostructured sensors, in *NATO Science for Peace and Security Series B: Physics and Biophysics*. ed. by J.P. Reithmaier, P. Paunovic, W. Kulisch, C. Popov, P. Petkov (Springer, Netherlands, 2011), pp.235–238
97. Q. Chen, N.B. Feng, X.H. Huang, Y. Yao, Y.R. Jin, W. Pan, D. Liu, Humidity-sensing properties of a BiOCl-coated quartz crystal microbalance. ACS Omega **5**, 18818 (2020)

Chapter 7
The Catalysts for an Environmental Remediation

7.1 Photocatalysis

7.1.1 Introduction to Photocatalysis

A presence of the harmful pollutant in the environment impacts significantly on health of humans and animals. The methods of environmental remediation include a degradation of pharmaceutical products from water, groundwater remediation, water disinfection, air purification, and reduction of carbon dioxide in air [1]. The toxic organic compounds can be eliminated from environment using adsorption [2], ozonation [3, 4], forward/reverse osmosis [5], ultrasonic degradation [6], electrochemical oxidation [7, 8], Fenton process [9], biodegradation [10], photolysis [11], and photocatalysis [12–14]. The advanced oxidation processes (AOPs) involve in situ generation of highly reactive radical species, which are able to degrade the contaminants in wastewater [15]. Among different AOPs techniques, the photocatalysis possesses significant advantages, like high efficiency and low-cost [16]. A removal of various harmful pollutants have been achieved so far by using of different photocatalysts, like carbon based materials, metal oxides, perovskite materials, metal–organic frameworks (MOFs), chalcogenides, and chalcohalides [12].

A general mechanism of the pollutant degradation using semiconductor as a photocatalyst is presented in Fig. 7.1. When the semiconductor photocatalyst is exposed to light with energy higher than its energy band gap (E_g), the electron (e^-) is excited by a absorbed photon and it jumps to the conduction band, leading to the hole (h^+) formation in the valance band. Then, the hydroxyl radical ($^{\bullet}OH$) and superoxide anion ($^{\bullet}O_2^-$) are generated according following reactions [17]

$$h^+ + H_2O \rightarrow {}^{\bullet}OH + H^+, \tag{7.1}$$

$$h^+ + OH^- \rightarrow {}^{\bullet}OH_{ad}, \tag{7.2}$$

K. Mistewicz, *Low-Dimensional Chalcohalide Nanomaterials*, NanoScience and Technology, https://doi.org/10.1007/978-3-031-25136-8_7

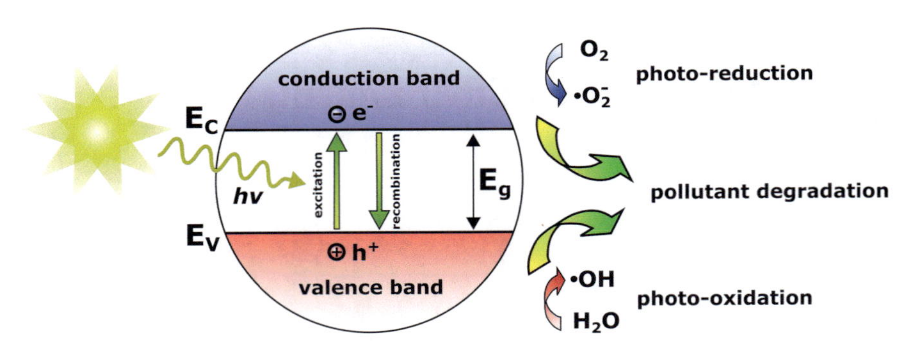

Fig. 7.1 A general mechanism of photocatalytic degradation of the pollutant using a semiconductor material as a photocatalyst. Used abbreviations and symbols: E_C—bottom of conduction energy band; E_g—energy band gap; E_V—top of valence energy band; e^-—electron; h^+—hole; $h\nu$—photon. Detailed description is provided in the text

$$e^- + O_2 \rightarrow {}^\bullet O_2^- . \quad (7.3)$$

Both radicals $^\bullet OH$ and $^\bullet O_2{}^-$ act as powerful oxidants responsible for a decomposition of the organic pollutant into the less harmful inorganic products [16]. In order to achieve a substantial efficiency of the photocatalytic process, a reduced recombination rate, high separation of photogenerated charges, large adsorption capacity, narrow energy band gap, and good stability of the photocatalyst are required. Moreover, the efficiency of photocatalysis is also affected by many additional factors, including pH of a solution, the ratio of photocatalyst mass to the initial mass of a dye, and intensity of the light used to irradiate a dye solution.

The kinetics of the heterogeneous photocatalytic removal of a dye or organic pollutant is commonly described by the Langmuir–Hinshelwood model [18, 19]

$$r = -\frac{dC}{dt} = \frac{k_r K_e C}{1 + K_e C}, \quad (7.4)$$

where r means the rate of a dye or pollutant degradation, C is its concentration after photocatalysis time (t), k_r denotes the rate constant expressed in mg L^{-1} min^{-1}, and K_e is the equilibrium constant for the adsorption of a dye or pollutant molecule on the catalyst surface at the reaction temperature [20]. When $K_e C \ll 1$, the relation (7.4) can be reduced to a first-order equation

$$r = -\frac{dC}{dt} = kC, \quad (7.5)$$

where $k = k_r K_e$ is the first-order rate constant expressed in min^{-1} units, known also as the apparent rate constant. The formula (7.5) can be integrated over a time into the following equation

$$C(t) = C_0 \exp(-kt), \tag{7.6}$$

where C_0 is initial concentration of a dye or organic pollutant after the completion of adsorption experiments performed in dark condition. Usually, a logarithm of the relative concentration (C/C_0) is presented as a linear function of a time [21–23]

$$\ln\left(\frac{C}{C_0}\right) = -kt. \tag{7.7}$$

The kinetics of the heterogeneous photocatalysis can also follow the second-order law [24, 25]

$$r = -\frac{dC}{dt} = k_2 C^2, \tag{7.8}$$

where k_2 is the second-order rate constant. In such case, time dependence of dye concentration is described by the equation

$$C(t) = \frac{C_0}{k_2 C_0 t + 1}. \tag{7.9}$$

The efficiency of the photocatalytic degradation of a dye or organic pollutant is defined by following formula [26]

$$\eta = \left(1 - \frac{C}{C_0}\right) \cdot 100\%. \tag{7.10}$$

When the reaction follows the first-order law (7.5), the Eqs. (7.6) and (7.10) can be combined into the relation

$$\eta = \left(1 - e^{-kt}\right) \cdot 100\%. \tag{7.11}$$

It is frequently observed [27–29] that the efficiency is reduced when the catalyst is reused in the subsequent cycles of the dye degradation. However, the calcination [30], UV irradiation [31], and chemical regeneration method [29] can be applied to remove the poisoning effect and restore the catalytic properties.

7.1.2 Antimony Chalcohalide Based Photocatalysts

Antimony sulfoiodide (SbSI) is a ferroelectric semiconductor with relatively narrow energy band gap with the values in the range from 1.8 eV [32] up to 1.96 eV [33] depending on the morphology of this material and method of its fabrication. The photocatalytic properties of SbSI were for the first time presented in Ref. [34]. The solution-based approach was used to prepare the SbSI microrods self-assembled into the “urchin” shaped microstructures. The mixture of $SbCl_3$, KI, thioacetamide, and glacial acetic acid was heated at temperature of 383 K and stirred for 2 h. The photocatalytic properties of SbSI “urchin” shaped microstructures were studied at room temperature using methyl orange (MO) as an organic dye. Before photocatalytic experiments, the suspension of SbSI in MO solution was agitated in dark condition for 2 h to obtain an equilibrium between the adsorption and desorption of MO on the SbSI surface. The SbSI-MO suspension was irradiated with visible light emitted by a solar simulator (1 Sun, 1.5G AM). A control experiment was performed to prove that methyl orange is almost no photodegraded in the absence of SbSI catalyst. The kinetics of the MO decomposition followed first-order law (Eqs. 7.5–7.7). A decrease of the reaction rate constant from 0.19 min^{-1} to 0.05 min^{-1} was observed with increase of initial MO concentration from 20 to 50 ppm [34]. This effect was explained as the result of the photocatalyst shielding from the incident light by the methyl orange. In order to reveal a mechanism of the photocatalytic MO degradation, ammonium oxalate (AO), sodium azide, (SA), benzoquinone (BQ), isopropanol (IPA) were used as the scavengers of the following reactive species: hole, singlet oxygen, superoxide, and hydroxyl radicals, respectively. The efficiency of MO photodegradation (Eq. 7.10) of 97% was achieved after 20 min of photocatalysis without any scavenger. The value of η was reduced to 17, 21, 54, 86%, when BQ, SA, AO, IPA were added to the SbSI-MO suspension, accordingly. It was concluded that $^{\bullet}O_2^-$ radicals and holes were played the significant roles in the photocatalytic degradation of methyl orange. The photodegradation of MO was repeated five times to examine the recyclability of the SbSI photocatalyst. The efficiency of MO photocatalytic removal was decreased from 97% (first cycle) to 93% (fifth cycle). Furthermore, the chemical composition and morphology of the SbSI “urchin” shaped microstructures remained almost unchanged after recyclability experiments. High photocatalytic activity of the SbSI microstructures was attributed to two factors. The electron–hole recombination rate was reduced due to the low dangling bonds on the surface of the SbSI microrods. A large static dielectric constant of SbSI resulted in enhanced separation of photogenerated charges.

Wang and co-workers demonstrated a top-down strategy of fabrication of the SbSI nanocrystals with desired dimensions. It involved ball milling of the bulk SbSI crystals and the size selective centrifugation [21]. The nanocrystals of SbSI exhibited excellent photocatalytic properties toward removal of methyl orange. The suspension of SbSI nanocrystals in MO aqueous solution was irradiated with visible light emitted by xenon lamp. The light intensity was approximately 400 mW/cm^2. When the SbSI nanocrystals with average size of 80 nm were used as photocatalysts, 99% of the methyl orange was degraded in extremely short time of 10 s. The first-order kinetics of the MO photodegradation was confirmed. The huge reaction rate constant of 25.2 min^{-1} was determined [21]. A reusability of the photocatalyst was proved by performing five cycles of the MO photodegradation using recycled the SbSI nanocrystals. The efficiency of MO decomposition was only slight decreased. Energy-dispersive X-ray spectroscopy (EDS) showed that the chemical composition of the SbSI nanocrystals was almost unchanged after the photocatalytic experiments. However, X-ray photoelectron spectroscopy (XPS) revealed a small reduction of iodine concentration. The different scavenging agents were used to study the mechanism of photocatalytic degradation of methyl orange in the presence of SbSI nanocrystals. Carotene and sodium azide were applied to scavenge the singlet oxygen. Moreover, the superoxide dismutase (SOD), isopropanol, and catalase were added to SbSI-MO suspension to trap the $^{\bullet}O_2^-$, $^{\bullet}OH$, and H_2O_2, respectively. It was concluded in Ref. [21] that the singlet oxygen played a key role in the high performance photocatalytic decomposition of methyl orange.

An application of the SbSI nanowires as photocatalysts toward fast and efficient removal of methyl orange from aqueous solution was shown for the first time in Ref. [35]. The UV light was used to carry out the photocatalytic experiments. A strong influence of the photocatalysis time on the absorption spectrum (Fig. 7.2a) and color (Fig. 7.2b) of MO solution was observed. The change of MO concentration under light illumination with the absence of SbSI photocatalyst was negligible. The photocatalytic removal of methyl orange attained high value of 95% after short time of 160 s. The kinetics of this reaction followed the first-order law. Thus, the theoretical dependence (7.6) was best fitted to the experimental data (Fig. 7.2b). It allowed to determine the reaction rate constant of 9(1) min^{-1}, which was higher than this parameter reported for SbSI microrods [34] and lower than the reaction rate constant achieved for MO photodegradation with SbSI nanocrystals [21]. The outstanding photocatalytic properties of SbSI can result from a ferroelectric polarization that diminishes the electron–hole recombination [36], steers the migration of photogenerated charge carriers [37], and induces their efficient separation [38–40].

Antimony sulfoiodide was also used for a photodegradation of other organic compounds, i.e. methylene blue (MB) [34], rhodamine B (RhB) [34], and Acid Blue 92 [41]. An overview of the photocatalytic performance of the SbSI

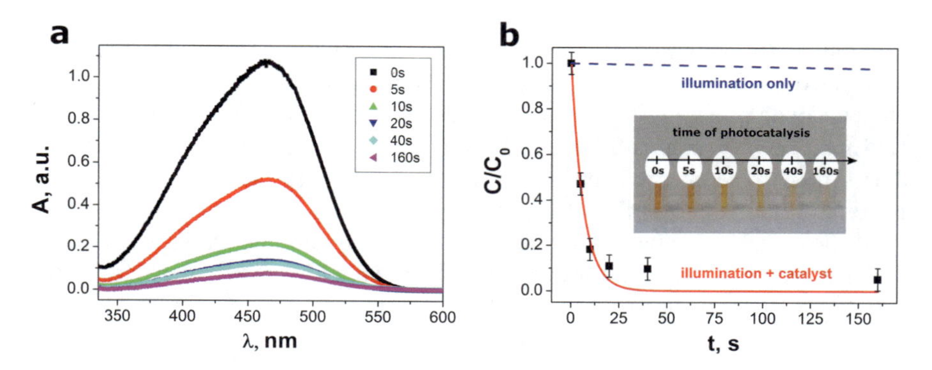

Fig. 7.2 The UV–Vis absorption spectra **a** and relative concentration **b** of the methyl orange aqueous solution after various times of UV illumination using SbSI nanowires as photocatalyst. Red solid line in graph **b** represents the best-fitted dependence (7.6). Blue dashed line in graph **b** refers to the control experiment performed under UV illumination without photocatalyst. The inset in graph **b** shows the color fading of the methyl orange solution with increasing photocatalysis time. Reprinted from Mistewicz et al. [35] under the terms of the Creative Commons Attribution 4.0 International License (CC BY 4.0). Copyright (2020) MDPI

Table 7.1 Photocatalytic removal of different dyes using antimony chalcohalides

Photocatalyst	Photocatalyst morphology	Dye	m_{ph}/m_d	Illumination	t_{ph}, min	η, %	k, min^{-1}	Reference
SbSI	Nanowires	AB92	12.5	Hg lamp (125 W)	90	90	0.023	[41]
	Microrods	MB		solar simulator (1.5G AM)	20	18		[34]
	Microrods	MO	50	solar simulator (1.5G AM)	20	97	0.19	[34]
	Nanocrystals	MO	33.3	Xe lamp (400 mW/cm^2)	0.17	99	25.2	[21]
	Nanowires	MO	33.3	UV lamp (300 W)	2.67	95	9	[35]
	Nanowires	MO	16.7	metal halide lamp (80 mW/cm^2)	10	97	0.366	[45]
	Microrods	MO	16.7	metal halide lamp (80 mW/cm^2)	10	78.5	0.149	[45]
	Microrods	RhB		solar simulator (1.5G AM)	20	61		[34]
SbSI@CNTs	Encapsulated SbSI in CNTs	AB92	12.5	Hg lamp (125 W)	50	95	0.069	[41]

Used abbreviations and symbols: AB92—Acid Blue 92; CNTs—carbon nanotubes; k—reaction rate constant; MB—methylene blue; MO—methyl orange; m_{ph}/m_d—the ratio of photocatalyst mass to the initial mass of a dye; RhB—rhodamine B; t_{ph}—time of photocatalysis; η—efficiency of a dye photodegradation

nano/microstructures with different morphologies is given in Table 7.1. It should be underlined that SbSI is the only one compound among antimony chalcohalides which photocatalytic properties have been examined so far. A special attention should be paid in the future to other interesting antimony chalcohalides, e.g. antimony selenoiodide (SbSeI). The energy band gap of this material is reported in the range from 1.68 eV [42] to 1.86 eV [43]. The narrow energy band gap of SbSeI should favor a good photocatalytic activity of this semiconductor. Furthermore, the high adsorption of a dye can be obtained in for one-dimensional SbSeI nanostructures due to their large surface to volume ratio [44].

7.1.3 Bismuth Chalcohalide Based Photocatalysts

The values of energy band gaps of the bismuth chalcohalides comprise a broad range from a few hundreds of meV up to several eV. The bismuth oxyhalides are the most frequently explored as photocatalysts among all bismuth chalcohalide compounds (Table 7.2). This family of materials is denoted as BiOX, where X refers to the elements from group 17 of the periodic table: F, Cl, Br, or I. The BiOX compounds are indirect band gap semiconductors [46]. They possess a tetragonal matlockite structure that consists of fluorite-like $[Bi_2O_2]$ layers sandwiched between double halogen layers [47]. An internal electric field exists between the $[Bi_2O_2]$ and halogen layers [46]. It supports a separation and migration of photogenerated electric carriers resulting in a high photocatalytic activity of the BiOX compounds. The bismuth oxybromide (BiOBr) has a relatively wide energy band gap of 2.64 eV [48]. It can be tuned by La doping [49] or modifying of the BiOBr surface with the polyvinyl pyrrolidone (PVP) surfactant [50]. Due to the high value of the energy band gap, the photocatalytic activity of the BiOBr under visible light is limited (Table 7.2). Usually, the BiOBr was used toward photocatalytic degradation of rhodamine B [51–53] and methyl orange [22, 54]. The bismuth oxychloride (BiOCl) has large energy band gap of approximately 3.3 eV [48]. Therefore, it can only respond to the ultraviolet light in the solar spectrum, what leads to low photocatalytic performance of this material (Table 7.2). However, the open layered crystal structure and indirect band gap of the BiOCl favor the separation of the photocarriers and efficient charge transfer [55, 56]. The BiOCl was frequently utilized for photocatalytic degradation of rhodamine B [57–59] and methyl orange [22, 60, 61]. The energy band gap of the bismuth oxyiodide (BiOI) is equal to 1.74 eV [62]. It can be tuned through iodine self-doping [63] and preparation of the composite of BiOI and reduced graphene oxide [64]. The BiOI was applied for photocatalytic degradation of methyl orange [22, 64–67], rhodamine B [57, 68, 69], and phenol [67, 70].

A comparative study of photocatalytic properties of three bismuth oxyhalides (BiOBr, BiOCl, and BiOI) was presented in Ref. [22]. These materials were synthesized from bismuth nitrate ($Bi(NO_3)_3{\bullet}5H_2O$) and potassium halides: KCl, KBr, and KI. The BiOBr grown into the microplates with round corners and variable lateral dimensions from 1 to 2 μm. The thickness of the microplates was approximately 10 nm. The BiOCl and BiOI had the forms of nanoneedles and nanoflakes. The effective surface areas of 13.6 m^2/g, 6.3 m^2/g, and 3.4 m^2/g were calculated for BiOCl, BiOI, and BiOBr, respectively. The prepared bismuth oxyhalides were used for a photocatalytic degradation of the methyl orange under different illumination conditions: UV radiation generated by the mercury lamp (Fig. 7.3a), visible light emitted by 100 W tungsten filament lamp (Fig. 7.3b), and natural sunlight (Fig. 7.3c). The best photocatalytic performance under UV irradiation was observed for the BiOCl. Among three investigated bismuth oxyhalides, the BiOBr exhibited the highest photocatalytic activity under visible light and natural sunlight exposure. A stability of the bismuth oxyhalide photocatalysts was examined by exposing them to UV radiation, visible light, and natural sunlight. A small decrease of the photoactivity (from 5 to 9%) was observed for the BiOCl and BiOBr. A significant reduction of the BiOI photocatalytic activity (21%) was noted when it was illuminated with UV radiation [22].

Table 7.2 Photocatalytic degradation of different dyes or pollutants using bismuth chalcohalides

Photocatalyst	Photocatalyst morphology	Dye or pollutant	m_{ph}/m_d	Illumination	t_{ph}, min	η, %	k, min^{-1}	Reference
BiOBr	Microspheres composed of nanosheets	IBF	32.3	Xe lamp (300 W)			0.0083	[71]
	Nanosheets	MB	80	Xe lamp (500 W)			0.0042	[54]
	Nanosheets	MO	80	Xe lamp (500 W)	60	42	0.0077	[54]
	Microplates	MO		UV light from Hg lamp			0.004	[22]
	Microflowers	NOR	25	Visible light	240	99.8	0.0346	[26]
	Microflowers	OFL	25	Visible light	240	94	0.0177	[26]
	Microplates	phenol	106	Visible light	240	42		[72]
	Nanosheets	RhB	40	Xe lamp (500 W)	60	100	0.1292	[54]
	Nanosheets and nanoplates	RhB	32.8	Xe lamp (500 W)				[57]
	Nanosheets	RhB	20	Metal halide lamp (400 W)	15	95	0.142	[51]
	Nanosheets assembled into microspheres	RhB	100	Visible light	80	100		[52]
	Microplates	RhB		Visible light (1.9 mW/cm^2)	1200	42.9		[53]
	Nano-roundels	RhB	80	Xe lamp (500 W)	120	99.2		[73]
Bi_3O_4Br	Nanorings	SA	50	Xe lamp (350 W)	300	95		[74]
$Bi_4O_5Br_2$	Nanosheets	resorcinol	83.3	Xe lamp (350 W)	180	93.8		[75]
BiOCl	Nanoflakes	IC		Solar simulator (36 mW/cm^2)	90	97		[76]
	Nanoplates	MO	200	Hg lamp (300 W)	10	100		[60]
	Nanosheets	MO	100	Xe lamp (500 W)	45	99	0.086	[61]
	Nanoneedles	MO		UV light from Hg lamp			0.0054	[22]
	Microspheres	NOR	25	Hg lamp (125 W)	240	92	0.018	[58]

(continued)

Table 7.2 (continued)

Photocatalyst	Photocatalyst morphology	Dye or pollutant	m_{ph}/m_d	Illumination	t_{ph}, min	η, %	k, min^{-1}	Reference
	Microspheres	OFL	25	Hg lamp (125 W)	80	93	0.0336	[58]
	Microspheres composed of nanoplates	phenol		UV lamp	120	31.3		[59]
	Nanosheets and nanoplates	RhB	38.4	Xe lamp (500 W)				[57]
	Microspheres	RhB	25	Hg lamp (125 W)	240	99	0.004	[58]
	Microspheres composed of nanoplates	RhB		UV lamp	120	91		[59]
Bi_3O_4Cl		MO	30	Hg lamp (500 W)	55	100	0.0523	[23]
	Nanosheets	MO	10	Visible light (80 mW/cm^2)	180	91		[77]
$Bi_{12}O_{17}Cl_2$	Microspheres composed of nanoplates	RhB	25	Xe lamp (300 W)	180	99	0.0212	[78]
BiOF	Nanoflakes	IC		Solar simulator (36 mW/cm^2)	180	92		[76]
BiOI	Nanoplates	MO	40	Halogen lamp (300 W)	180	42		[65]
	Microspheres	MO	40	Halogen lamp (300 W)	180	97		[65]
	Nanoplates	MO	100	Halogen lamp (500 W)	180	27	0.0014	[66]
	Nanoplate microspheres	MO	100	Halogen lamp (500 W)	180	78	0.0085	[66]
	Nanoplates	MO	100	Xe lamp (500 W)	60	97		[67]
	Nanoflakes	MO		UV light from Hg lamp			0.0035	[22]
	Nanosheets	MO	40	Xe lamp (300 W)			0.002	[64]
	Microspheres composed of nanoplatelets	phenol	40	Xe lamp (1000 W)	240	90.3		[70]
	Nanoplates	phenol	50	Xe lamp (500 W)	240	55		[67]

(continued)

Table 7.2 (continued)

Photocatalyst	Photocatalyst morphology	Dye or pollutant	m_{ph}/m_d	Illumination	t_{ph}, min	η, %	k, min^{-1}	Reference
	Nanoplates and microsheets	RhB	28.4	Xe lamp (500 W)				[57]
	Microspheres	RhB	76.9	UV lamp	180	90	0.0092	[68]
	Microspheres composed of nanosheets	RhB	25	Sunlight	80	100		[69]
$Bi_4O_5I_2$	Nanosheets	phenol	40	Xe lamp (300 W)	60	84.2	0.028	[79]
Bi_5O_7I	Nanosheets	phenol	40	Xe lamp (300 W)	60	90.1	0.035	[79]
$Bi_{19}S_{27}Br_3$	Microfabrics constructed by nanofibers	RhB		Visible light	70	95		[80]

Used abbreviations and symbols: IBF—ibuprofen; IC—indigo carmine; k—reaction rate constant; MB—methylene blue; MO—methyl orange; m_{ph}/m_d—the ratio of photocatalyst mass to the initial mass of a dye or pollutant; NOR—norfloxacin; OFL—ofloxacin; RhB—rhodamine B; SA—salicylic acid; t_{ph}—time of photocatalysis; η—efficiency of a dye photodegradation

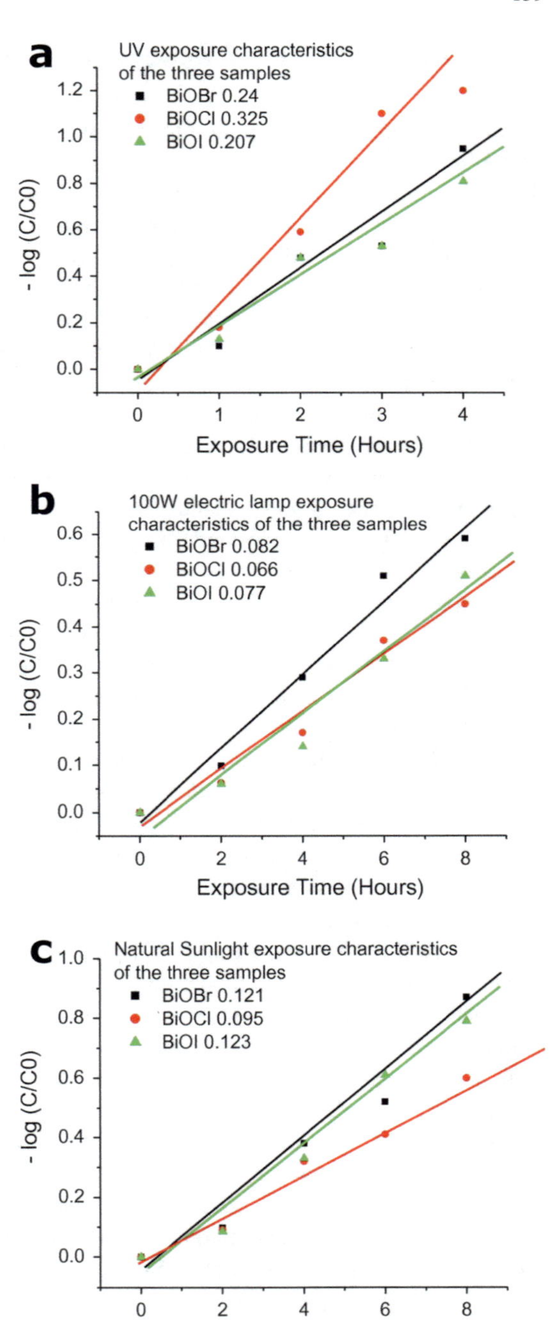

Fig. 7.3 A comparison of the kinetics of methyl orange degradation under **a** UV radiation, **b** visible light emitted by 100 W electric lamp, and **c** natural sunlight exposure using the BiOBr, BiOCl, and BiOI as the photocatalysts. The solid lines represent theoretical formula (7.7). Adapted from Sharma et al. [22] under the terms of the Creative Commons Attribution 4.0 International License (CC BY 4.0). Copyright (2015) Taylor & Francis Group

7.1.4 Heterostructured and Composite Photocatalysts

Various strategies can be applied to improve the photocatalytic performance, such as doping, vacancies formation, dye photosensitization, using co-catalysts, fabrication of composites and heterostructures. The last mentioned method is an effective approach to overcome the drawbacks of a single-phase photocatalysis [81]. Figure 7.4 presents different heterojunction systems commonly used in photocatalysis. In the case of the type I heterostructure (straddling gap), the photogenerated electron and holes are transferred from the semiconductor with higher energy band gap to this with lower energy band gap (Fig. 7.4a). Since type I heterojunction is not able to separate spatially charge carries, they are accumulated and their recombination rate is high [82]. It results in low photocatalytic performance of this type of heterojunction. The electrons and holes in the type II heterojunction (Fig. 7.4b) flow in the opposite directions, i.e. to the semiconductor with a lower conduction band level and to the semiconductor with higher valence band level, respectively. It leads to the spatial separation of the photogenerated charge carriers and reduces the possibility of the electron–hole recombination. It was shown in Ref. [83] that type I heterostructure photocatalyst can be transformed into the type II heterojunction via energy band engineering. The *pn* junction (Fig. 7.4c) has a similar band alignment as the type II heterostructure. However, an additional internal electric field exists at the *pn* interface, what supports separation of the photogenerated charge carriers [84]. The band structure and charge carriers transfer in the Z-scheme photocatalyst is depicted in Fig. 7.4d. The electrons of the semiconductor B recombine with the holes of semiconductor A. In such case, strong reduction and oxidation potentials are obtained along with a spatial separation of oxidation and reduction locations [85]. Therefore, this Z-scheme band configuration allows to achieve a high photocatalytic performance.

Table 7.3 presents an overview of the photocatalyst based on the chalcohalide composites and heterostructures. Generally, they can be divided into three main groups: type II heterostructures [86, 87], *pn* junctions [88–90], and Z-scheme heterojunctions [91, 92]. The chalcohalide hetersotructures were demonstrated as suitable materials for photocatalytic removal of different organic pollutants, including rhodamine B [93–95], methyl orange [86, 96], methylene blue [97, 98], tetracycline [92, 99], phenol [100], and p-nitrophenol [101]. One can see that the bismuth oxyhalides [102] are the most frequently used to construct heterojunction photocatalysts (Table 7.3). The best photocatalytic activity among chalcohalide composites and heterostructures was achieved for ternary hierarchical photocatalyst composed of $BiOBr/Bi_2S_3$ heterostructured composite and cerium based metal–organic framework (CeMOF) [103]. The remarkable photocatalytic performance was also reported for $Bi_4O_5I_2/BiOCl$ Z-scheme heterojunction [92], $Bi_2W_{0.4}Mo_{0.6}O_6$-BiOCl [104] and $Bi_2O_2CO_3$-BiOI [88] *pn* junctions.

A fabrication and characterization of the bismuth oxyiodide/activated carbon (BiOI/C) composite photocatalyst was described in Ref. [94]. The BiOI nanosheets were grown on the activated carbon surface leading to formation of the hierarchical

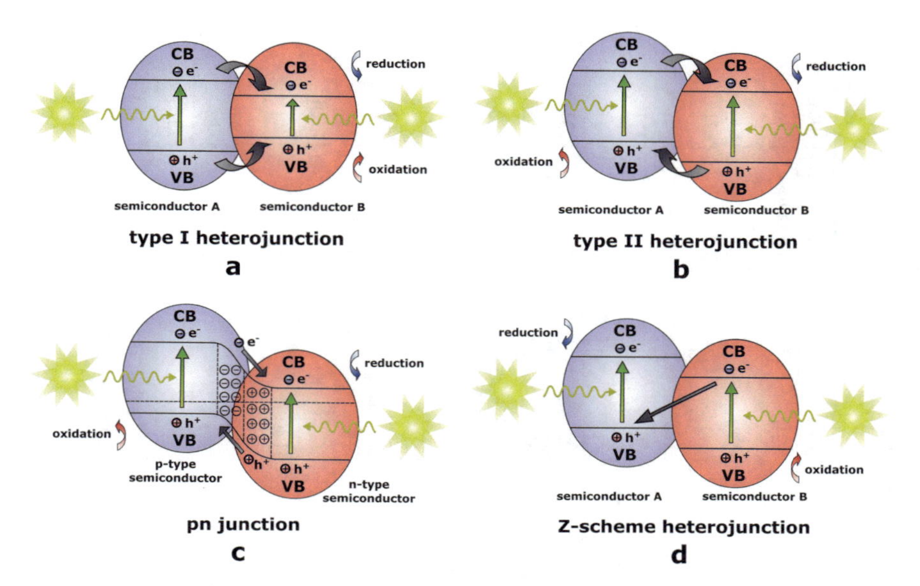

Fig. 7.4 The band structure and charge carries transfer of different heterojunction systems commonly used in photocatalysis: **a** type I heterostructure (straddling gap), **b** type II heterostructure (staggered gap), **c** *pn* junction, and **d** Z-scheme heterojunction. Used abbreviations and symbols: CB—conduction band; e^-—electron; h^+—hole; VB—valence band. Detailed description is provided in the text

structure composite with large specific surface area. It was applied for a photocatalytic degradation of rhodamine B under visible light irradiation emitted by the 500 W xenon lamp. The efficiency of RhB removal using BiOI/C as photocatalyst attained 95% in 120 min (Fig. 7.5a). It was over three times higher than RhB photodegradation efficiency obtained with pure BiOI. An investigation of the long-term cyclic photocatalytic performance of BiOI/C revealed that decrease of the RhB removal efficiency after five cycles was lower than 5% (Fig. 7.5b). It proved that the BiOI/C photocatalyst is stable and it can be successfully recycled. In order to establish the photocatalytic mechanism of the RhB photodegradation, the edta-disodium (EDTA-2Na), benzoquinone, and isopropyl alcohol were used as trapping agents (Fig. 7.5c). It was found that holes and superoxide anions mainly contributed to the photocatalytic degradation of rhodamine B. The active carbon played a role of the absorption center responsible for a higher visible light harvesting ability of the BiOI/C photocatalyst in comparison to pristine BiOI (Fig. 7.5d).

An ultrasonic-assisted method was used in Ref. [120] to prepare the composite of nitrogen-carbon dots (N-CDs) and three-dimensional hierarchical BiOBr. The N-CD/BiOBr composites with different mass ratios of N-CDs to BiOBr (0.005, 0.01, 0.015, 0.02, 0.03, and 0.04) were fabricated and labeled as BC-0.5, BC-1, BC-1.5, BC-2, BC-3, and BC-4, respectively. These materials were tested toward rhodamine B degradation under a visible light illumination (Fig. 7.6a, b). A large RhB removal of 89.3% was achieved after 50 min of irradiation in the presence of BC-1 photocatalyst.

Table 7.3 Photocatalytic removal of various dyes or pollutants using heterostructured and composite chalcohalides

Photocatalyst	Type of junction	Dye or pollutant	m_{ph}/m_d	Illumination	t_{ph}, min	η, %	k, min^{-1}	Reference
Ag/AgBr/BiOBr		TC	100	Xe lamp (300 W)	24	100	0.2	[105]
Ag/Bi_3O_4Cl		TC	50	Xe lamp (250 W)	120	94.2	0.0232	[106]
AgBr-BiOBr	type II	RhB	50	Xe lamp (300 W)	10	80		[107]
$AgBr/Bi_3O_4Br$	type II	RhB	100	Xe lamp (350 W)			0.032	[87]
$Ag_3PO_4/BiOBr$		MB		Xe lamp (500 W)	120	95.2	0.0260	[98]
$BaTiO_3/BiOI$		MO	153	Xe lamp (300 W)	60	85.5		[37]
BiOBr/AgBr/CNFs		PNP		Xe lamp (500 W)	210	99		[101]
$BiOBr/BiFeWO_6$		RhB	209	Xe lamp (300 W)	90	96.3	0.031	[93]
$Bi_{24}O_{31}Br_{10}/BiOI$	type I	RhB	52.2	Hg lamp (300 W)	30	90	0.1091	[108]
$BiOBr/Bi_2S_3@CeMOF$		4NP		Visible light	30	99.7	3.91	[103]
$BiOBr/Bi_2WO_6$	*pn*	phenol		Xe lamp (1000 W)				[109]
BiOBr/CNT		phenol	106	Visible light				[72]
$BiOBr\text{-}C_3N_4$	*pn*	RhB	25	Xe lamp (300 W)	40	100		[90]
BiOBr/RGO		MB	93.8	LED lamp (5 W)	75	96.4	0.0392	[97]
$BiOCl/CeO_2$	type II	RhB	100	Xe lamp (245 W)	60	98.8	0.074	[110]
$BiOCl/TiO_2$	type II	MO	50	Xe lamp (300 W)	180	95		[86]
$Bi_2O_2CO_3\text{-}BiOI$	*pn*	BPA	50	Xe lamp (300 W)			0.3219	[88]
		MO	50	Xe lamp (300 W)			0.2335	[88]
BiOF/BiOI		DP	15	Halogen lamp (200 W)			0.0018	[111]
BiOI/C		RhB	20	Xe lamp (500 W)	120	95		[94]
$BiOI\text{-}TiO_2$		BPA	25	Halogen lamp (250 W)	90	82.5	0.241	[112]

(continued)

Table 7.3 (continued)

Photocatalyst	Type of junction	Dye or pollutant	m_{ph}/m_d	Illumination	t_{ph}, min	η, %	k, min^{-1}	Reference
BiOI/ZnSn(OH)$_6$		phenol	26.7	UV lamp			0.079	[62]
$Bi_4O_5I_2$/BiOCl	Z-scheme	TC	100	Sunlight	60	84	0.4563	[92]
$Bi_4O_5I_2$-Bi_5O_7I/Ni	type II	RhB		Xe lamp (500 W)	300	99	0.0143	[95]
BiSI/Bi_2WO_6/g-C_3N_4	Z-scheme	TC	30	Hg lamp	30	90	0.0655	[91]
BiSI/MoS_2		CV	100	Hg lamp (250 W)	250	90	0.006	[113]
$Bi_2W_{0.4}Mo_{0.6}O_6$-BiOCl	*pn*	RhB	50	Xe lamp (300 W)	7.5	99.9	0.345	[104]
Bi_2WO_6-BiOCl		OTC	50	Xe lamp (42 mW/cm^2)	80	98.6	0.0451	[100]
		phenol	50	Xe lamp (42 mW/cm^2)	300	93.4	0.0098	[100]
Bi_2WO_6-BiSI	Z-scheme	TC	20	Hg lamp	30	93	0.0828	[99]
$BiVO_4$/BiOBr	*pn*	MB	50	Xe lamp (500 W)			0.0065	[89]
$BiVO_4$/BiOCl	*pn*	MB	50	Xe lamp (500 W)			0.0038	[89]
$BiVO_4$/BiOF	*pn*	MB	50	Xe lamp (500 W)			0.0207	[89]
$BiVO_4$/BiOI	*pn*	MB	50	Xe lamp (500 W)			0.0167	[89]
CdS/BiOBr		RhB	50	Halide lamp (250 W)	60	97	0.043	[114]
Co_3O_4/$Bi_4O_5I_2$/Bi_5O_7I	C-scheme	BB	53.3	LED lamp (60 W)	180	91.6	0.0132	[115]
		PNP	32	LED lamp (60 W)	120	99.3	0.0199	[115]
CSs-BiOI/$BiOIO_3$		gaseous Hg^0		LED lamp		92.9		[116]
$FeVO_4$@BiOCl	*pn*	PNP	5	UV lamp	60	90	0.0403	[117]
g-C_3N_4/BiOBr		BPA	20	Xe lamp (400 W)	120	96.6	0.0275	[118]
MWCNT/BiOBr		phenol	120	Xe lamp (500 W)			0.0014	[119]
N-CDs/BiOBr		RhB	20	Solar simulator	50	89.3	0.0444	[120]

(continued)

Table 7.3 (continued)

Photocatalyst	Type of junction	Dye or pollutant	m_{ph}/m_d	Illumination	t_{ph}, min	η, %	k, min^{-1}	Reference
PANI/$Bi_4O_5Br_2$		CIP	40	Visible light	30	96	0.106	[121]
		TC	20	Visible light	10	73	0.0935	[121]
TiO_2/CdTe/BiOI	*pn*	MO	560	Xe lamp (500 W)	90	99.7		[96]

Used abbreviations and symbols: 4NP—4-nitrophenol; BB—brilliant black; BPA—bisphenol A; CeMOF—cerium based metal–organic frameworks; CIP—ciprofloxacin; CNFs—carbon nanofibers; CNT—carbon nanotube; CSs—carbon spheres; CV—crystal violet; DP—diclofenac potassium; g-C_3N_4—graphitized carbon nitride; k—reaction rate constant; MB—methylene blue; MO—methyl orange; m_{ph}/m_d—the ratio of photocatalyst mass to the initial mass of a dye or pollutant; MWCNT—multiwalled carbon nanotube; N-CDs—nitrogen-doped carbon dots; OTC—oxytetracycline; PNP—p-nitrophenol; RGO—reduced graphene oxide; RhB—rhodamine B; TC—tetracycline; t_{ph}—time of photocatalysis; η—efficiency of a dye photodegradation

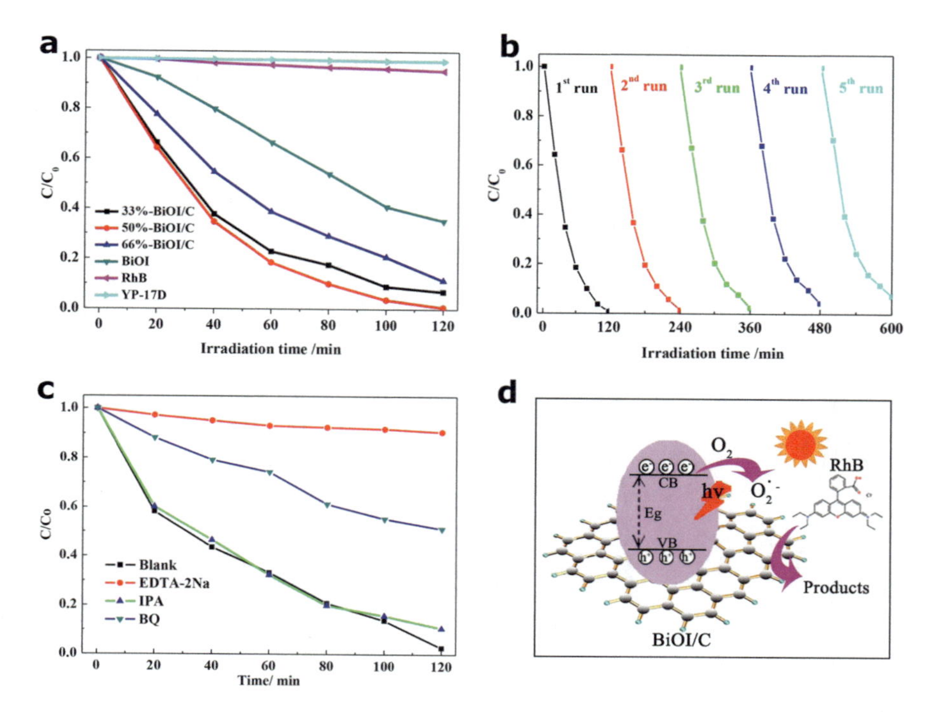

Fig. 7.5 The kinetics **a** of removal of rhodamine B under visible-light (500 W Xe lamp, $\lambda > 420$ nm) without any photocatalyst (RhB) and using different materials as photocatalysts: commercial activated carbon (YP-17D), BiOI, and BiOI/C composites with various molar percentages of activated carbon (33%-BiOI/C, 50%-BiOI/C, 66%-BiOI/C). The cycling degradation **b** of rhodamine B under visible-light irradiation using 50%-BiOI/C as photocatalyst. Trapping experiments **c** of active species in the presence of various agents: edta-disodium (EDTA-2Na), isopropyl alcohol (IPA), and p-benzoquinone (BQ). The scheme **d** of photocatalytic degradation of rhodamine B with BiOI/C. Adapted from Hou et al. [94] under the terms of the Creative Commons Attribution 4.0 International License (CC BY 4.0). Copyright (2017) Springer Nature

A reusability of the BC-1 composite was examined (Fig. 7.6c, d). The photocatalytic activity and chemical composition of the BC-1 composite did not change significantly after five cycles of the RhB photodegradation.

Guo and co-workers developed the facile hydro/solvothermal fabrication of the heterojunction composite of the Bi_2WO_6 nanoparticles and two-dimensional layered BiOCl [100]. This material was applied to photocatalytic decomposition of oxytetracycline (OTC) and phenol under the light emitted by 500 W xenon lamp. The cyclic photodegradation experiments were conducted to investigate stability of the Bi_2WO_6–BiOCl heterojunction composite. A small decrease of the photocatalytic activity was observed in the case of OTC (Fig. 7.7a) and phenol (Fig. 7.7b) degradation after four cycles. The X-ray diffraction (Fig. 7.7c) and scanning electron microscopy (Fig. 7.7d) confirmed no apparent differences between fresh and used Bi_2WO_6–BiOCl composites. It proved good stability and recyclability of

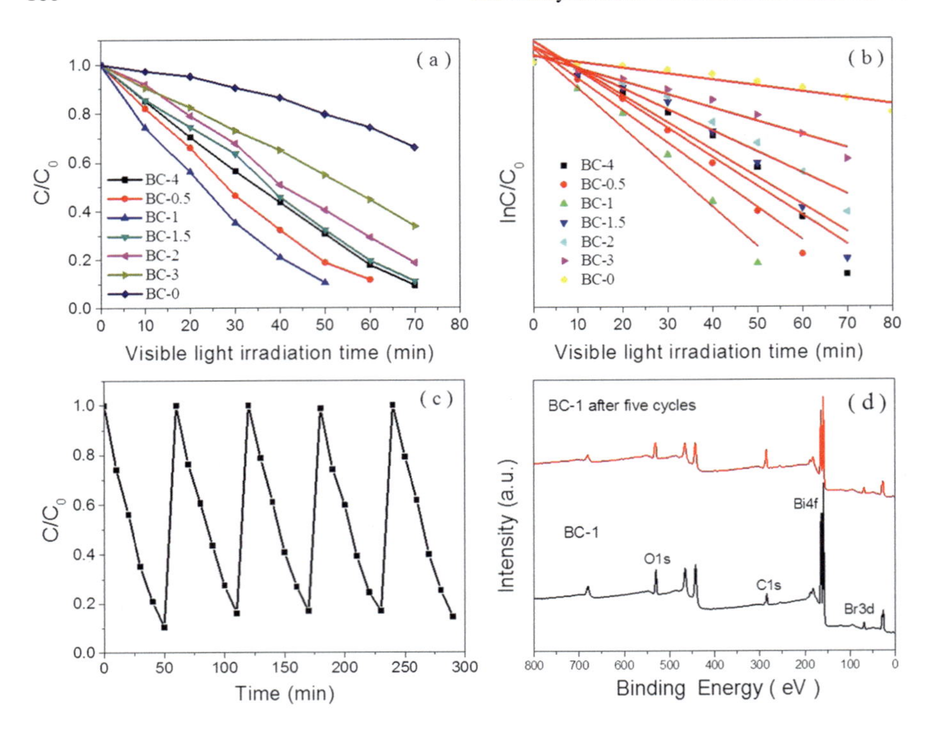

Fig. 7.6 Photocatalytic degradation **a**, **b** of rhodamine B using nanocomposites of nitrogen-doped carbon dots (N-CD) and BiOBr nanoplates. The cycling degradation **c** of rhodamine B in a presence of BC-1 photocatalyst. The XPS patterns **d** of BC-1 before and after five cycles of photocatalysis. The BC-0, BC-0.5, BC-1, BC-1.5, BC-2, BC-3, and BC-4 labels refer to different mass ratios of N-CDs to BiOBr: 0, 0.005, 0.01, 0.015, 0.02, 0.03, and 0.04, respectively. The solid lines in **b** represent theoretical formula (7.7). Reprinted from Zhang et al. [120] under the terms of the Creative Commons Attribution 4.0 International License (CC BY 4.0). Copyright (2017) Springer Nature

the Bi_2WO_6–BiOCl composite as promising photocatalyst for an environmental remediation.

7.2 Piezo- and Sonocatalysis

The piezocatalytic properties of the SbSI were for the first time revealed in Ref. [35]. The one-dimensional SbSI nanowires were synthesized using a simple and fast sonochemical method. The purification procedure was carried out to clean the surfaces of the SbSI nanowires and increase their piezocatalytic activity. Methyl orange was chosen as a common dye for the piezocatalytic tests. The suspension of the SbSI nanowires in MO aqueous solution was stirred in the dark for 20 min

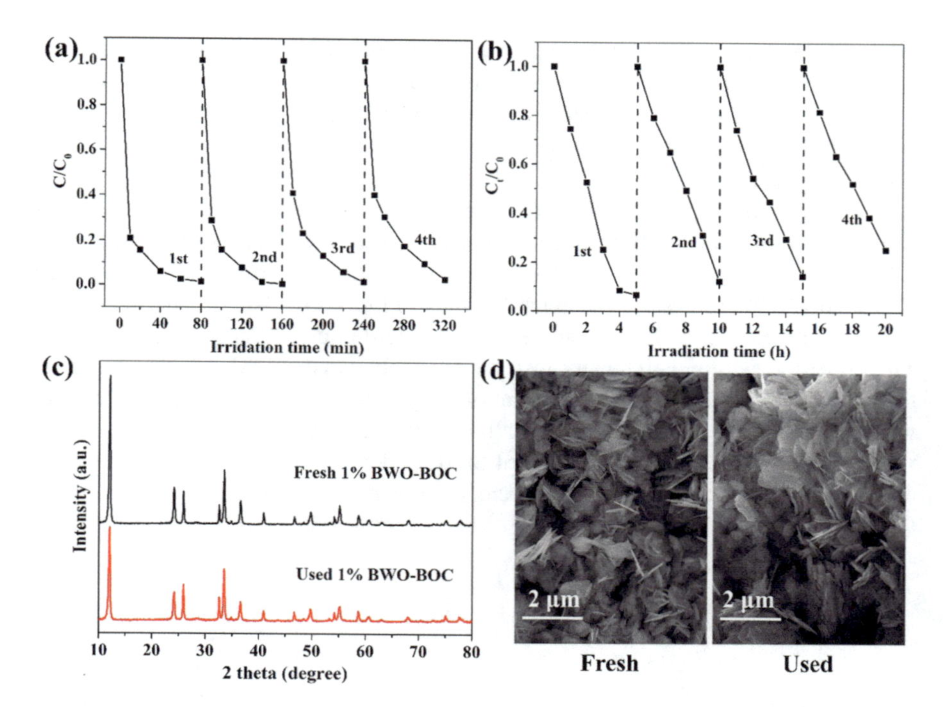

Fig. 7.7 The recycling tests of the Bi_2WO_6–BiOCl heterostructure for photocatalytic degradation of **a** oxytetracycline and **b** phenol under simulated sunlight irradiation. The **c** XRD patterns and **d** SEM micrographs of the fresh and used Bi_2WO_6–BiOCl for phenol degradation. Reprinted from Guo et al. [100] under the terms of the Creative Commons Attribution 4.0 International License (CC BY 4.0). Copyright (2020) Springer Nature

before photocatalytic experiments. It was done in order to achieve an adsorption–desorption equilibrium between the MO solution and the SbSI piezocatalyst. An ultrasonic vibration of the SbSI-MO suspension was provided by the two various ultrasonic reactors, i.e. the VCX-750 processor with working frequency of 20 kHz and maximum power of 750 W and the Sonic-6 ultrasonic bath with operating frequency of 40 kHz and maximum power of 480 W. Figure 7.8a shows an apparent discoloration of the MO aqueous solution after a very short time of piezocatalysis (40 s). The photocatalytic degradation of methyl orange was monitored by measuring its optical transmittance spectrum (Fig. 7.8b). The kinetics of the MO photodegradation was described by the first-order law. Thus, the theoretical dependence (7.7) was best fitted to the experimental data (Fig. 7.8c). It allowed to determine a large reaction rate constant of 7.6 min^{-1} [35]. When MO solution was subjected to ultrasonic vibration without using piezocatalyst, the MO concentration did not change (Fig. 7.8c). It confirmed that the SbSI nanowires played a crucial role in a degradation of methyl orange under ultrasonic irradiation. The efficiency of the MO removal was calculated using formula (7.10) and presented in Fig. 7.8d as a function of the piezocatalysis time. A decrease of the photocatalytic activity of SbSI nanowires was observed with the increase of the initial MO concentration to the mass of SbSI piezocatalyst. It was ascribed to the shielding of the piezocatalyst by the methyl orange molecules [34]. The efficiency of the MO degradation was higher when more intense ultrasounds were applied. The acoustic power of ultrasound influences the generation of hot spots and affects the efficiency of MO degradation [122].

A mechanism of the methyl orange degradation using SbSI nanowires as piezocatalysts was provided in Ref. [35]. When high power ultrasounds was applied to the SbSI-MO suspension (Fig. 7.9a), the acoustic cavitation occurred [123]. It involved a formation and collapse of the microbubbles in the MO solution. The huge local pressure of the order of 10^8 Pa can be generated within the collapsing bubble [124]. Therefore, the extremely intensive forces were applied to the SbSI nanowires leading to their lateral or longitudinal deformation (Fig. 7.9b). The piezoelectric potential was created in bent, compressed or stretched SbSI nanowires. It led to a tilt of the valence and conduction energy bands of SbSI, as depicted in Fig. 7.9c. The electron and holes were separated under a piezoelectric polarization and moved toward opposite surfaces of SbSI nanowire. Then, they interacted with the molecules adsorbed on

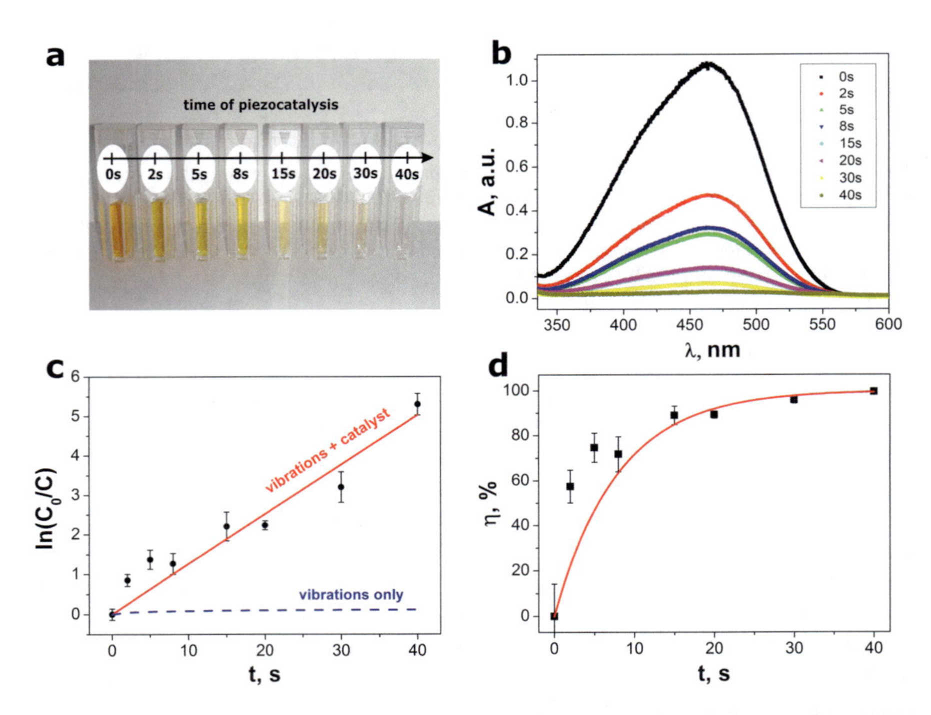

Fig. 7.8 A photograph **a**, kinetics **b**, **c**, and efficiency **d** of piezocatalytic methyl orange degradation using the SbSI nanowires as piezocatalysts. Blue dashed line in graph **c** represents data registered in control experiment performed under ultrasonic vibration without SbSI piezocatalyst. Red lines in graphs **c** and **d** show the best fitted dependences (7.7) and (7.11), respectively. Adapted from Mistewicz et al. [35] under the terms of the Creative Commons Attribution 4.0 International License (CC BY 4.0). Copyright (2020) MDPI

the SbSI surface resulting in formation of free radicals. According to reaction (7.3), the electrons were donated to the adsorbed oxygen molecules leading to formation of superoxide anions. Moreover, the hydroxyl radicals were created through an interaction of holes with water molecules (Eq. 7.1). The hydroxyl radicals, holes, and superoxide anions were recognized in Ref. [45] as main species responsible for a degradation of methyl orange using SbSI piezocatalyst. The methyl orange was decomposed into intermediate products. Finally, they were mineralized into the inorganic compounds, including CO_2 and H_2O [35].

The piezocatalytic properties of the microrods and nanowires of SbSI were also examined in Ref. [45]. They were applied for a piezocatalytic degradation of methyl orange in aqueous solution. The large of the reaction rate constants of 0.124 and 0.055 min^{-1} were determined for nanowires and microrods of SbSI, respectively. These values are much higher than the reaction rate constants reported for the piezocatalysts based on bismuth chalcohalide compounds (Table 7.4), and other piezoelectric materials, e.g. $BaTiO_3$ [125, 126], $SrTiO_3$ [127], polyvinylidene fluoride

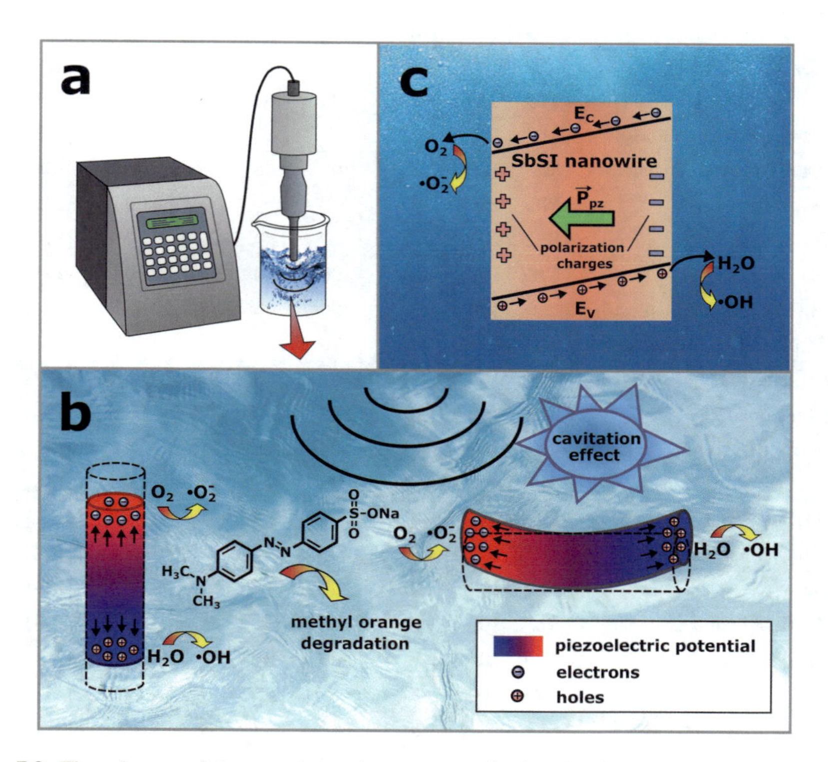

Fig. 7.9 The schemes of the experimental setup **a**, mechanism **b** of methyl orange degradation using SbSI nanowires as piezocatalyts, and energy band diagram **c** of the SbSI nanowire under a piezoelectric potential. Used abbreviations and symbols: E_C—the conduction energy band; E_V—the valence energy band; P_{pz}—piezoelectric polarization. A detailed description is provided in the text. Reprinted from Mistewicz et al. [35] under the terms of the Creative Commons Attribution 4.0 International License (CC BY 4.0). Copyright (2020) MDPI

(PVDF) [128], and lead zirconate titanate (PZT) [129]. The excellent piezocatalytic properties of one-dimensional SbSI [35, 45] can originate from its high surface to volume ratio, piezoelectric modulus, electromechanical coupling coefficient [130], and flexibility supported by one-dimensional morphology.

The bismuth chalcohalide piezocatalysts were applied mainly for degradation of methyl orange and rhodamine B (Table 7.4). The BiOBr [131, 132], BiOCl [131, 135, 136], BiOI [138], and its derivatives (Bi_5O_7I [122], $PbBiO_2I$ [139]) were the most explored as piezocatalysts among the bismuth chalcohalide compounds. However, the catalytic performance of aforementioned piezocatalysts (Table 7.4) is poor in comparison to antimony (Table 7.1) and bismuth (Table 7.2) chalcohalide photocatalysts. Therefore, the recent effort towards increasing piezocatalytic activity of the

Table 7.4 The piezo/sonocatalytic degradation of various dyes using chalcohalide materials

Piezocatalyst	Dye	m_p/m_d	f, kHz	P_U, W	t_p, min	η, %	k, min^{-1}	Reference
BiOBr nanoplates	MO	20	40	120			0.004	[131]
BiOBr nanoflakes	RhB	100	40	120			0.006	[132]
$BiOBr/BaTiO_3$ composite	MO	20	40	120	60	54	0.012	[131]
$Bi_2MoO_6/BiOBr$ composite	MV	50					0.010	[133]
TCQDs/BiOBr composite	MV	100	35	100	48	53.6	0.008	[134]
BiOCl nanoplates	MO	20	40	120			0.003	[131]
BiOCl microspheres	RhB	20	40	120	96	26	0.003	[135]
BiOCl nanosheets	CBZ	100	40	100	30	22.4	0.009	[136]
Bi/BiOCl nanocomposite	CBZ	100	40	100	30	75.1	0.051	[136]
$BiOCl/BaTiO_3$ composite	MO	20	40	120			0.007	[131]
$BiOCl/NaNbO_3$ heterojunction	RhB	200	40	50	100	9.9	0.0005	[137]
BiOI nanoparticles	MB	100	40	80	180	85	0.062	[138]
	MO	100	40	80	180	61		[138]
	RhB	100	40	80	180	91		[138]
Bi_5O_7I nanorods	MO	100	40	60			0.007	[122]
Ag/Bi_5O_7I nanocomposite	MO	100	40	60			0.033	[122]
$PbBiO_2I$	RhB	100	40	120			0.002	[139]
$Ag/PbBiO_2I$ nanocomposite	RhB	100	40	120			0.017	[139]
SbSI nanowires	MO	100	40	480	0.75	84.0	2.0	[35]
	MO	200	40	480	0.75	97.9	5.1	[35]
	MO	200	20	750	0.67	99.5	7.6	[35]
SbSI nanowires	MO	16.7	40	180	10	69.0	0.124	[45]
SbSI microrods	MO	16.7	40	180	10	43.9	0.055	[45]

Used abbreviations and symbols: CBZ—carbamazepine; f—frequency of the ultrasounds; k—reaction rate constant; MB—methylene blue; MO—methyl orange; m_p/m_d—the ratio of piezocatalyst mass to the initial mass of a dye; MV—methyl violet; P_U—power of the ultrasonic reactor; RhB—rhodamine B; TCQDs—Ti_3C_2 quantum dots; t_p—time of piezocatalysis; η—efficiency of a dye degradation bismuth chalcohalides have been focused on construction of the heterojunctions [137, 140, 141] and application of the synergetic piezo/photocatalytic effect [131, 132].

Further characterization of the chalcohalide piezocatalysts requires an additional attention. For instance, the new methods and devices are needed to proper determination of the acoustic power in the ultrasonic reactors [142]. It is expected that the piezocatalytic activity of the chalcohalide materials depends on the various factors, including frequency of ultrasounds, acoustic power, and temperature of the reaction. An influence of these parameters on the piezocatalytic efficiency has to be examined in the future. Such investigations should provide a new insight into the mechanism of the organic compounds degradation in a presence of the chalcohalide piezocatalysts.

References

1. S. Nagendran, P. Sivasubramanian, J.-H. Chang, S.-Y. Shen, A.K. Nayak, M. Kumar, Fundamentals of environmental remediation techniques in *Bismuth-Based Materials for Environmental Remediation* (2022), pp. 1–22
2. M. Nageeb, Adsorption technique for the removal of organic pollutants from water and wastewater, in *Organic Pollutants—Monitoring, Risk and Treatment* (IntechOpen, Rijeka, 2013), Ch. 7
3. S. Lim, J.L. Shi, U. von Gunten, D.L. McCurry, Ozonation of organic compounds in water and wastewater: a critical review. Water Res. **213**, 118053 (2022)
4. U. Hübner, U. von Gunten, M. Jekel, Evaluation of the persistence of transformation products from ozonation of trace organic compounds—a critical review. Water Res. **68**, 150 (2015)
5. M. Xie, L.D. Nghiem, W.E. Price, M. Elimelech, Comparison of the removal of hydrophobic trace organic contaminants by forward osmosis and reverse osmosis. Water Res. **46**, 2683 (2012)
6. J. Wang, Z. Wang, C.L.Z. Vieira, J.M. Wolfson, G. Pingtian, S. Huang, Review on the treatment of organic pollutants in water by ultrasonic technology. Ultrason. Sonochem. **55**, 273 (2019)
7. D. Ghime, P. Ghosh, Removal of organic compounds found in the wastewater through electrochemical advanced oxidation processes: a review. Russ. J. Electrochem. **55**, 591 (2019)
8. M. Shestakova, M. Sillanpää, Electrode materials used for electrochemical oxidation of organic compounds in wastewater. Rev. Environ. Sci. Biotechnol. **16**, 223 (2017)
9. A. Shokri, M.S. Fard, A critical review in fenton-like approach for the removal of pollutants in the aqueous environment. Environ. Challenges **7**, 100534 (2022)
10. S.S. Chan, K.S. Khoo, K.W. Chew, T.C. Ling, P.L. Show, Recent advances biodegradation and biosorption of organic compounds from wastewater: microalgae-bacteria consortium—a review. Bioresour. Technol. **344**, 126159 (2022)
11. I.S. Kang, J. Xi, H.Y. Hu, Photolysis and photooxidation of typical gaseous VOCs by UV irradiation: removal performance and mechanisms. Front. Environ. Sci. Eng. **12**, 8 (2018)
12. P.M. Rajaitha, S. Hajra, M. Sahu, K. Mistewicz, B. Toroń, R. Abolhassani, S. Panda, Y.K. Mishra, H.J. Kim, Unraveling highly efficient nanomaterial photocatalyst for pollutant removal: a comprehensive review and future progress. Mater. Today Chem. **23**, 100692 (2022)
13. M. Saeed, M. Muneer, A. ul Haq, N. Akram, Photocatalysis: an effective tool for photodegradation of dyes—a review. Environ. Sci. Pollut. Res. **29**, 293 (2022)
14. M. Melchionna, P. Fornasiero, Updates on the roadmap for photocatalysis. ACS Catal. **10**, 5493 (2020)
15. R.K. Gautam, M.C. Chattopadhyaya, Advanced oxidation process–based nanomaterials for the remediation of recalcitrant pollutants, in *Nanomaterials for Wastewater Remediation*, ed.

by R.K. Gautam, M.C.B.T.-N. for W. R. Chattopadhyaya (Butterworth-Heinemann, Boston, 2016), pp. 33–48
16. R.S. Pedanekar, S.K. Shaikh, K.Y. Rajpure, Thin film photocatalysis for environmental remediation: a status review. Curr. Appl. Phys. **20**, 931 (2020)
17. M.Y. Ghaly, T.S. Jamil, I.E. El-Seesy, E.R. Souaya, R.A. Nasr, Treatment of highly polluted paper mill wastewater by solar photocatalytic oxidation with synthesized nano TiO_2. Chem. Eng. J. **168**, 446 (2011)
18. K. Vasanth Kumar, K. Porkodi, A. Selvaganapathi, Constrain in solving Langmuir-Hinshelwood kinetic expression for the photocatalytic degradation of auramine O aqueous solutions by ZnO catalyst. Dye. Pigment. **75**, 246 (2007)
19. Y. Nosaka, A.Y. Nosaka, Langmuir-Hinshelwood and light-intensity dependence analyses of photocatalytic oxidation rates by two-dimensional-ladder kinetic simulation. J. Phys. Chem. C **122**, 28748 (2018)
20. N.G. Asenjo, R. Santamaría, C. Blanco, M. Granda, P. Álvarez, R. Menéndez, Correct use of the Langmuir-Hinshelwood equation for proving the absence of a synergy effect in the photocatalytic degradation of phenol on a suspended mixture of titania and activated carbon. Carbon N. Y. **55**, 62 (2013)
21. C. Wang et al., SbSI nanocrystals: an excellent visible light photocatalyst with efficient generation of singlet oxygen. ACS Sustain. Chem. Eng. **6**, 12166 (2018)
22. I.D. Sharma, G.K. Tripathi, V.K. Sharma, S.N. Tripathi, R. Kurchania, C. Kant, A.K. Sharma, K.K. Saini, One-pot synthesis of three bismuth oxyhalides (BiOCl, BiOBr, BiOI) and their photocatalytic properties in three different exposure conditions. Cogent Chem. **1**, 1076371 (2015)
23. L. Xinping, H. Tao, H. Fuqiang, W. Wendeng, S. Jianlin, Photocatalytic activity of a bi-based oxychloride Bi_3O_4Cl. J. Phys. Chem. B **110**, 24629 (2006)
24. W. Gao, C. Ran, M. Wang, L. Li, Z. Sun, X. Yao, The role of reduction extent of graphene oxide in the photocatalytic performance of Ag/AgX (X=Cl, Br)/RGO composites and the pseudo-second-order kinetics reaction nature of the Ag/AgBr system. Phys. Chem. Chem. Phys. **18**, 18219 (2016)
25. L. Ernawati, A.W. Yusariarta, A.D. Laksono, R.A. Wahyuono, H. Widiyandari, R. Rebeka, V. Sitompul, Kinetic studies of methylene blue degradation using $CaTiO_3$ photocatalyst from chicken eggshells. J. Phys. Conf. Ser. **1726**, 12017 (2021)
26. T. Senasu, T. Chankhanittha, K. Hemavibool, S. Nanan, Solvothermal synthesis of BiOBr photocatalyst with an assistant of PVP for visible-light-driven photocatalytic degradation of fluoroquinolone antibiotics. Catal. Today **384–386**, 209 (2022)
27. L. Thi Thanh Nhi, L. Van Thuan, D. My Uyen, M.H. Nguyen, V.T. Thu, D.Q. Khieu, L.H. Sinh, Facile fabrication of highly flexible and floatable Cu_2O/RGO on vietnamese traditional paper toward high-performance solar-light-driven photocatalytic degradation of ciprofloxacin antibiotic. RSC Adv. **10**, 16330 (2020)
28. F. Ghasemy-Piranloo, S. Dadashian, F. Bavarsiha, $Fe_3O_4/SiO_2/TiO_2$–Ag cubes with core/shell/shell nano-structure: synthesis, characterization and efficient photo-catalytic for phenol degradation. J. Mater. Sci. Mater. Electron. **30**, 12757 (2019)
29. K.S. Ranjith, R.T. Rajendra Kumar, Regeneration of an efficient, solar active hierarchical ZnO flower photocatalyst for repeatable usage: controlled desorption of poisoned species from active catalytic sites. RSC Adv. **7**, 4983 (2017)
30. N. Miranda-García, S. Suárez, M.I. Maldonado, S. Malato, B. Sánchez, Regeneration approaches for TiO_2 immobilized photocatalyst used in the elimination of emerging contaminants in water. Catal. Today **230**, 27 (2014)
31. Z. Chen, Y. Peng, J. Chen, C. Wang, H. Yin, H. Wang, C. You, J. Li, Performance and mechanism of photocatalytic toluene degradation and catalyst regeneration by thermal/UV treatment. Environ. Sci. Technol. **54**, 14465 (2020)
32. G. Chen, W. Li, Y. Yu, Q. Yang, Fast and low-temperature synthesis of one-dimensional (1D) single-crystalline SbSI microrod for high performance photodetector. RSC Adv. **5**, 21859 (2015)

33. Y.C. Choi, E. Hwang, D.H. Kim, Controlled growth of SbSI thin films from amorphous Sb_2S_3 for low-temperature solution processed chalcohalide solar cells. APL Mater. **6**, 121108 (2018)
34. M. Tamilselvan, A.J. Bhattacharyya, Antimony sulphoiodide (SbSI), a narrow band-gap non-oxide ternary semiconductor with efficient photocatalytic activity. RSC Adv. **6**, 105980 (2016)
35. K. Mistewicz, M. Kępińska, M. Nowak, A. Sasiela, M. Zubko, D. Stróż, Fast and efficient piezo/photocatalytic removal of methyl orange using SbSI nanowires. Materials **13**, 4803 (2020)
36. M.A. Khan, M.A. Nadeem, H. Idriss, Ferroelectric polarization effect on surface chemistry and photo-catalytic activity: a review. Surf. Sci. Rep. **71**, 1 (2016)
37. H. Huang, S. Tu, X. Du, Y. Zhang, Ferroelectric spontaneous polarization steering charge carriers migration for promoting photocatalysis and molecular oxygen activation. J. Colloid Interface Sci. **509**, 113 (2018)
38. Z. Zhang, C. Zou, S. Yang, Z. Yang, Y. Yang, Ferroelectric polarization effect promoting the bulk charge separation for enhance the efficiency of photocatalytic degradation. Chem. Eng. J. **410**, 128430 (2021)
39. S. Tu, Y. Zhang, A.H. Reshak, S. Auluck, L. Ye, X. Han, T. Ma, H. Huang, Ferroelectric polarization promoted bulk charge separation for highly efficient CO_2 photoreduction of $SrBi_4Ti_4O_{15}$. Nano Energy **56**, 840 (2019)
40. Y. Liu et al., Internal-field-enhanced charge separation in a single-domain ferroelectric $PbTiO_3$ photocatalyst. Adv. Mater. **32**, 1906513 (2020)
41. M. Tasviri, Z. Sajadi-Hezave, SbSI nanowires and CNTs encapsulated with SbSI as photocatalysts with high visible-light driven photoactivity. Mol. Catal. **436**, 174 (2017)
42. Y.C. Choi, K.W. Jung, One-step solution deposition of antimony selenoiodide films via precursor engineering for lead-free solar cell applications. Nanomaterials **11**, 3206 (2021)
43. K.T. Butler, S. McKechnie, P. Azarhoosh, M. Van Schilfgaarde, D.O. Scanlon, A. Walsh, Quasi-particle electronic band structure and alignment of the V-VI-VII semiconductors SbSI, SbSBr, and SbSeI for solar cells. Appl. Phys. Lett. **108**, 112103 (2016)
44. M. Nowak, B. Kauch, P. Szperlich, M. Jesionek, M. Kepińska, Bober, J. Szala, G. Moskal, T. Rzychoń, D. Stróz, Sonochemical preparation of SbSeI gel. Ultrason. Sonochem. **16**, 546 (2009)
45. R. Wang, Y. Wang, N. Zhang, S. Lin, Y. He, Y. Yan, K. Hu, H. Sun, X. Liu, Synergetic piezo-photocatalytic effect in SbSI for highly efficient degradation of methyl orange. Ceram. Int. **48**, 31818 (2022)
46. Y. Yang, C. Zhang, C. Lai, G. Zeng, D. Huang, M. Cheng, J. Wang, F. Chen, C. Zhou, W. Xiong, BiOX (X=Cl, Br, I) photocatalytic nanomaterials: applications for fuels and environmental management. Adv. Colloid Interface Sci. **254**, 76 (2018)
47. M. Arumugam, M.Y. Choi, Recent progress on bismuth oxyiodide (BiOI) photocatalyst for environmental remediation. J. Ind. Eng. Chem. **81**, 237 (2020)
48. Y. Li, H. Jiang, X. Wang, X. Hong, B. Liang, Recent advances in bismuth oxyhalide photocatalysts for degradation of organic pollutants in wastewater. RSC Adv. **11**, 26855 (2021)
49. Q. Fan, X. Chen, F. Chen, J. Tian, C. Yu, C. Liao, Regulating the stability and bandgap structure of BiOBr during thermo-transformation via La doping. Appl. Surf. Sci. **481**, 564 (2019)
50. B. Zhang, M. Zhang, L. Zhang, P.A. Bingham, W. Li, S. Kubuki, PVP surfactant-modified flower-like BiOBr with tunable bandgap structure for efficient photocatalytic decontamination of pollutants. Appl. Surf. Sci. **530**, 147233 (2020)
51. Y. Mi, H. Li, Y. Zhang, N. Du, W. Hou, Synthesis and photocatalytic activity of BiOBr nanosheets with tunable crystal facets and sizes. Catal. Sci. Technol. **8**, 2588 (2018)
52. X. Wang, F. Zhang, Y. Yang, Y. Zhang, L. Liu, W. Lei, Controllable synthesis and photocatalytic activity of nano-BiOBr photocatalyst. J. Nanomater. **2020**, 1013075 (2020)
53. Z. Geng, L. Zhang, J. Wang, Y. Yu, G. Zhang, X. Cheng, D. Wang, Biobr precursor solutions modified cement paste: the photocatalytic performance and effects. Crystals **11**, 969 (2021)
54. D. Zhang, J. Li, Q. Wang, Q. Wu, High 001 facets dominated BiOBr lamellas: facile hydrolysis preparation and selective visible-light photocatalytic activity. J. Mater. Chem. A **1**, 8622 (2013)

55. W.W. Liu, R.F. Peng, Recent advances of bismuth oxychloride photocatalytic material: property, preparation, and performance enhancement. J. Electron. Sci. Technol. **18**, 119 (2020)
56. L. Wang, Y. Liu, G. Chen, M. Zhang, X. Yang, R. Chen, Y. Cheng, Bismuth oxychloride nanomaterials fighting for human health: from photodegradation to biomedical applications. Crystals **12**, 491 (2022)
57. G. Li, F. Qin, R. Wang, S. Xiao, H. Sun, R. Chen, BiOX (X=Cl, Br, I) nanostructures: mannitol-mediated microwave synthesis, visible light photocatalytic performance, and Cr(VI) removal capacity. J. Colloid Interface Sci. **409**, 43 (2013)
58. T. Senasu, T. Narenuch, K. Wannakam, T. Chankhanittha, S. Nanan, Solvothermally grown BiOCl catalyst for photodegradation of cationic dye and fluoroquinolone-based antibiotics. J. Mater. Sci. Mater. Electron. **31**, 9685 (2020)
59. E. Bárdos, V.A. Márta, S. Fodor, E.Z. Kedves, K. Hernadi, Z. Pap, Hydrothermal crystallization of bismuth oxychlorides (BiOCl) using different shape control reagents. Materials **14**, 2261 (2021)
60. K.L. Zhang, C.M. Liu, F.Q. Huang, C. Zheng, W.D. Wang, Study of the electronic structure and photocatalytic activity of the BiOCl photocatalyst. Appl. Catal. B Environ. **68**, 125 (2006)
61. J. Jiang, K. Zhao, X. Xiao, L. Zhang, Synthesis and facet-dependent photoreactivity of BiOCl single-crystalline nanosheets. J. Am. Chem. Soc. **134**, 4473 (2012)
62. H. Li, Y. Cui, W. Hong, B. Xu, Enhanced photocatalytic activities of BiOI/ZnSn$(OH)_6$ composites towards the degradation of phenol and photocatalytic H_2 Production. Chem. Eng. J. **228**, 1110 (2013)
63. X. Zhang, L. Zhang, Electronic and band structure tuning of ternary semiconductor photocatalysts by self doping: the case of BiOI. J. Phys. Chem. C **114**, 18198 (2010)
64. X. Ren, J. Yao, L. Cai, J. Li, X. Cao, Y. Zhang, B. Wang, Y. Wei, Band gap engineering of BiOI: via oxygen vacancies induced by graphene for improved photocatalysis. New J. Chem. **43**, 1523 (2019)
65. J. Xia, S. Yin, H. Li, H. Xu, L. Xu, Q. Zhang, Enhanced photocatalytic activity of bismuth oxyiodine (BiOI) porous microspheres synthesized via reactable ionic liquid-assisted solvothermal method. Colloids Surfaces A Physicochem. Eng. Asp. **387**, 23 (2011)
66. Y. Wang, K. Deng, L. Zhang, Visible light photocatalysis of BiOI and its photocatalytic activity enhancement by in situ ionic liquid modification. J. Phys. Chem. C **115**, 14300 (2011)
67. Y. Li, J. Wang, H. Yao, L. Dang, Z. Li, Efficient decomposition of organic compounds and reaction mechanism with BiOI photocatalyst under visible light irradiation. J. Mol. Catal. A Chem. **334**, 116 (2011)
68. Z. Kása, E. Orbán, Z. Pap, I. Ábrahám, K. Magyari, S. Garg, K. Hernadi, Innovative and cost-efficient BiOI immobilization technique on ceramic paper—total coverage and high photocatalytic activity. Nanomaterials **10**, 1959 (2020)
69. T. Narenuch, T. Senasu, T. Chankhanittha, S. Nanan, Sunlight-active BiOI photocatalyst as an efficient adsorbent for the removal of organic dyes and antibiotics from aqueous solutions. Molecules **26**, 5624 (2021)
70. X. Xiao, W. De Zhang, Facile synthesis of nanostructured BiOI microspheres with high visible light-induced photocatalytic activity. J. Mater. Chem. **20**, 5866 (2010)
71. Y. Liu, Z. Hu, J.C. Yu, Photocatalytic degradation of ibuprofen on S-doped BiOBr. Chemosphere **278**, 130376 (2021)
72. N. Sharma, Z. Pap, S. Garg, K. Hernádi, Hydrothermal synthesis of BiOBr and BiOBr/CNT composites, their photocatalytic activity and the importance of early $Bi_6O_6(OH)_3(NO_3)_3 \cdot 1.5H_2O$ formation. Appl. Surf. Sci. **495**, 143536 (2019)
73. Y. Zhang, P. Cao, X. Zhu, B. Li, Y. He, P. Song, R. Wang, Facile construction of BiOBr ultra-thin nano-roundels for dramatically enhancing photocatalytic activity. J. Environ. Manage. **299**, 113636 (2021)
74. X. Xiong, T. Zhou, X. Liu, S. Ding, J. Hu, Surfactant-mediated synthesis of single-crystalline Bi_3O_4Br nanorings with enhanced photocatalytic activity. J. Mater. Chem. A **5**, 15706 (2017)
75. M. Li, Y. Cui, Y. Jin, H. Li, Facile hydrolysis synthesis of $Bi_4O_5Br_2$ photocatalyst with excellent visible light photocatalytic performance for the degradation of resorcinol. RSC Adv. **6**, 47545 (2016)

76. L.S. Gómez-Velázquez, A. Hernández-Gordillo, M.J. Robinson, V.J. Leppert, S.E. Rodil, M. Bizarro, The bismuth oxyhalide family: thin film synthesis and periodic properties. Dalt. Trans. **47**, 12459 (2018)
77. B. Xu, Y. Gao, Y. Li, S. Liu, D. Lv, S. Zhao, H. Gao, G. Yang, N. Li, L. Ge, Synthesis of Bi_3O_4Cl nanosheets with oxygen vacancies: the effect of defect states on photocatalytic performance. Appl. Surf. Sci. **507**, 144806 (2020)
78. F. Wu, F. Chang, J. Zheng, M. Jiao, B. Deng, X. Hu, X. Liu, Synthesis and photocatalytic performance of $Bi_{12}O_{17}Cl_2$ semiconductors calcined at different temperatures. J. Inorg. Organomet. Polym. Mater. **28**, 721 (2018)
79. C. Liu, X.J. Wang, Room temperature synthesis of $Bi_4O_5I_2$ and Bi_5O_7I ultrathin nanosheets with a high visible light photocatalytic performance. Dalt. Trans. **45**, 7720 (2016)
80. C. Deng, H. Guan, X. Tian, Novel $Bi_{19}S_{27}Br_3$ superstructures: facile microwave-assisted aqueous synthesis and their visible light photocatalytic performance. Mater. Lett. **108**, 17 (2013)
81. T. Wu, X. Liu, Y. Liu, M. Cheng, Z. Liu, G. Zeng, B. Shao, Q. Liang, W. Zhang, Q. He, Application of QD-MOF composites for photocatalysis: energy production and environmental remediation. Coord. Chem. Rev. **403**, 213097 (2020)
82. X. He, T. Kai, P. Ding, Heterojunction photocatalysts for degradation of the tetracycline antibiotic: a review. Environ. Chem. Lett. **19**, 4563 (2021)
83. X. Jia, J. Cao, H. Lin, M. Zhang, X. Guo, S. Chen, Transforming type-I to type-II heterostructure photocatalyst via energy band engineering: a case study of I-BiOCl/I-BiOBr. Appl. Catal. B Environ. **204**, 505 (2017)
84. G.M. Tomboc, B.T. Gadisa, J. Joo, H. Kim, K. Lee, Hollow structured metal sulfides for photocatalytic hydrogen generation. ChemNanoMat **6**, 850 (2020)
85. U. Ghosh, A. Pal, Graphitic carbon nitride based Z scheme photocatalysts: design considerations, synthesis, characterization and applications. J. Ind. Eng. Chem. **79**, 383 (2019)
86. X. Qu, X. Zhao, M. Liu, Z. Gao, D. Yang, L. Shi, Y. Tang, H. Song, $BiOCl/TiO_2$ composite photocatalysts synthesized by the sol-gel method for enhanced visible-light catalytic activity toward methyl orange. J. Mater. Res. **35**, 3067 (2020)
87. J. Yu, M. Cui, X. Liu, Q. Chen, Y. Wu, Y. He, Preparation of novel $AgBr/Bi_3O_4Br$ hybrid with high photocatalytic activity via in situ ion exchange method. Mater. Lett. **193**, 73 (2017)
88. Y. Peng, H. Qian, N. Zhao, Y. Li, Synthesis of a novel 1D/2D $Bi_2O_2CO_3$–BiOI heterostructure and its enhanced photocatalytic activity. Catalysts **11**, 1284 (2021)
89. H. Razavi-Khosroshahi, S. Mohammadzadeh, M. Hojamberdiev, S. Kitano, M. Yamauchi, M. Fuji, $BiVO_4$/BiOX (X=F, Cl, Br, I) heterojunctions for degrading organic dye under visible light. Adv. Powder Technol. **30**, 1290 (2019)
90. J. Fu, Y. Tian, B. Chang, F. Xi, X. Dong, BiOBr-carbon nitride heterojunctions: synthesis, enhanced activity and photocatalytic mechanism. J. Mater. Chem. **22**, 21159 (2012)
91. R. Zhang, K. Zeng, A novel flower-like dual Z-scheme $BiSI/Bi_2WO_6/g\text{-}C_3N_4$ photocatalyst has excellent photocatalytic activity for the degradation of organic pollutants under visible light. Diam. Relat. Mater. **115**, 108343 (2021)
92. J. Liu, Y. Li, L. Huang, C. Wang, L. Yang, J. Liu, C. Huang, Y. Song, Fabrication of novel narrow/wide band gap $Bi_4O_5I_2$/BiOCl heterojunction with high antibacterial and degradation efficiency under LED and sunlight. Appl. Surf. Sci. **567**, 150713 (2021)
93. C. Lu, W. Wu, H. Zhou, In situ fabrication of $BiOBr/BiFeWO_6$ heterojunction with excellent photodegradation activity under visible light. J. Solid State Chem. **303**, 122465 (2021)
94. J. Hou, K. Jiang, M. Shen, R. Wei, X. Wu, F. Idrees, C. Cao, Micro and nano hierachical structures of BiOI/activated carbon for efficient visible-light-photocatalytic reactions. Sci. Rep. **7**, 11665 (2017)
95. S. Wang, W. An, Y. Guan, Z. Li, D. Mao, T. Xu, H. Wang, Z. Hu, S. Yan, $Bi_4O_5I_2\text{-}Bi_5O_7I$/Ni foam constructed in-situ accelerating interfacial carrier transfer for efficient photocatalysis. Appl. Surf. Sci. **564**, 150485 (2021)
96. X. Qu, M. Liu, L. Li, C. Wang, C. Zeng, J. Liu, L. Shi, F. Du, Fabrication of CdTe QDs/BiOI-promoted TiO_2 hollow microspheres with superior photocatalytic performance under simulated sunlight. Nanoscale Res. Lett. **14**, 50 (2019)

97. K.Y. Shih, Y.L. Kuan, E.R. Wang, One-step microwave-assisted synthesis and visible-light photocatalytic activity enhancement of BiOBr/RGO nanocomposites for degradation of methylene blue. Materials **14**, 4577 (2021)
98. Z.K. Cui, F.L. Zhang, Z. Zheng, W.J. Fa, B.J. Huang, Preparation and characterisation of Ag_3PO_4/BiOBr composites with enhanced visible light driven photocatalytic performance. Mater. Technol. **29**, 214 (2014)
99. R. Zhang, K. Zeng, T. Zhang, Enhanced visible-light-driven photocatalytic activity of Bi_2WO_6-BiSI Z-scheme heterojunction photocatalysts for tetracycline degradation. Int. J. Environ. Anal. Chem. **102**, 7084 (2020)
100. M. Guo, Z. Zhou, S. Yan, P. Zhou, F. Miao, S. Liang, J. Wang, X. Cui, Bi_2WO_6–BiOCl heterostructure with enhanced photocatalytic activity for efficient degradation of oxytetracycline. Sci. Rep. **10**, 18401 (2020)
101. Q. Huang, G. Jiang, H. Chen, L. Li, Y. Liu, Z. Tong, W. Chen, Hierarchical nanostructures of BiOBr/AgBr on electrospun carbon nanofibers with enhanced photocatalytic activity. MRS Commun. **6**, 61 (2016)
102. K. Sridharan, S. Shenoy, S. G. Kumar, C. Terashima, A. Fujishima, S. Pitchaimuthu, Advanced two-dimensional heterojunction photocatalysts of stoichiometric and non-stoichiometric bismuth oxyhalides with graphitic carbon nitride for sustainable energy and environmental applications. Catalysts **11**, 426 (2021)
103. E.M. El-Fawal, Visible light-driven BiOBr/Bi_2S_3@CeMOF heterostructured hybrid with extremely efficient photocatalytic reduction performance of nitrophenols: modeling and optimization. ChemistrySelect **6**, 6904 (2021)
104. Q.W. Cao, Y.F. Zheng, X.C. Song, The enhanced visible light photocatalytic activity of $Bi_2WxMo_{1-x}O_6$-BiOCl heterojunctions with adjustable energy band. Ceram. Int. **42**, 14533 (2016)
105. W. Zhu, J. Song, X. Wang, Y. Lu, G. Hu, J. Yang, The fast degradation for tetracycline over the Ag/AgBr/BiOBr photocatalyst under visible light. J. Mater. Sci. Mater. Electron. **32**, 26465 (2021)
106. E. Jiang, X. Liu, H. Che, C. Liu, H. Dong, G. Che, Visible-light-driven Ag/Bi_3O_4Cl nanocomposite photocatalyst with enhanced photocatalytic activity for degradation of tetracycline. RSC Adv. **8**, 37200 (2018)
107. L. Kong, Z. Jiang, H.H. Lai, R.J. Nicholls, T. Xiao, M.O. Jones, P.P. Edwards, Unusual reactivity of visible-light-responsive AgBr-BiOBr heterojunction photocatalysts. J. Catal. **293**, 116 (2012)
108. X. Lou, J. Shang, L. Wang, H. Feng, W. Hao, T. Wang, Y. Du, Enhanced photocatalytic activity of $Bi_{24}O_{31}Br_{10}$: constructing heterojunction with BiOI. J. Mater. Sci. Technol. **33**, 281 (2017)
109. A. Pancielejko, J. Łuczak, W. Lisowski, A. Zaleska-Medynska, P. Mazierski, Novel two-step synthesis method of thin film heterojunction of BiOBr/Bi_2WO_6 with improved visible-light-driven photocatalytic activity. Appl. Surf. Sci. **569**, 151082 (2021)
110. Y. Zhang, Q. Shao, H. Jiang, L. Liu, M. Wu, J. Lin, J. Zhang, S. Wu, M. Dong, Z. Guo, One-step co-precipitation synthesis of novel BiOCl/CeO_2 composites with enhanced photodegradation of rhodamine B. Inorg. Chem. Front. **7**, 1345 (2020)
111. A. Alzamly et al., Construction of BiOF/BiOI nanocomposites with tunable band gaps as efficient visible-light photocatalysts. J. Photochem. Photobiol. A Chem. **375**, 30 (2019)
112. Y. Chen, X. Xu, J. Fang, G. Zhou, Z. Liu, S. Wu, W. Xu, J. Chu, X. Zhu, Synthesis of BiOI-TiO_2 composite nanoparticles by microemulsion method and study on their photocatalytic activities. Sci. World J. **2014**, 647040 (2014)
113. S. Bargozideh, M. Tasviri, Construction of a novel BiSI/MoS_2 nanocomposite with enhanced visible-light driven photocatalytic performance. New J. Chem. **42**, 18236 (2018)
114. B. Cui, W. An, L. Liu, J. Hu, Y. Liang, Synthesis of CdS/BiOBr composite and its enhanced photocatalyticdegradation for rhodamine. Appl. Surf. Sci. **319**, 298 (2014)
115. M.E. Malefane, Co_3O_4/$Bi_4O_5I_2$/Bi_5O_7I C-scheme heterojunction for degradation of organic pollutants by light-emitting diode irradiation. ACS Omega **5**, 26829 (2020)

116. J. Wu et al., Hydrothermal synthesis of carbon spheres—BiOI/$BiOIO_3$ heterojunctions for photocatalytic removal of gaseous Hg_0 under visible light. Chem. Eng. J. **304**, 533 (2016)
117. G. Eshaq, S. Wang, H. Sun, M. Sillanpää, Core/shell FeVO4@BiOCl heterojunction as a durable heterogeneous fenton catalyst for the efficient sonophotocatalytic degradation of p-nitrophenol. Sep. Purif. Technol. **231**, 115915 (2020)
118. J. Wu, Y. Xie, Y. Ling, Y. Dong, J. Li, S. Li, J. Zhao, Synthesis of flower-like g-C_3N_4/BiOBr and enhancement of the activity for the degradation of bisphenol a under visible light irradiation. Frontiers in Chemistry **7**, 49 (2019)
119. D. Liu, J. Xie, Y. Xia, Improved photocatalytic activity of MWCNT/BiOBr composite synthesized via interfacial covalent bonding linkage. Chem. Phys. Lett. **729**, 42 (2019)
120. Y. Zhang, M. Park, H.Y. Kim, B. Ding, S.J. Park, A facile ultrasonic-assisted fabrication of nitrogen-doped carbon dots/BiOBr up-conversion nanocomposites for visible light photocatalytic enhancements. Sci. Rep. **7**, 45086 (2017)
121. Y. Xu, Y. Ma, H. Ji, S. Huang, M. Xie, Y. Zhao, H. Xu, H. Li, Enhanced long-wavelength light utilization with polyaniline/bismuth-rich bismuth oxyhalide composite towards photocatalytic degradation of antibiotics. J. Colloid Interface Sci. **537**, 101 (2019)
122. L. Chen, W. Zhang, J. Wang, X. Li, Y. Li, X. Hu, L. Zhao, Y. Wu, Y. He, High piezo/photocatalytic efficiency of Ag/Bi_5O_7I nanocomposite using mechanical and solar energy for N_2 fixation and methyl orange degradation. Green Energy Environ. **8**, 283 (2023)
123. K. Yasui, Fundamentals of acoustic cavitation and sonochemistry, in *Theoretical and Experimental Sonochemistry Involving Inorganic Systems*. ed. by M. Ashokkumar (Springer, Netherlands, Dordrecht, 2011), pp.1–29
124. S. Merouani, O. Hamdaoui, Y. Rezgui, M. Guemini, Theoretical estimation of the temperature and pressure within collapsing acoustical bubbles. Ultrason. Sonochem. **21**, 53 (2014)
125. C. Jin, D. Liu, J. Hu, Y. Wang, Q. Zhang, L. Lv, F. Zhuge, The role of microstructure in piezocatalytic degradation of organic dye pollutants in wastewater. Nano Energy **59**, 372 (2019)
126. X. Wang, X. Gao, M. Li, S. Chen, J. Sheng, J. Yu, Synthesis of flexible $BaTiO_3$ nanofibers for efficient vibration-driven piezocatalysis. Ceram. Int. **47**, 25416 (2021)
127. J. Ling, K. Wang, Z. Wang, H. Huang, G. Zhang, Enhanced piezoelectric-induced catalysis of $SrTiO_3$ nanocrystal with well-defined facets under ultrasonic vibration. Ultrason. Sonochem. **61**, 104819 (2020)
128. W. Ma, B. Yao, W. Zhang, Y. He, Y. Yu, J. Niu, Fabrication of PVDF-based piezocatalytic active membrane with enhanced oxytetracycline degradation efficiency through embedding few-layer E-MoS_2 nanosheets. Chem. Eng. J. **415**, 129000 (2021)
129. Y. Feng, L. Ling, Y. Wang, Z. Xu, F. Cao, H. Li, Z. Bian, Engineering spherical lead zirconate titanate to explore the essence of piezo-catalysis. Nano Energy **40**, 481 (2017)
130. K. Hamano, T. Nakamura, Y. Ishibashi, T. Ooyane, Piezoelectric property of SbSI single crystal. J. Phys. Soc. Japan **20**, 1886 (1965)
131. X. Zhou, F. Yan, S. Wu, B. Shen, H. Zeng, J. Zhai, Remarkable piezophoto coupling catalysis behavior of BiOX/$BaTiO_3$ (X=Cl, Br, $Cl_{0.166}$ $Br_{0.834}$) piezoelectric composites. Small **16**, 2001573 (2020)
132. H. Lei, H. Zhang, Y. Zou, X. Dong, Y. Jia, F. Wang, Synergetic photocatalysis/piezocatalysis of bismuth oxybromide for degradation of organic pollutants. J. Alloys Compd. **809**, 151840 (2019)
133. Z. Yao, H. Sun, S. Xiao, Y. Hu, X. Liu, Y. Zhang, Synergetic piezo-photocatalytic effect in a Bi_2MoO_6/BiOBr composite for decomposing organic pollutants. Appl. Surf. Sci. **560**, 150037 (2021)
134. Z. Yao, H. Sun, S. Xiao, Y. Hu, X. Liu, Y. Zhang, Ti_3C_2 quantum dots modified on BiOBr surface for sewage disposal: the induction of the piezo-phototronic effect from edge to whole. Appl. Surf. Sci. **599**, 153911 (2022)
135. M. Ismail, Z. Wu, L. Zhang, J. Ma, Y. Jia, Y. Hu, Y. Wang, High-efficient synergy of piezocatalysis and photocatalysis in bismuth oxychloride nanomaterial for dye decomposition. Chemosphere **228**, 212 (2019)

136. K. Fan, C. Yu, S. Cheng, S. Lan, M. Zhu, Metallic Bi self-deposited BiOCl promoted piezocatalytic removal of carbamazepine. Surfaces and Interfaces **26**, 101335 (2021)
137. L. Li, W. Cao, J. Yao, W. Liu, F. Li, C. Wang, Synergistic piezo-photocatalysis of BiOCl/$NaNbO_3$ heterojunction piezoelectric composite for high-efficient organic pollutant degradation. Nanomaterials **12**, 353 (2022)
138. L. Song, S. Zhang, Q. Wei, Porous BiOI sonocatalysts: hydrothermal synthesis, characterization, sonocatalytic, and kinetic properties. Ind. Eng. Chem. Res. **51**, 1193 (2012)
139. Z. Li, Q. Zhang, L. Wang, J. Yang, Y. Wu, Y. He, Novel application of Ag/$PbBiO_2I$ nanocomposite in piezocatalytic degradation of rhodamine B via harvesting ultrasonic vibration energy. Ultrason. Sonochem. **78**, 105729 (2021)
140. Q. Zhao, L. Hao, F. Li, T. Liu, Y. He, J. Yang, Y. Zhang, Y. Lu, Piezo/photocatalytic activity of flexible BiOCl-BiOI films immobilized on SUS_{304} wire mesh. J. Water Process Eng. **42**, 102105 (2021)
141. C. Zhang, W. Fei, H. Wang, N. Li, D. Chen, Q. Xu, H. Li, J. He, J. Lu, P-n heterojunction of BiOI/ZnO nanorod arrays for piezo-photocatalytic degradation of bisphenol A in water. J. Hazard. Mater. **399**, 123109 (2020)
142. K. Mistewicz, M. Jesionek, H.J. Kim, S. Hajra, M. Kozioł, Ł Chrobok, X. Wang, Nanogenerator for determination of acoustic power in ultrasonic reactors. Ultrason. Sonochem. **78**, 105718 (2021)

Chapter 8
Conclusions and Future Prospects

8.1 A Summary on the Recent Development of Low-Dimensional Chalcohalides

A family of the chalcohalide materials comprises a very wide range of different compounds. They contain both chalcogen and halogen atoms. The chalcohalides exist in the forms of binary, ternary, quaternary, pentanary, and even more complex materials. A powerful group of these compounds are pnictogen chalcohalides. They are denoted as V-VI-VII [1–3] or $A^{V}B^{VI}C^{VII}$ [4–6] or $A^{15}B^{16}C^{17}$ [7] compounds, where A^{15} represents the element from group 15 (As, Sb, Bi), B^{16} means the element from group 16 (O, S, Se, Te), and C^{17} is the element from group 17 (F, Cl, Br, I). In general, the chemical formula of the pnictogen chalcohalides is $A_xB_yC_z$, where x, y, z determine the material stoichiometry. Some of these compounds have been known from the beginning of the nineteenth century, when antimony sulfoiodide (SbSI) was synthesized for the first time [8]. However, the increased interest in the bulk crystals of ternary chalcohalides was observed in the 60 s of the 20th Century after the photoconductivity [9] and ferroelectricity [10, 11] were revealed in these materials. Further studies demonstrated other interesting properties of the pnictogen chalcohalides, such as electromechanical [12], piezoelectric [5, 13, 14], pyroelectric [15, 16], electrocaloric [17], thermoelectric [18–21], photovoltaic [1, 22–24], ferroelectric-photovoltaic [25, 26], photoelectrochemical [2, 27, 28], photocatalytic [29–31], and piezocatalytic properties [32, 33]. The numerous pnictogen chalcohalides exhibit one-dimensional crystal structure. They consist of the double chains which are held together by weak van der Waals forces. Highly anisotropic morphology of these materials leads to a large internal electric field which supports the charge carriers separation. This property of the pnictogen chalcohalides is crucial for photovoltaic, photodetector, and photocatalytic applications. The recent investigations have shown Rashba type spin splitting in bismuth tellurohalides [34–38]. The chalcohalides have been considered as the potential "defect-tolerant" semiconductors [39–41]. The ferroelectric characteristic combined with photosensitive

K. Mistewicz, *Low-Dimensional Chalcohalide Nanomaterials*, NanoScience and Technology, https://doi.org/10.1007/978-3-031-25136-8_8

property, known as photoferroelectricity [42, 43], seems to be one of the most important feature of the chalcohalide materials. Figure 8.1 presents the number of papers per year published on selected ternary chalcohalides: SbSI, antimony selenoiodide (SbSeI), bismuth sulfoiodide (BiSI), bismuth oxyiodide (BiOI), bismuth oxychloride (BiOCl), and bismuth oxybromide (BiOBr). It should be underlined that the bismuth oxyhalides have received a gained attention over the last decade especially due to excellent photocatalytic properties [44–47].

The methods of fabrication of the low-dimensional chalcohalide compounds can be divided into two main groups: "top-down" and "bottom-up" approaches. The first one includes mechanical milling of bulk crystals [29, 48, 49] and liquid-phase exfoliation [50–52]. These techniques are facile and do not require use of the complex equipment. The vapor phase growth [53, 54], hydrothermal synthesis [55, 56], solvothermal method [57], synthesis under ultrasonic irradiation [16, 58], microwave synthesis [59], laser/heat-induced crystallization [60], electrospinning [61], successive ionic layer adsorption and reaction [62] belong to the "bottom-up" methods of chalcohalide nanostructures fabrication. It should be noted that hydrothermal and solvothermal are versatile methods which allow to obtain a plenty of different bare chalcohalide nanomaterials as well as their composites or heterostructures. The sonochemical technique was demonstrated to be suitable for preparation of one-dimensional nanostructures of SbSI type compounds. This method is a low-cost, simple, and it can be utilized successfully used on a large scale. The details of all aforementioned methods of chalcohalide nanomaterials fabrication were provided in the Chap. 2 of this book. The applied chemical procedures, used reagents, and basic properties of the prepared products were reviewed.

The recent achievements in technologies for incorporation of low-dimensional chalcohalides into the functional devices were summarized in Chap. 3. The solution processing of the thin films [63, 64] was discussed as a method commonly used for preparation of the solar cells and photodetectors. The spin-coating deposition of polymer composites [5, 65] was demonstrated as a convenient way to construct

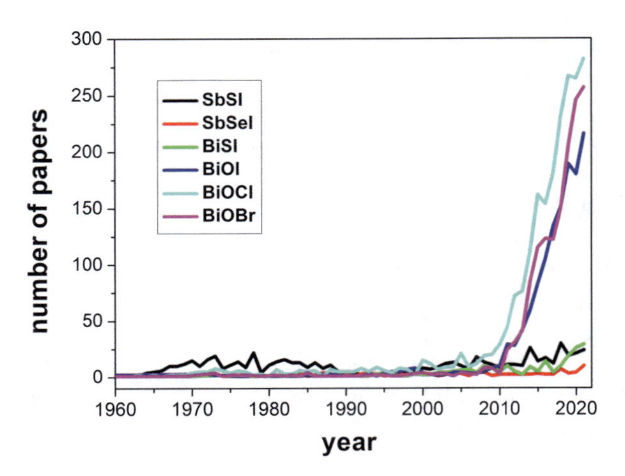

Fig. 8.1 The number of papers published per year on selected antimony and bismuth chalcohalides according to Scopus database

mechanical energy harvesters and photovoltaic devices. The drop-casting of the inorganic nanocomposites [66] allowed for the fabrication of a thin film pyroelectric nanogenerator. The room temperature compression of the chalcohalide nanowires under a high pressure [16, 67] was applied to obtain humidity sensors as well as piezoelectric, pyroelectric, and triboelectric nanogenerators. The pressure assisted sintering of bismuth chalcohalides [68, 69] was developed to prepare the porous photoelectrodes and radiation detectors. The alignment of the SbSI nanowires in an electric field [70, 71] was frequently utilized for fabrication of gas sensors and photodetectors.

The overview of the devices for energy harvesting and storage based on the chalcohalide nanomaterials was given in Chap. 4. A conversion of the mechanical energy into the electric output was presented using piezoelectric [67, 72] and triboelectric [73, 74] nanogenerators. The piezoelectric nanogenerators were used as the sensors for a detection of low [5] and high [75] frequency vibrations. The SbSI and SbSeI nanowires were confirmed to be suitable for use in the piezoelectric sensors and mechanical energy harvesters due to their outstanding piezoelectric and electromechanical properties as well as one-dimensional morphology that enables high flexibility. The nanowires of SbSI and SbSeI were also successfully used in the pyroelectric generators that converted the thermal fluctuations into the electric energy [16, 66]. The bismuth [76–78] and antimony chalcohalides [79] have been recently examined as electrode materials for the supercapacitors. The remarkable electrochemical performance and high stability of the these compounds proved their great potential for supercapacitor applications.

The photovoltaic and photodetection properties of low-dimensional chalcohalide materials were discussed comprehensively in Chap. 5. The chalcohalide compounds have been intensively investigated as the promising light or radiation inorganic absorbers due to relatively narrow energy band gaps, high charge carrier mobilities, low effective masses of the charge carriers, and a high level of defect tolerance. The ferroelectric-photovoltaic effect in the SbSI [25, 80, 81] was presented. The solar cells were categorized into two main groups: antimony [65, 82, 83] and bismuth [84–86] chalcohalide based photovoltaic devices. Fabrication methods and photovoltaic figures of merit of the chalcohalide solar cells were reviewed. The applications of low-dimensional chalcohalides in photodetectors [87–89] and ionizing radiation detectors [69, 90] were presented.

The chalcohalide nanomaterials were found to be suitable for use in gas sensors (Chap. 6). These devices were distinguished into four main types: conductometric [70, 91, 92], photoconductive [92, 93], impedance [94, 95], and quartz crystal microbalance [96, 97] sensors. These devices were examined taking into consideration fundamental features of gas sensors: sensitivity, detection limit, response kinetics, its hysteresis, stability, repeatability, reversibility, and selectivity against the interfering gases. The mechanisms of the different gases detection were discussed.

Numerous antimony [29, 98, 99] and bismuth [47, 100, 101] chalcohalides were applied as efficient photocatalyst for degradation of dyes or hazardous organic compounds (Chap. 7). The excellent photocatalytic properties of SbSI were attributed to the ferroelectric polarization. It was responsible for reduction of charge carriers recombination and enhancement of their separation. The chalcohalide heterostructures [102, 103] and composites [104, 105] were investigated as promising materials with high photocatalytic activities. The heterostructures were distinguished into different groups: type I [106], type II [107], *pn* [108], and Z-scheme [103] junctions. The photocatalytic performances and mechanisms of the dye photodegradation were analyzed. The low-dimensional chalcohalides were also studied as new piezocatalyst for decomposition of toxic compounds [32, 98, 109].

8.2 Author Contribution to Progress in Chalcohalide Nanomaterials

A scientific activity of Dr. Krystian Mistewicz, author of this book, is focused on fabrication, characterization, and possible applications of ternary chalcohalide nanowires in energy harvesting, sensing, and environmental remediation (Fig. 8.2). He developed a novel method of the SbSI nanowires alignment and their ultrasonic bonding to the microelectrodes [110]. A formation of reliable connections between nanowires and microelectrodes was confirmed by using atomic force microscopy (AFM), scanning electron microscopy (SEM), and direct current (DC) electrical measurements. The ultrasonic processing technology allows to prepare an array of a few SbSI nanowires suitable for use in conductometric gas nanosensors [110]. He proposed a simple method for determination of electrical conductivity type of the SbSI nanowires [71]. The influence of hydrogen and oxygen adsorption on electric response of an array of a few SbSI nanowires was studied. The character of this response proved the p-type electrical conductivity of SbSI nanowires. Mistewicz and coworkers examined the SbSI xerogel and an array of a few SbSI nanowires as conductometric and photoconductive humidity nanosensors [92, 111]. The differences in operation of these devices were analyzed. The humidity sensing mechanism was proposed taking into account proton-transfer process and Grotthuss chain reaction. A first report on detection of ammonia gas using an array of a few SbSI nanowires as gas sensor was provided in Ref. [70]. The nanosensor response was registered at operating temperatures below and above Curie temperature of SbSI. It was found that ferroelectric properties of the SbSI nanosensor were crucial for its large sensitivity [70]. Furthermore, the SbSI nanosensor exhibited short term response, good stability, reversibility, and remarkable selectivity to ammonia against other interfering gases. The ammonia sensing mechanism was explained. In 2018, Mistewicz reviewed the recent achievements in exploration of ferroelectric nanomaterials (including SbSI nanowires) for their application in photodetectors, ionizing radiation detectors, biosensors, piezoelectric, pyroelectric, and piezoresistive sensors of mechanothermal signals [112]. The first humidity sensor based on the SbSeI nanowires was shown in Ref. [95]. The influence of temperature and relative humidity on the impedance characteristics of SbSeI nanosensor was examined. The impedance response of the nanosensor displayed a low hysteresis, excellent repeatability, and long-term stability. An analysis of the Nyquist characteristics revealed a presence of the relaxation processes with a non-Debye type nature [95]. An equivalent electric model of SbSeI xerogel was proposed. The humidity sensing mechanism was elaborated with reference to a proton hopping.

Mistewicz and coworkers investigated a ferroelectric-photovoltaic effect in SbSI nanowires for the first time [25]. The photoelectric response of SbSI nanowires was studied as a function of light intensity. The dependence of optical power density on a short circuit photocurrent density followed the Glass law. The photovoltage and photocurrent were found to be switchable through applying a poling electric field. The open circuit photovoltage of 119 mV was generated in the SbSI nanowires under monochromatic light illumination with wavelength of 488 nm and optical

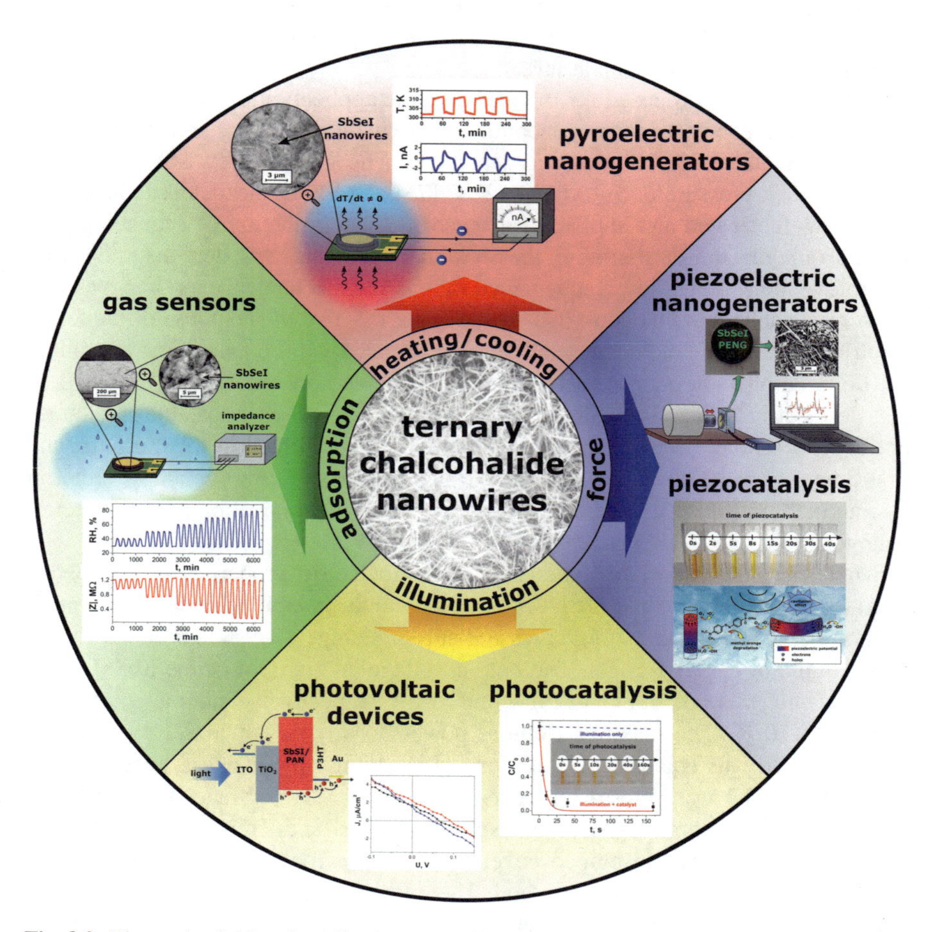

Fig. 8.2 The main fields of applications of ternary chalcohalide nanowires. Reprinted from Mistewicz [113] under the terms of the Creative Commons Attribution International License (CC BY). Copyright (2021) IAAM

power density of 127 mW/cm^2. An application of the composite of SbSI nanowires and polyacrylonitrile (PAN) in photovoltaic device was presented in Ref. [65]. This device was fabricated via simple and fast method which involved of spin-coating of the SbSI-PAN nanocomposite on indium tin oxide substrate. This technology did not require high temperature treatment in contrast to solution processing commonly used for preparation of chalcohalide based solar cells. The titanium dioxide (TiO_2) and poly(3-hexylthiophene) (P3HT) were chosen as an electron and hole transporting layers, respectively. The developed photovoltaic device exhibited open circuit voltage of 69(13) mV and short circuit current density of 1.84(20) μA/cm^2 under a white light standard illumination with power density of 100 mW/cm^2. For the first time, Mistewicz provided the figures of merit for photodetector based on the SbSI nanowires (Sect. 5.3 in Chap. 5). The responsivity, noise equivalent power, detectivity, external quantum efficiency, response and recovery times were determined for SbSI nanowires and compared to these parameters reported for photodetectors based on other chalcohalide nanomaterials.

Mistewicz et al. proposed for the first time the application of SbSI nanowires in ferroelectric/piezoelectric nanogenerator for mechanical energy harvesting [72]. A large electric field of $3 \cdot 10^7$ V/m was generated in the array of a few SbSI nanowires under a shock pressure of 5.9 MPa. In Ref. [75], the SbSeI piezoelectric nanogenerator was fabricated and applied as a sensor for determination of acoustic power in ultrasonic reactor. Furthermore, the SbSeI nanogenerator was demonstrated to be suitable for detection of low frequency vibrations and mechanical energy harvesting. Mistewicz showed in this book (Chap. 4, Sect. 4.1.2) that the response of the ferroelectric/piezoelectric nanogenerator can be enhanced via the nanowires alignment and elimination of the contacts between them. Mistewicz proved in Sect. 3.7 (Chap. 3) that the electric field assisted alignment of the SbSI nanowires is scalable and it does not depend on the electrode distance. Moreover, he presented in this book for the first time a preparation of the SbSI/PVDF nanocomposite film via simple solution casting method (Sect. 3.3 in Chap. 3) and presented its application in piezoelectric nanogenerator (Sect. 4.1.2 in Chap. 4).

Mistewicz and associates examined the pyroelectric properties of the SbSeI nanowires for the first time [16]. A new, facile, and scalable method of compression of the nanowires under high pressure at room temperature was used to prepare bulk sample of SbSeI xerogel. The output voltage and current of the SbSeI pyroelectric nanogenerator attained 12 mV and 11 nA, respectively, under a temperature change of 10 K. The maximum output power density of the SbSeI pyroelectric nanogenerator was equal to 0.59 μW/m^2 [16]. A high pyroelectric coefficient of 44(5) nC/(cm^{2}K) was determined at temperature of 327 K. Mistewicz developed fabrication of nanocomposite of the SbSI nanowires and TiO_2 nanoparticles for the first time [66]. He developed novel SbSI-TiO_2 pyroelectric nanogenerator by applying a simple drop-casting method. The pyroelectric current increased linearly with an increase of the temperature change rate. The pyroelectric current was highly correlated with the temperature fluctuations. The large pyroelectric coefficient of 264 nC/(cm^2 K) was calculated for the SbSI-TiO_2 nanocomposite. The maximum power density of the $SbSI\text{-}TiO_2$ nanogenerator reached 8.39(2) $\mu W/m^2$ when temperature change was equal to 62.5 mK/s. The experiments, described above, proved a great potential of antimony chalcohalide nanowires for a converting of thermal fluctuations into an electrical energy.

Mistewicz and coworkers investigated piezocatalytic properties of the SbSI for the first time [98]. Methyl orange was used as a typical dye for the piezo- and photocatalytic tests. The high degradation efficiencies of 99.5% and 95% were obtained after 40 s and 160 s of ultrasonic vibration and UV illumination, respectively. The large reaction rate constants of 7.6 min^{-1} and 9 min^{-1} were determined for piezo- and photocatalytic decomposition of methyl orange, accordingly [98]. This research opened a way to future application of the SbSI nanowires in a purification of wastewater.

8.3 The Main Challenges

The fundamental optical, electrical, and photoelectrical properties of many chalcohalide compounds still remain unknown. Over 161,000 ternary metal chalcohalides were selected in Ref. [114] from possible 32 million compositions as promising candidate photoactive semiconductors using the computer-aided hierarchical screening procedure. It allowed to predict the crystal structures and basic optoelectronic properties of four novel chalcohalides: $Sn_5S_4Cl_2$, Sn_4SF_6, $Cd_5S_4Cl_2$, Cd_4SF_6. The effective masses of charge carriers of the antimony and bismuth chalcohalides were computed in Ref. [115]. Thermoelectric properties of antimony chalcohalides were simulated theoretically in Ref. [18, 21]. These results have not been verified in an experimental way. Similarly, the piezoelectric properties of many different chalcohalide materials have not been investigated so far.

The ferroelectric behavior of the chalcohalide compounds is essential for their possible use in ferroelectric memories [116] and ferroelectric solar cells [117]. However, the Curie temperatures of the pnictogen chalcohalides are low (see Table 1.3 in Chap. 1). It strongly limits their potential applications. Moreover, an explanation of the size effect in ferroelectric nanomaterials [118] is still not clear and it has to be examined in future. In order to visualize the ferroelectric domains and measure polarization switching properties at the nanoscale, the chalcohalide nanostructures can be investigated using piezoresponse force microscopy [119–121].

The SbSI and SbSeI xerogels are complex systems composed of large number of randomly oriented nanowires. The piezoelectric or pyroelectric responses of the xerogel was averaged over the volume of a bulk sample [16, 67, 75]. Thus, determined figures of merit do not represent the exact parameters describing property of an individual nanowire. Moreover, a packing factor achieved for high pressure compression of the SbSI and SbSeI nanowires did not exceed 50% [73, 95, 122]. It means that the voids existed between the bundles of the nanowires. In the case of polymer composites [5, 65], an interpretation of the experimental data is even more complicated. The polymer matrix influenced the mechanical, optical, and electrical properties of the composite. A significant effort has to be done toward a construction of the piezoelectric and pyroelectric energy harvesters or sensors from a single nanowire.

Numerous reports [5, 67, 72–75] proved that chalcohalide nanomaterials are suitable for use in piezoelectric and triboelectric nanogenerators. However, the exact mechanism of the charge carriers generation in such devices is still not fully understood. A deep insight into the relation between mechanical excitation and generated electric output is needed to proper understanding of working principle of these devices and their future improvement. Another challenge is a relatively low efficiency of energy harvesting. The nanogenerators, prepared via high compression of chalcohalide nanowires, exhibited poor flexibility and hardness [67]. The electrical impedances of chalcohalide composites are usually large and they do not correspond to the electrical loads commonly used in the electronic circuits. Additional experiments should be conducted toward selecting the optimal polymer matrix and tailoring the concentration of chalcohalide nanowires in the nanocomposite. It will result in amplification of the output power of the nanogenerator. Temperature dependence of piezoelectric response has to be determined in order to find the best operating conditions of the chalcohalide nanogenerator.

The ferroelectric-photovoltaic effect in SbSI is attractive for cutting-edge optically rewritable [123] or readable [124] ferroelectric memories. Nevertheless, the short circuit photocurrent and open circuit photovoltage generated by ferroelectric-photovoltaic cell based on the SbSI nanowires [25] are too small for commercial applications. Until now, the highest power conversion efficiency of the chalcohalide based solar cells has attained approximately 6% [83]. This value is much lower than power conversion efficiencies reported for perovskite solar cells [125]. The application of the low-dimensional chalcohalide compounds in the detectors of ionizing radiation is at the initial stage. In order to improve the detector performance, the sensitivity, stability, repeatability, and kinetics of its response have to be studied in detail.

A majority of the conductometric or photoconductive gas nanosensors were fabricated from the SbSI nanowires. Thus, the attention should be paid to other chalcohalide nanomaterials as possible sensing materials. Despite intensive investigations of the SbSI based gas sensors, the character of the gas adsorption on the SbSI surface at relatively low temperatures is controversial. Further research is required to determine whether physi- or chemisorption occurs.

The antimony chalcohalides seem to be overlooked photocatalytic materials. Antimony sulfoiodide is the only one compound from this group of materials which photocatalytic properties have been investigated so far. The characterization of chalcohalide piezocatalysts needs additional studies. It is supposed that different factors (excitation frequency, acoustic power, reaction temperature) influence the efficiency of the piezocatalysis. The influence of mentioned parameters on piezocatalytic degradation of various organic pollutants has to be determined in the future.

8.4 Future Trends and Outlooks

A precise tailoring of the optoelectronic properties of the chalcohalide semiconductor material is needed to construct high performance devices. It can be achieved in several possible manners. The first one is a band gap engineering [126]. An adjusting the concentrations of the elements in the quaternary chalcohalide compound may result in a continuous change of its energy band gap. The low temperature synthesis of the $BiSBr_{1-x}I_x$ was presented in Ref. [127]. This method allowed to tune the energy band gap of the $BiSBr_{1-x}I_x$ by changing the ratio between iodine and bromine. A rise of selenium doping in the $BiS_{1-x}Se_xI$ from $x = 0$ to $x = 0.6$ led to reduction of energy band gap from 1.63 to 1.48 eV [128]. Similarly, it was demonstrated in Ref. [129] that the increase of selenium concentration in the $SbS_{1-x}Se_xI$ nanowires from $x = 0$ to $x = 1$ resulted in a linear decrease of energy band gap by 0.21 eV. Another approach toward band gap engineering is a material doping [130]. The next strategy toward improving the optoelectronic properties is a modulation of the charge carriers mobility. It was shown in Ref. [131] that the reorganization of the BiOCl surface led to increase of the carriers mobility and expanding visible-light absorption edge. The charge carrier concentration may be modulated via the bulk doping [132]. Aforementioned methods can be considered in the future to fabricate new optoelectronic components from chalcohalide nanomaterials.

The Curie temperature of the chalcohalide compound can be tuned through a change of chemical composition [133, 134] and an applying of a strain to the ferroelectric material [135]. The first mentioned approach requires use of advanced chemical procedures and equipment. The strain engineering seems to be more convenient and effective way to enhance the Curie temperature of a ferroelectric chalcohalide material. In this method, the flexoelectric effect [136, 137] can also occur. According to Ref. [135], the Curie temperature of SbSI can increased by over 60 K under a small strain of 3%. It was demonstrated in Ref. [26] that the lone-pair electrons in one-dimensional SbSI were reconfigured under a pressure of 14 GPa leading to a symmetry-breaking and significant rise of the photocurrent. Recently developed compression of the chalcohalide nanomaterials under a high pressure [16, 122] should be useful in preparation of photoferroelectric devices operating at elevated temperatures.

Further increase of the interest in fabrication of the hybrid nanogenerators is anticipated. The nanogenerator that can harvest different types of energy (mechanical, thermal, solar) is expected to be more effective than device which recovers only one specific energy. Moreover, it will be interesting to construct a piezoelectric, pyroelectric, thermoelectric, or triboelectric nanogenerator combined with a supercapacitor for a simultaneous storage of scavenged energy. Such device can be prepared from one chalcohalide material. The pnictogen chalcohalides are highly anisotropic materials which exhibit the largest ferroelectric, pyroelectric, and piezoelectric properties along their lengths. The experimental results, presented in Chap. 4 (Sect. 4.1.2), proved that the nanowires alignment and elimination of the contacts between separate nanowires are crucial factors for achieving the large amplitude of the ferroelectric/piezoelectric nanogenerator response. Therefore, the efficiency of the energy conversion may be enhanced via the alignment of the chalcohalide nanowires in the polymer matrix.

A growing interest in application of the pnictogen chalcohalide compounds in the supercapacitors [77, 79] is predicted. The incorporation of the chalcohalide nanostructures into the carbon based material [76] is an innovative method for preparing the supercapacitors with high specific capacity and improved cycling stability.

The magnitude of the ferroelectric-photovoltaic effect can be enlarged through photo-sensitizing of ferroelectric film. It was proved in Ref. [138] that a selection of an appropriate substrate for ferroelectric thin film leads to decrease of energy band gap of a ferroelectric material and favors electron injection in the substrate/electrode. Moreover, the open circuit photovoltage can be enhanced by increasing the distance between electrodes [139]. In the case of photovoltaic device constructed from SbSI nanowires, its short circuit photocurrent may be increased due to an improvement of electrical contacts between separate nanowires [25]. The future strategies toward increasing the power conversion efficiencies of the chalcohalide based solar cells involve defect engineering [140, 141], band alignment against the properly chosen electron/hole transporting layers [142], application of additional interlayers [83, 143, 144], dimensional reduction for tuning the electronic structure of the chalcohalide material, panchromatic photon extraction by hole transporting material [145], bandgap engineering via transition-metal or chalcogenide doping [22] and adjusting of chemical composition [127]. A fabrication of tandem solar cells is a new interesting trend in photovoltaics [146, 147]. Such tandem photovoltaic devices can be prepared by combining traditional silicon junctions with wide bandgap chalcohalide semiconductors [1]. The chalcohalide compounds with heavy-Z cations can be considered for use in the betavoltaic cells [148, 149]. It is expected that these devices will be fabricated and examined in a near future.

Several strategies can be considered for a future development of gas sensors based on chalcohalide nanomaterials. The examination of the gas sensor response in the broad range of temperature should result in finding the optimal operating temperature. A reversibility of the sensor response may be ameliorated by exposing the device to the UV light. The new photoconductive gas sensors can be constructed from chalcohalide materials other than SbSI, e.g. SbSeI, BiSI, BiSBr, BiOI, or BiSCl. These semiconductors exhibit photovoltaic properties and possess the relatively low energy bang gaps. Therefore, they are promising for use in novel photovoltaic self-powered gas sensors [150–152]. The technologies, described in Chap. 3, can be applied in order to prepare high quality sensing layers. The solution processing, spin-coating deposition, drop-casting, high pressure compression of nanostructures at room temperature, and pressure assisted sintering seem to be attractive for fabrication of conductometric, photoconductive, impedance, and quartz crystal sensors. A combining the different chalcohalide nanomaterials into one device will enable to construct the matrix of gas sensors. Such device should show a better sensitivity and selectivity in comparison to a conventional gas detector based on one sensing material.

Antimony and bismuth chalcohalides possess a great potential for photo/piezocatalytic removal of organic pollutants from aqueous solutions.

One can expect that future research in this field will be focused on improving the efficiency, stability, and recyclability of these catalysts. The catalytic performance can be boosted via defect engineering [153, 154], elemental doping [155, 156], control of nanostructures size [157], interfacing with carbon materials [158–160], deposition of noble metals on catalyst surface [161], and heterojunction formation [104, 108, 162]. The energy band engineering allows to transform type-I into type-II heterostructure photocatalyst [163]. The hydro/solvothermal method is a powerful tool for preparation of the photocatalyst based on the pristine chalcohalide materials and their heterostructures [164–166]. However, a special attention should be paid to sonochemical synthesis as promising approach toward obtaining novel heterostructured and composite chalcohalides [66]. Such experiments are in a progress. Their results will be presented in near future. It is predicted that a possible simultaneous pyro-, piezo-, and photocatalysis should yield much higher efficiency than single photocatalysis. It is anticipated that bare chalcohalide nanomaterials [167] and their heterostructures [168] will be explored in future toward efficient photocatalytic hydrogen production by water decomposition.

Emerging trends in research of chalcohalide materials refer to their potential use in spintronics [38] and Kitaev spin liquid [169]. Another new interesting direction in investigation of chalcohalide nanoparticles is related with inverse “guest–host” phenomenon in liquid crystals [170]. In this effect, the ferroelectric nanoparticles reorient and hold liquid crystal molecules in a direction of nanoparticles orientation [170]. It is achieved despite an external electric field tries to align a liquid crystal in an orthogonal direction. Finally, it should be concluded that a rapid growth of the interest in fabrication, characterization, and applications of the low-dimensional chalcohalide nanomaterials is expected in the coming years.

References

1. F. Palazon, Metal chalcohalides: next generation photovoltaic materials? Sol. RRL **6**, 2100829 (2022)
2. D. Quarta et al., Colloidal bismuth chalcohalide nanocrystals. Angew. Chemie - Int. Ed. **61**, e202201747 (2022)
3. K.T. Butler, S. McKechnie, P. Azarhoosh, M. Van Schilfgaarde, D.O. Scanlon, A. Walsh, Quasi-particle electronic band structure and alignment of the V-VI-VII semiconductors SbSI, SbSBr, and SbSeI for solar cells. Appl. Phys. Lett. **108**, 112103 (2016)
4. S. Palaz, O. Oltulu, A.M. Mamedov, E. Ozbay, AVBVICVII ferroelectrics as novel materials for phononic crystals. Ferroelectrics **511**, 12 (2017)
5. Y. Purusothaman, N.R. Alluri, A. Chandrasekhar, S.J. Kim, Photoactive piezoelectric energy harvester driven by antimony sulfoiodide (SbSI): A $A_V B_{VI} C_{VII}$ class ferroelectric-semiconductor compound. Nano Energy **50**, 256 (2018)
6. R. Bai, B. Xiao, F. Li, X. Liu, S. Xi, M. Zhu, W. Jie, B. Bin Zhang, Y. Xu, Growth of bismuth- and antimony-based chalcohalide single crystals by the physical vapor transport method. CrystEngComm **24**, 1094 (2022)
7. P. Szperlich, Piezoelectric $A^{15}B^{16}C^{17}$ compounds and their nanocomposites for energy harvesting and sensors: a review. Materials **14**, 6973 (2021)

8. H. Garot, D'un Produit Résultant de l'action Réciproque Du Sulfure d'antimoine et de l'iode. J. Pharm **10**, 511 (1824)
9. R. Nitsche, W.J. Merz, Photoconduction in ternary V-VI-VII compounds. J. Phys. Chem. Solids **13**, 154 (1960)
10. E. Fatuzzo, G. Harbeke, W.J. Merz, R. Nitsche, H. Roetschi, W. Ruppel, Ferroelectricity in SbSI. Phys. Rev. **127**, 2036 (1962)
11. R. Nitsche, H. Roetschi, P. Wild, New ferroelectric V. VI. VII compounds of the SbSI type. Appl. Phys. Lett. **4**, 210 (1964)
12. A.A. Grekov, E.D. Rogach, On the Influence of Illumination on the Electromechanical Properties of SbSI single crystal. Ferroelectrics **6**, 87 (1973)
13. K. Hamano, T. Nakamura, Y. Ishibashi, T. Ooyane, Piezoelectric property of SbSI single crystal. J. Phys. Soc. Japan **20**, 1886 (1965)
14. Y. Poratt, R.Y. Ting, The piezoelectric and dielectric properties of $SbSI(Sb_2S_3)_X$ composites. Ferroelectrics **87**, 155 (1988)
15. W.A. Smith, J.P. Doughertyt, L.E. Cross, Pyroelectricity in SbSI. Ferroelectrics **33**, 3 (1981)
16. K. Mistewicz, M. Jesionek, M. Nowak, M. Kozioł, SbSeI pyroelectric nanogenerator for a low temperature waste heat recovery. Nano Energy **64**, 103906 (2019)
17. M.A. Hamad, Detecting giant electrocaloric properties of ferroelectric SbSI at room temperature. J. Adv. Dielectr. **03**, 1350008 (2013)
18. B. Peng, K. Xu, H. Zhang, Z. Ning, H. Shao, G. Ni, J. Li, Y. Zhu, H. Zhu, C.M. Soukoulis, 1D SbSeI, SbSI, and SbSBr with high stability and novel properties for microelectronic, optoelectronic, and thermoelectric applications. Adv. Theory Simulations **1**, 1700005 (2018)
19. S.D. Guo, H.C. Li, Monolayer enhanced thermoelectric properties compared with bulk for BiTeBr. Comput. Mater. Sci. **139**, 361 (2017)
20. Y. He, J. Zhou, First-principles study on the ultralow lattice thermal conductivity of BiSeI. Phys. B Condens. Matter **646**, 414278 (2022)
21. W. Khan, S. Hussain, J. Minar, S. Azam, Electronic and thermoelectric properties of ternary chalcohalide semiconductors: first principles study. J. Electron. Mater. **47**, 1131 (2018)
22. S. J. Adjogri and E. L. Meyer, *Chalcogenide Perovskites and Perovskite-Based Chalcohalide as Photoabsorbers: A Study of Their Properties, and Potential Photovoltaic Applications*, Materials.
23. D.R. Akopov, A.I. Rodin, A.A. Grekov, Anomalous photovoltaic effect in $A^VB^{VI}C^{VII}$ Ferroelectrics. Ferroelectrics **26**, 855 (1980)
24. D.R. Akopov, A.I. Rodin, A.A. Grekov, Photovoltaic effects in ferroelectrics of the AVBVICVII type. Ferroelectrics **174**, 1 (1995)
25. K. Mistewicz, M. Nowak, D. Stróż, A ferroelectric-photovoltaic effect in SbSI nanowires. Nanomaterials **9**, 580 (2019)
26. T. Liu et al., Pressure-enhanced photocurrent in one-dimensional SbSI via lone-pair electron reconfiguration. Materials **15**, 3845 (2022)
27. G. Peng, H. Lu, Y. Liu, D. Fan, The construction of a single-crystalline SbSI Nanorod array-WO_3 heterostructure photoanode for high PEC performance. Chem. Commun. **57**, 335 (2021)
28. P. Kwolek, K. Pilarczyk, T. Tokarski, J. Mech, J. Irzmański, K. Szaciłowski, Photoelectrochemistry of N-type antimony sulfoiodide nanowires. Nanotechnology **26**, 105710 (2015)
29. C. Wang et al., SbSI nanocrystals: an excellent visible light photocatalyst with efficient generation of singlet oxygen. ACS Sustain. Chem. Eng. **6**, 12166 (2018)
30. W.W. Liu, R.F. Peng, Recent advances of bismuth oxychloride photocatalytic material: property, preparation, and performance enhancement. J. Electron. Sci. Technol. **18**, 119 (2020)
31. Y. Li, H. Jiang, X. Wang, X. Hong, B. Liang, Recent advances in bismuth oxyhalide photocatalysts for degradation of organic pollutants in wastewater. RSC Adv. **11**, 26855 (2021)
32. R. Wang, Y. Wang, N. Zhang, S. Lin, Y. He, Y. Yan, K. Hu, H. Sun, X. Liu, Synergetic piezo-photocatalytic effect in SbSI for highly efficient degradation of methyl orange. Ceram. Int. **48**, 31818 (2022)

33. L. Li, W. Cao, J. Yao, W. Liu, F. Li, C. Wang, Synergistic piezo-photocatalysis of BiOCl/$NaNbO_3$ heterojunction piezoelectric composite for high-efficient organic pollutant degradation. Nanomaterials **12**, 353 (2022)
34. K. Ishizaka et al., Giant Rashba-Type spin splitting in bulk BiTeI. Nat. Mater. **10**, 521 (2011)
35. A. Akrap, J. Teyssier, A. Magrez, P. Bugnon, H. Berger, A.B. Kuzmenko, D. Van Der Marel, Optical properties of BiTeBr and BiTeCl. Phys. Rev. B—Condens. Matter Mater. Phys. **90**, 35201 (2014)
36. Z. Tajkov, D. Visontai, L. Oroszlány, J. Koltai, Topological phase diagram of BiTeX-graphene hybrid structures. Applied Sciences **9**, 4330 (2019)
37. Y. Qi et al., Topological quantum phase transition and superconductivity induced by pressure in the bismuth tellurohalide BiTeI. Adv. Mater. **29**, 1605965 (2017)
38. A. Bafekry, S. Karbasizadeh, C. Stampfl, M. Faraji, D.M. Hoat, I.A. Sarsari, S.A.H. Feghhi, M. Ghergherehchi, Two-dimensional janus semiconductor BiTeCl and BiTeBr monolayers: a first-principles study on their tunable electronic propertiesviaan electric field and mechanical strain. Phys. Chem. Chem. Phys. **23**, 15216 (2021)
39. R.E. Brandt et al., Searching for "defect-tolerant" photovoltaic materials: combined theoretical and experimental screening. Chem. Mater. **29**, 4667 (2017)
40. S.R. Kavanagh, C.N. Savory, D.O. Scanlon, A. Walsh, Hidden spontaneous polarisation in the chalcohalide photovoltaic absorber $Sn_2SbS_2I_3$. Mater. Horizons **8**, 2709 (2021)
41. R.E. Brandt, V. Stevanović, D.S. Ginley, T. Buonassisi, Identifying defect-tolerant semiconductors with high minority-carrier lifetimes: beyond hybrid lead halide perovskites. MRS Commun. **5**, 265 (2015)
42. M. Nowak, Photoferroelectric nanowires, in nanowires science and technology (IntechOpen, Rijeka, 2010), Ch. 13
43. L. Belyaev, V. Fridkin, A. Grekov, N. Kosonogov, A. Rodin, Photoferroelectric effects in $A_VB_{VI}C_{VII}$ and $BaTiO_3$-type ferroelectrics. J. Phys. **C2**, 123 (1972)
44. Y.N. Teja, K. Gayathri, C. Ningaraju, A. Murali, M. Sakar, Oxyhalides-based photocatalysts: the case of bismuth oxyhalides, in *Photocatalytic Systems by Design: Materials, Mechanisms and Applications*, edited by M. Sakar, R.G. Balakrishna, T.-O.B.T.-P.S. by D. Do (Elsevier, 2021), pp. 441–474
45. X. Jin, L. Ye, H. Xie, G. Chen, Bismuth-rich bismuth oxyhalides for environmental and energy photocatalysis. Coord. Chem. Rev. **349**, 84 (2017)
46. L. Wang, L. Wang, Y. Du, X. Xu, S.X. Dou, Progress and perspectives of bismuth oxyhalides in catalytic applications. Mater. Today Phys. **16**, 100294 (2021)
47. Y. Yang, C. Zhang, C. Lai, G. Zeng, D. Huang, M. Cheng, J. Wang, F. Chen, C. Zhou, W. Xiong, BiOX ($X = Cl$, Br, I) photocatalytic nanomaterials: applications for fuels and environmental management. Adv. Colloid Interface Sci. **254**, 76 (2018)
48. A.V. Gomonnai, I.M. Voynarovych, A.M. Solomon, Y.M. Azhniuk, A.A. Kikineshi, V.P. Pinzenik, M. Kis-Varga, L. Daroczy, V.V. Lopushansky, X-ray diffraction and Raman scattering in SbSI nanocrystals. Mater. Res. Bull. **38**, 1767 (2003)
49. S.Z.M. Murtaza, P. Vaqueiro, Rapid synthesis of chalcohalides by ball milling: preparation and characterisation of BiSI and BiSeI. J. Solid State Chem. **291**, 121625 (2020)
50. H. Yu, H. Huang, K. Xu, W. Hao, Y. Guo, S. Wang, X. Shen, S. Pan, Y. Zhang, Liquid-phase exfoliation into monolayered BiOBr nanosheets for photocatalytic oxidation and reduction. ACS Sustain. Chem. Eng. **5**, 10499 (2017)
51. C. Wang et al., Nonlinear optical response of SbSI nanorods dominated with direct band gaps. J. Phys. Chem. C **125**, 15441 (2021)
52. Y. Shi, J. Li, C. Mao, S. Liu, X. Wang, X. Liu, S. Zhao, X. Liu, Y. Huang, L. Zhang, Van Der Waals gap-rich BiOCl atomic layers realizing efficient, pure-water CO_2-to-CO photocatalysis. Nat. Commun. **12**, 5923 (2021)
53. K.C. Gödel, U. Steiner, Thin film synthesis of SbSI micro-crystals for self-powered photodetectors with rapid time response. Nanoscale **8**, 15920 (2016)
54. D. Hajra, R. Sailus, M. Blei, K. Yumigeta, Y. Shen, S. Tongay, Epitaxial synthesis of highly oriented 2D Janus Rashba semiconductor BiTeCl and BiTeBr layers. ACS Nano **14**, 15626 (2020)

55. I. Cho, B.K. Min, S.W. Joo, Y. Sohn, One-dimensional single crystalline antimony sulfur iodide, SbSI. Mater. Lett. **86**, 132 (2012)
56. Y. Zhang, P. Cao, X. Zhu, B. Li, Y. He, P. Song, R. Wang, Facile construction of BiOBr ultra-thin nano-roundels for dramatically enhancing photocatalytic activity. J. Environ. Manage. **299**, 113636 (2021)
57. T. Senasu, T. Narenuch, K. Wannakam, T. Chankhanittha, S. Nanan, Solvothermally grown BiOCl catalyst for photodegradation of cationic dye and fluoroquinolone-based antibiotics. J. Mater. Sci. Mater. Electron. **31**, 9685 (2020)
58. M. Nowak et al., Fabrication and characterization of SbSI gel for humidity sensors. Sensors Actuators, A Phys. **210**, 119 (2014)
59. G. Li, F. Qin, R. Wang, S. Xiao, H. Sun, R. Chen, BiOX (X=Cl, Br, I) nanostructures: mannitol-mediated microwave synthesis, visible light photocatalytic performance, and Cr(VI) removal capacity. J. Colloid Interface Sci. **409**, 43 (2013)
60. D.I. Kaynts, A.P. Shpak, V.M. Rubish, O.A. Mykaylo, O.G. Guranich, P.P. Shtets, P.P. Guranich, Formation of ferroelectric nanostructures in $(As_2S_3)100–x(SbSI)x$ glassy matrix. Ferroelectrics **371**, 28 (2008)
61. V.J. Babu, R.S.R. Bhavatharini, S. Ramakrishna, Electrospun BiOI nano/microtectonic plate-like structure synthesis and UV-light assisted photodegradation of ARS dye. RSC Adv. **4**, 19251 (2014)
62. A.A. Abuelwafa, R.M. Matiur, A.A. Putri, T. Soga, Synthesis, structure, and optical properties of the nanocrystalline bismuth oxyiodide (BiOI) for optoelectronic application. Opt. Mater. (Amst). **109**, 110413 (2020)
63. Y.C. Choi, E. Hwang, D.H. Kim, Controlled growth of SbSI thin films from amorphous Sb_2S_3 for low-temperature solution processed chalcohalide solar cells. APL Mater. **6**, 121108 (2018)
64. K.W. Jung, Y.C. Choi, Compositional engineering of antimony chalcoiodides via a two-step solution process for solar cell applications. ACS Appl. Energy Mater. **5**, 5348 (2021)
65. K. Mistewicz et al., A simple route for manufacture of photovoltaic devices based on chalcohalide nanowires. Appl. Surf. Sci. **517**, 146138 (2020)
66. K. Mistewicz, Pyroelectric nanogenerator based on an SbSI-TiO_2 nanocomposite. Sensors **22**, 69 (2022)
67. B. Toroń, K. Mistewicz, M. Jesionek, M. Kozioł, D. Stróż, M. Zubko, Nanogenerator for dynamic stimuli detection and mechanical energy harvesting based on compressed SbSeI nanowires. Energy **212**, 118717 (2020)
68. N. Nishimura, H. Suzuki, M. Higashi, R. Abe, A pressure-assisted low temperature sintering of particulate bismuth chalcohalides BiSX ($X = Br$, I) for fabricating efficient photoelectrodes with porous structures. J. Photochem. Photobiol. A Chem. **413**, 113264 (2021)
69. M.M. Frutos, M.E.P. Barthaburu, L. Fornaro, I. Aguiar, Bismuth chalcohalide-based nanocomposite for application in ionising radiation detectors. Nanotechnology **31**, 225710 (2020)
70. K. Mistewicz, M. Nowak, D. Stróż, A. Guiseppi-Elie, Ferroelectric SbSI nanowires for ammonia detection at a low temperature. Talanta **189**, 225 (2018)
71. K. Mistewicz, M. Nowak, A. Starczewska, M. Jesionek, T. Rzychoń, R. Wrzalik, A. Guiseppi-Elie, Determination of electrical conductivity type of SbSI nanowires. Mater. Lett. **182**, 78 (2016)
72. K. Mistewicz, M. Nowak, D. Stróz, R. Paszkiewicz, SbSI Nanowires for ferroelectric generators operating under shock pressure. Mater. Lett. **180**, 15 (2016)
73. B. Toroń, K. Mistewicz, M. Jesionek, M. Kozioł, M. Zubko, D. Stróż, A New hybrid piezo/triboelectric SbSeI nanogenerator. Energy **238**, 122048 (2022)
74. Z. Yu, H. Yang, N. Soin, L. Chen, N. Black, K. Xu, P.K. Sharma, C. Tsonos, A. Kumar, J. Luo, Bismuth oxyhalide based photo-enhanced triboelectric nanogenerators. Nano Energy **89**, 106419 (2021)
75. K. Mistewicz, M. Jesionek, H.J. Kim, S. Hajra, M. Kozioł, Ł Chrobok, X. Wang, Nanogenerator for determination of acoustic power in ultrasonic reactors. Ultrason. Sonochem. **78**, 105718 (2021)

76. H. Sun, X. Xiao, V. Celorrio, Z. Guo, Y. Hu, C. Kirk, N. Robertson, A novel method to synthesize BiSI uniformly coated with RGO by chemical bonding and its application as a supercapacitor electrode material. J. Mater. Chem. A **9**, 15452 (2021)
77. A.K. Pathak, A.C. Mohan, S.K. Batabyal, Bismuth sulfoiodide (BiSI) for photo-chargeable charge storage device. Appl. Phys. A Mater. Sci. Process. **128**, 298 (2022)
78. H. Sun, G. Yang, J. Chen, C. Kirk, N. Robertson, Facile synthesis of BiSI and $Bi_{13}S_{18}I_2$ as stable electrode materials for supercapacitor applications. J. Mater. Chem. C **8**, 13253 (2020)
79. S. Manoharan, D. Kesavan, P. Pazhamalai, K. Krishnamoorthy, S.J. Kim, Ultrasound irradiation mediated preparation of antimony sulfoiodide (SbSI) nanorods as a high-capacity electrode for electrochemical supercapacitors. Mater. Chem. Front. **5**, 2303 (2021)
80. M. Nakamura, H. Hatada, Y. Kaneko, N. Ogawa, Y. Tokura, M. Kawasaki, Impact of electrodes on the extraction of shift current from a ferroelectric semiconductor SbSI. Appl. Phys. Lett. **113**, 232901 (2018)
81. M. Sotome et al., Spectral dynamics of shift current in ferroelectric semiconductor SbSI. Proc. Natl. Acad. Sci. U. S. A. **116**, 1929 (2019)
82. R. Nie, M. Hu, A.M. Risqi, Z. Li, S. Il Seok, Efficient and stable antimony selenoiodide solar cells. Adv. Sci. **8**, 2003172 (2021)
83. R. Nie, S. Il Seok, Efficient antimony-based solar cells by enhanced charge transfer. Small Methods **4**, 1900698 (2020)
84. Y.C. Choi, E. Hwang, Controlled growth of BiSI nanorod-based films through a two-step solution process for solar cell applications. Nanomaterials **9**, 1650 (2019)
85. S. Li, L. Xu, X. Kong, T. Kusunose, N. Tsurumach, Q. Feng, Enhanced photovoltaic performance of BiSCl solar cells through nanorod array. Chemsuschem **14**, 3351 (2021)
86. D. Tiwari, F. Cardoso-Delgado, D. Alibhai, M. Mombrú, D.J. Fermín, Photovoltaic performance of phase-pure orthorhombic BiSI thin-films. ACS Appl. Energy Mater. **2**, 3878 (2019)
87. G. Chen, W. Li, Y. Yu, Q. Yang, Fast and low-temperature synthesis of one-dimensional (1D) single-crystalline SbSI microrod for high performance photodetector. RSC Adv. **5**, 21859 (2015)
88. S. Farooq, T. Feeney, J.O. Mendes, V. Krishnamurthi, S. Walia, E. Della Gaspera, J. van Embden, High gain solution-processed carbon-free BiSI chalcohalide thin film photodetectors. Adv. Funct. Mater. **31**, 2104788 (2021)
89. L. Sun, C. Wang, L. Xu, J. Wang, X. Liu, X. Chen, G.C. Yi, SbSI whisker/PbI_2 flake mixed-dimensional van Der Waals heterostructure for photodetection. CrystEngComm **21**, 3779 (2019)
90. I. Aguiar, M. Mombrú, M.P. Barthaburu, H.B. Pereira, L. Fornaro, Influence of solvothermal synthesis conditions in BiSI nanostructures for application in ionizing radiation detectors. Mater. Res. Express **3**, 25012 (2016)
91. M. Devetak, B. Berčič, M. Uplaznik, A. Mrzel, D. Mihailovic, $Mo_6S_3I_6$ Nanowire Network Vapor Pressure Chemisensors. Chem. Mater. **20**, 1773 (2008)
92. K. Mistewicz, M. Nowak, R. Paszkiewicz, A. Guiseppi-Elie, SbSI nanosensors: from gel to single nanowire devices. Nanoscale Res. Lett. **12**, 97 (2017)
93. M. Nowak, K. Mistewicz, A. Nowrot, P. Szperlich, M. Jesionek, A. Starczewska, Transient characteristics and negative photoconductivity of SbSI humidity sensor. Sensors Actuators, A Phys. **210**, 32 (2014)
94. A. Starczewska, M. Nowak, P. Szperlich, B. Toroń, K. Mistewicz, D. Stróz, J. Szala, Influence of humidity on impedance of SbSI gel. Sensors Actuators, A Phys. **183**, 34 (2012)
95. K. Mistewicz, A. Starczewska, M. Jesionek, M. Nowak, M. Kozioł, D. Stróż, Humidity dependent impedance characteristics of SbSeI nanowires. Appl. Surf. Sci. **513**, 145859 (2020)
96. Q. Chen, N.B. Feng, X.H. Huang, Y. Yao, Y.R. Jin, W. Pan, D. Liu, Humidity-sensing properties of a BiOCl-coated quartz crystal microbalance. ACS Omega **5**, 18818 (2020)
97. B. Monchev, D. Filenko, N. Nikolov, C. Popov, T. Ivanov, P. Petkov, I.W. Rangelow, Investigation of the sorption properties of thin Ge-S-AgI films deposited on cantilever-based gas sensor. Appl. Phys. A Mater. Sci. Process. **87**, 31 (2007)

98. K. Mistewicz, M. Kępińska, M. Nowak, A. Sasiela, M. Zubko, D. Stróż, Fast and efficient piezo/photocatalytic removal of methyl orange using SbSI nanowires. Materials **13**, 4803 (2020)
99. M. Tasviri, Z. Sajadi-Hezave, SbSI nanowires and CNTs encapsulated with SbSI as photocatalysts with high visible-light driven photoactivity. Mol. Catal. **436**, 174 (2017)
100. L. Wang, Y. Liu, G. Chen, M. Zhang, X. Yang, R. Chen, Y. Cheng, Bismuth oxychloride nanomaterials fighting for human health: from photodegradation to biomedical applications. Crystals **12**, 491 (2022)
101. X. Wang, F. Zhang, Y. Yang, Y. Zhang, L. Liu, W. Lei, Controllable synthesis and photocatalytic activity of nano-BiOBr photocatalyst. J. Nanomater. **2020**, 1013075 (2020)
102. R. Zhang, K. Zeng, A novel flower-like dual Z-scheme BiSI/Bi_2WO_6/g-C_3N_4 photocatalyst has excellent photocatalytic activity for the degradation of organic pollutants under visible light. Diam. Relat. Mater. **115**, 108343 (2021)
103. J. Liu, Y. Li, L. Huang, C. Wang, L. Yang, J. Liu, C. Huang, Y. Song, Fabrication of novel narrow/wide band gap $Bi_4O_5I_2$/BiOCl heterojunction with high antibacterial and degradation efficiency under LED and sunlight. Appl. Surf. Sci. **567**, 150713 (2021)
104. X. Qu, X. Zhao, M. Liu, Z. Gao, D. Yang, L. Shi, Y. Tang, H. Song, BiOCl/TiO_2 composite photocatalysts synthesized by the sol-gel method for enhanced visible-light catalytic activity toward methyl orange. J. Mater. Res. **35**, 3067 (2020)
105. S. Bargozideh, M. Tasviri, Construction of a novel BiSI/MoS_2 nanocomposite with enhanced visible-light driven photocatalytic performance. New J. Chem. **42**, 18236 (2018)
106. X. Lou, J. Shang, L. Wang, H. Feng, W. Hao, T. Wang, Y. Du, Enhanced photocatalytic activity of $Bi_{24}O_{31}Br_{10}$: constructing heterojunction with BiOI. J. Mater. Sci. Technol. **33**, 281 (2017)
107. J. Yu, M. Cui, X. Liu, Q. Chen, Y. Wu, Y. He, Preparation of novel AgBr/Bi_3O_4Br hybrid with high photocatalytic activity via in situ ion exchange method. Mater. Lett. **193**, 73 (2017)
108. H. Razavi-Khosroshahi, S. Mohammadzadeh, M. Hojamberdiev, S. Kitano, M. Yamauchi, M. Fuji, $BiVO_4$/BiOX (X = F, Cl, Br, I) heterojunctions for degrading organic dye under visible light. Adv. Powder Technol. **30**, 1290 (2019)
109. L. Chen, W. Zhang, J. Wang, X. Li, Y. Li, X. Hu, L. Zhao, Y. Wu, Y. He, High piezo/photocatalytic efficiency of Ag/Bi_5O_7I nanocomposite using mechanical and solar energy for N_2 fixation and methyl orange degradation. Green Energy Environ. **8**, 283 (2023)
110. K. Mistewicz, M. Nowak, R. Wrzalik, J. Śleziona, J. Wieczorek, A. Guiseppi-Elie, Ultrasonic processing of SbSI nanowires for their application to gas sensors. Ultrasonics **69**, 67 (2016)
111. K. Mistewicz, M. Nowak, Prevention of food spoilage using nanoscale sensors, in *Nanobiosensors*, edited by A.M.B.T.-N. Grumezescu (Academic Press, 2017), pp. 245–288
112. K. Mistewicz, Recent advances in ferroelectric nanosensors: toward sensitive detection of gas, mechanothermal signals, and radiation. J. Nanomater. **2018**, 2651056 (2018)
113. K. Mistewicz, Ternary chalcohalide nanowires for energy harvesting, sensing, and environmental remediation. Video Proc. Adv. Mater. **2**, (2021)
114. D.W. Davies, K.T. Butler, J.M. Skelton, C. Xie, A.R. Oganov, A. Walsh, Computer-aided design of metal chalcohalide semiconductors: from chemical composition to crystal structure. Chem. Sci. **9**, 1022 (2018)
115. Z. Ran, X. Wang, Y. Li, D. Yang, X.G. Zhao, K. Biswas, D.J. Singh, L. Zhang, Bismuth and antimony-based oxyhalides and chalcohalides as potential optoelectronic materials. Npj Comput. Mater. **4**, 14 (2018)
116. S. Surthi, S. Kotru, R.K. Pandey, SbSI films for ferroelectric memory applications. Integr. Ferroelectr. **48**, 263 (2002)
117. K.T. Butler, J.M. Frost, A. Walsh, Ferroelectric materials for solar energy conversion: photoferroics revisited. Energy Environ. Sci. **8**, 838 (2015)
118. A. Rüdiger, R. Waser, Size effects in nanoscale ferroelectrics. J. Alloys Compd. **449**, 2 (2008)
119. D. Denning, J. Guyonnet, B.J. Rodriguez, Applications of piezoresponse force microscopy in materials research: from inorganic ferroelectrics to biopiezoelectrics and beyond. Int. Mater. Rev. **61**, 46 (2016)

120. S. Martin, B. Gautier, N. Baboux, A. Gruverman, A. Carretero-Genevrier, M. Gich, A. Gomez, Characterizing ferroelectricity with an atomic force microscopy: an all-around technique, in *NanoScience and Technology*. ed. by U. Celano (Springer International Publishing, Cham, 2019), pp.173–203
121. A. Gruverman, M. Alexe, D. Meier, Piezoresponse force microscopy and nanoferroic phenomena. Nat. Commun. **10**, 1661 (2019)
122. A. Starczewska, K. Mistewicz, M. Kozioł, M. Zubko, D. Stróż, J. Dec, Interfacial polarization phenomena in compressed nanowires of SbSI. Materials **15**, 1543 (2022)
123. D. Kundys, A. Cascales, A.S. Makhort, H. Majjad, F. Chevrier, B. Doudin, A. Fedrizzi, B. Kundys, Optically rewritable memory in a graphene-ferroelectric-photovoltaic heterostructure. Phys. Rev. Appl. **13**, 64034 (2020)
124. R. Guo, L. You, Y. Zhou, Z.S. Lim, X. Zou, L. Chen, R. Ramesh, J. Wang, Non-volatile memory based on the ferroelectric. Nat. Commun. **4**, 1990 (2013)
125. J.Y. Kim, J.W. Lee, H.S. Jung, H. Shin, N.G. Park, High-efficiency perovskite solar cells. Chem. Rev. **120**, 7867 (2020)
126. Z. Liu, R. Mi, G. Ji, Y. Liu, P. Fu, S. Hu, B. Xia, Z. Xiao, Bandgap engineering and thermodynamic stability of oxyhalide and chalcohalide antiperovskites. Ceram. Int. **47**, 32634 (2021)
127. H. Kunioku, M. Higashi, R. Abe, Lowerature synthesis of bismuth chalcohalides: candidate photovoltaic materials with easily, continuously controllable band gap. Sci. Rep. **6**, 32664 (2016)
128. N.T. Hahn, A.J.E. Rettie, S.K. Beal, R.R. Fullon, C.B. Mullins, N-BiSI thin films: selenium doping and solar cell behavior. J. Phys. Chem. C **116**, 24878 (2012)
129. M. Nowak, B. Kauch, P. Szperlich, D. Stróz, J. Szala, T. Rzychoń, Bober, B. Toroń, A. Nowrot, Sonochemical Preparation of $SbS_{1-x}Se_xI$ Nanowires. Ultrason. Sonochem. **17**, 487 (2010)
130. M. Li, J. Li, C. Guo, L. Zhang, Doping bismuth oxyhalides with indium: a DFT calculations on tuning electronic and optical properties. Chem. Phys. Lett. **705**, 31 (2018)
131. S. Wu, W. Sun, J. Sun, Z.D. Hood, S.Z. Yang, L. Sun, P.R.C. Kent, M.F. Chisholm, Surface reorganization leads to enhanced photocatalytic activity in defective BiOCl. Chem. Mater. **30**, 5128 (2018)
132. E. Wlaźlak et al., Heavy pnictogen chalcohalides: the synthesis, structure and properties of these rediscovered semiconductors. Chem. Commun. **54**, 12133 (2018)
133. A. Audzijonis, L. Žigas, R. Sereika, R. Žaltauskas, Origin of ferroelectric phase transition in $SbSCl_xI_{1-x}$ mixed crystals. Int. J. Mod. Phys. B **28**, 1450209 (2014)
134. R. Sereika, R. Zaltauskas, V. Lapeika, S. Stanionytė, R. Juškenas, Structural changes in chlorine-substituted SbSI. J. Appl. Phys. **126**, 114101 (2019)
135. Y. Wang, Y. Hu, Z. Chen, Y. Guo, D. Wang, E.A. Wertz, J. Shi, Effect of strain on the curie temperature and band structure of low-dimensional SbSI. Appl. Phys. Lett. **112**, 183104 (2018)
136. C. Liu, J. Wang, G. Xu, M. Kamlah, T.Y. Zhang, An isogeometric approach to flexoelectric effect in ferroelectric materials. Int. J. Solids Struct. **162**, 198 (2019)
137. Y. Xia, Y. Ji, Y. Liu, L. Wu, Y. Yang, Controllable piezo-flexoelectric effect in ferroelectric $Ba_{0.7}Sr_{0.3}TiO_3$ materials for harvesting vibration energy. ACS Appl. Mater. Interfaces **14**, 36763 (2022)
138. J.J. Plata, J.A. Suárez, S. Cuesta-López, A.M. Márquez, J.F. Sanz, Photo-sensitizing thin-film ferroelectric oxides using materials databases and high-throughput calculations. J. Mater. Chem. A **7**, 27323 (2019)
139. S.Y. Yang et al., Above-bandgap voltages from ferroelectric photovoltaic devices. Nat. Nanotechnol. **5**, 143 (2010)
140. A.M. Ganose, S. Matsumoto, J. Buckeridge, D.O. Scanlon, Defect engineering of earth-abundant solar absorbers BiSI and BiSeI. Chem. Mater. **30**, 3827 (2018)
141. H. Shi, W. Ming, M.H. Du, Bismuth chalcohalides and oxyhalides as optoelectronic materials. Phys. Rev. B **93**, 104108 (2016)

142. A.M. Ganose, K.T. Butler, A. Walsh, D.O. Scanlon, Relativistic electronic structure and band alignment of BiSI and BiSeI: candidate photovoltaic materials. J. Mater. Chem. A **4**, 2060 (2016)
143. Y.C. Choi, K.W. Jung, Recent progress in fabrication of antimony/bismuth chalcohalides for lead-free solar cell applications. Nanomaterials **10**, 2284 (2020)
144. B. Yoo, D. Ding, J.M. Marin-Beloqui, L. Lanzetta, X. Bu, T. Rath, S.A. Haque, Improved charge separation and photovoltaic performance of BiI_3 absorber layers by use of an in situ formed BiSI interlayer. ACS Appl. Energy Mater. **2**, 7056 (2019)
145. J.A. Chang, S.H. Im, Y.H. Lee, H.J. Kim, C.S. Lim, J.H. Heo, S. Il Seok, Panchromatic photon-harvesting by hole-conducting materials in inorganic-organic heterojunction sensitized-solar cell through the formation of nanostructured electron channels. Nano Lett. **12**, 1863 (2012)
146. Y. Cheng, L. Ding, Perovskite/Si tandem solar cells: fundamentals advances, challenges, and novel applications. SusMat **1**, 324 (2021)
147. Z. Fang et al., Perovskite-based tandem solar cells. Sci. Bull. **66**, 621 (2021)
148. A.A. Krasnov, S.A. Legotin, Advances in the development of betavoltaic power sources (a review). Instruments Exp. Tech. **63**, 437 (2020)
149. G. Li, C. Zhao, Y. Liu, J. Ren, Z. Zhang, H. Di, W. Jiang, J. Mei, Y. Zhao, High-performance perovskite betavoltaics employing high-crystallinity $MAPbBr_3$ films. ACS Omega **6**, 20015 (2021)
150. Y. Niu, J. Zeng, X. Liu, J. Li, Q. Wang, H. Li, N.F. de Rooij, Y. Wang, G. Zhou, A photovoltaic self-powered gas sensor based on all-dry transferred MoS_2/GaSe heterojunction for Ppb-level NO_2 sensing at room temperature. Adv. Sci. **8**, 2100472 (2021)
151. Y. Kim et al., 2D transition metal dichalcogenide heterostructures for p- and n-type photovoltaic self-powered gas sensor. Adv. Funct. Mater. **30**, 2003360 (2020)
152. X.L. Liu, Y. Zhao, W.J. Wang, S.X. Ma, X.J. Ning, L. Zhao, J. Zhuang, Photovoltaic self-powered gas sensing: a review. IEEE Sens. J. **21**, 5628 (2021)
153. B. Xu, Y. Gao, Y. Li, S. Liu, D. Lv, S. Zhao, H. Gao, G. Yang, N. Li, L. Ge, Synthesis of Bi_3O_4Cl nanosheets with oxygen vacancies: the effect of defect states on photocatalytic performance. Appl. Surf. Sci. **507**, 144806 (2020)
154. J. Di et al., Defect-tailoring mediated electron-hole separation in single-unit-cell Bi_3O_4Br nanosheets for boosting photocatalytic hydrogen evolution and nitrogen fixation. Adv. Mater. **31**, 1807576 (2019)
155. Q. Geng, H. Xie, W. Cui, J. Sheng, X. Tong, Y. Sun, J. Li, Z. Wang, F. Dong, Optimizing the electronic structure of BiOBr nanosheets via combined Ba doping and oxygen vacancies for promoted photocatalysis. J. Phys. Chem. C **125**, 8597 (2021)
156. Y. Liu, Z. Hu, J.C. Yu, Photocatalytic degradation of ibuprofen on S-doped BiOBr. Chemosphere **278**, 130376 (2021)
157. Y. Mi, H. Li, Y. Zhang, N. Du, W. Hou, Synthesis and photocatalytic activity of BiOBr nanosheets with tunable crystal facets and sizes. Catal. Sci. Technol. **8**, 2588 (2018)
158. J. Wu et al., Hydrothermal synthesis of carbon spheres—BiOI/$BiOIO_3$ heterojunctions for photocatalytic removal of gaseous Hg_0 under visible light. Chem. Eng. J. **304**, 533 (2016)
159. D. Liu, J. Xie, Y. Xia, Improved photocatalytic activity of MWCNT/BiOBr composite synthesized via interfacial covalent bonding linkage. Chem. Phys. Lett. **729**, 42 (2019)
160. K.Y. Shih, Y.L. Kuan, E.R. Wang, one-step microwave-assisted synthesis and visible-light photocatalytic activity enhancement of BiOBr/RGO nanocomposites for degradation of methylene blue. Materials **14**, 4577 (2021)
161. X. Lv, F.L. Y. Lam, X. Hu, A review on bismuth oxyhalide (BiOX, X=Cl, Br, I) based photocatalysts for wastewater remediation. Frontiers in Catalysis **2**, 839072 (2022)
162. A. Alzamly et al., Construction of BiOF/BiOI nanocomposites with tunable band gaps as efficient visible-light photocatalysts. J. Photochem. Photobiol. A Chem. **375**, 30 (2019)
163. X. Jia, J. Cao, H. Lin, M. Zhang, X. Guo, S. Chen, Transforming type-I to type-II heterostructure photocatalyst via energy band engineering: a case study of I-BiOCl/I-BiOBr. Appl. Catal. B Environ. **204**, 505 (2017)

164. M. Guo, Z. Zhou, S. Yan, P. Zhou, F. Miao, S. Liang, J. Wang, X. Cui, Bi_2WO_6–BiOCl heterostructure with enhanced photocatalytic activity for efficient degradation of oxytetracycline. Sci. Rep. **10**, 18401 (2020)
165. G. Eshaq, S. Wang, H. Sun, M. Sillanpää, Core/shell $FeVO_4$@BiOCl heterojunction as a durable heterogeneous fenton catalyst for the efficient sonophotocatalytic degradation of p-nitrophenol. Sep. Purif. Technol. **231**, 115915 (2020)
166. R. Zhang, K. Zeng, T. Zhang, Enhanced visible-light-driven photocatalytic activity of Bi_2WO_6-BiSI Z-scheme heterojunction photocatalysts for tetracycline degradation. Int. J. Environ. Anal. Chem. **102**, 7084 (2022)
167. Z. Chen, Z. Wu, Z. Song, X. Zhang, H. Yang, Q. Jiang, T. Zhou, N. Liu, J. Hu, Crucial effect of halogen on the photocatalytic hydrogen evolution for $Bi_{19}X_3S_{27}$ (X=Cl, Br) nanomaterials. Ind. Eng. Chem. Res. **58**, 22958 (2019)
168. C. Zhou, R. Wang, C. Jiang, J. Chen, G. Wang, Dynamically optimized multi-interface novel BiSI-promoted redox sites spatially separated n-p-n double heterojunctions BiSI/MoS_2/CdS for hydrogen evolution. Ind. Eng. Chem. Res. **58**, 7844 (2019)
169. J. Ji, M. Sun, Y. Cai, Y. Wang, Y. Sun, W. Ren, Z. Zhang, F. Jin, Q. Zhang, Rare-earth chalcohalides: a family of van Der Waals layered kitaev spin liquid candidates. Chinese Phys. Lett. **38**, 47502 (2021)
170. Y. Garbovskiy, A.V. Emelyanenko, A. Glushchenko, Inverse "guest-host" effect: ferroelectric nanoparticles mediated switching of nematic liquid crystals. Nanoscale **12**, 16438 (2020)